AF595521

Measurement Methods in the Operation of Ships and Offshore Facilities

Measurement Methods in the Operation of Ships and Offshore Facilities

Editors

Leszek Chybowski
Arkadiusz Tomczak
Maciej Kozak

MDPI • Basel • Beijing • Wuhan • Barcelona • Belgrade • Manchester • Tokyo • Cluj • Tianjin

Editors
Leszek Chybowski
Maritime University of Szczecin
Poland

Arkadiusz Tomczak
Maritime University of Szczecin
Poland

Maciej Kozak
Maritime University of Szczecin
Poland

Editorial Office
MDPI
St. Alban-Anlage 66
4052 Basel, Switzerland

This is a reprint of articles from the Special Issue published online in the open access journal *Sensors* (ISSN 1424-8220) (available at: https://www.mdpi.com/journal/sensors/special_issues/Measurement_Ships_Offshore).

For citation purposes, cite each article independently as indicated on the article page online and as indicated below:

LastName, A.A.; LastName, B.B.; LastName, C.C. Article Title. *Journal Name* **Year**, *Volume Number*, Page Range.

ISBN 978-3-0365-2309-5 (Hbk)
ISBN 978-3-0365-2310-1 (PDF)

Contents

About the Editors

Leszek Chybowski is a professor at the Maritime University of Szczecin. He is editor-in-chef and director of the Maritime University of Szczecin Press and director of the Doctoral School of the Maritime University of Szczecin. He is a researcher and inventor, marine engineer, certified reliability professional, certified TRIZ specialist, graduate of the Polish Ministry of Science and Higher Education programme, and one of the Top 500 Innovators at Stanford University. He is an author/co-author of over 200 papers, chapters and monographs. The scope of his research includes reductions in the machinery–environment interference, improvements in machinery effectiveness, reliability and safety, and inventics and innovation management.

Arkadiusz Tomczak is a professor at the Maritime University of Szczecin with over 10 years of industrial experience including research and lecturing in the field of offshore–hydrographic surveillance, subsea mining, subsea and surface positioning systems, and ship and ROV navigation. The time he devoted to offshore industry led him to a Master's in Mariner and Hydrographer cat. A diploma. In 2013–2015, he was an editor-in-chief of the European Journal of Navigation, a joint journal of European navigation institutes. At present, he is involved in the national submarine search program and manages their first expedition, which is due to be held in June 2021 in the North Sea.

Maciej Kozak graduated from Szczecin University of Technology, Faculty of Electrical Engineering, where he received a Master of Science degree in Electrical Engineering in 1999. In 1999, he started work at Szczecin Maritime University, where he has been working intermittently to date. At that time, he obtained an international diploma of ship's electro-technical officer ETO. In 2011, he received a PhD degree in technical sciences at the Faculty of Electrical Engineering of Poznan University of Technology. From 2008, he was Deputy Director of the Institute of Electrical Engineering and Ship Automation of the Maritime University of Szczecin, and from 2015 to 2019, he was Head of the Institute. From 01.10.2019, he has held the position of Dean of the newly established Faculty of Mechatronics and Electrical Engineering at the Maritime Academy in Szczecin. He is a long-time member of the Association of Polish Electrical Engineers SEP; in 2010–2014, he was a member of the Board of the Szczecin Branch of the SEP; he has been a long-term and active member of IEEE. He is the author of over 20 publications in recognized scientific journals, administrator of the Green Energy Laboratory, and a manager of research and development projects. Lastly, he is involved in a project covering a ship's DC power grid using Energy Storage Sources.

Editorial

Measurement Methods in the Operation of Ships and Offshore Facilities

Leszek Chybowski [1,*], Arkadiusz Tomczak [2] and Maciej Kozak [3]

1 Faculty of Marine Engineering, Maritime University of Szczecin, ul. Willowa 2, 71-656 Szczecin, Poland
2 Faculty of Navigation, Maritime University of Szczecin, ul. Wały Chrobrego 1-2, 70-500 Szczecin, Poland; a.tomczak@am.szczecin.pl
3 Faculty of Mechatronics and Electrical Engineering, Maritime University of Szczecin, ul. Willowa 2, 71-656 Szczecin, Poland; m.kozak@am.szczecin.pl
* Correspondence: l.chybowski@am.szczecin.pl; Tel.: +48-91-4809-412

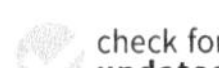

Citation: Chybowski, L.; Tomczak, A.; Kozak, M. Measurement Methods in the Operation of Ships and Offshore Facilities. *Sensors* **2021**, *21*, 2159. https://doi.org/10.3390/s21062159

Received: 11 March 2021
Accepted: 18 March 2021
Published: 19 March 2021

1. Introduction

The development of modern measurement methods for ship systems has occurred due to economic changes and increasingly stringent environmental requirements. At the same time, the specificity of ship systems and the conditions in which they work produce very strict requirements for the reliability and accuracy of the measuring systems on ships and offshore facilities. The digitization of many of the processes carried out on ships, oil rigs, platforms, etc., increases their complexity and security requirements. This book is dedicated to research concerning the measurement methods and condition monitoring of marine systems.

This collection presents the results of research related to the development of modern systems on ships and offshore facilities, with particular emphasis on measuring and assessing processes occurring in ship propulsion systems, ship navigation systems, maritime communications, maritime safety and alarm systems, marine cargo handling equipment, offshore technological systems, etc.

This book addresses all types of sensors and measurement systems designed for ships and offshore facilities. It provides an advanced forum for the science and technology of sensors and measuring systems. Regarding marine systems such as ships and offshore facilities, the scope of the book includes mostly topics associated with physical sensors, remote sensors, smart/intelligent sensors, sensor devices, sensor technologies and applications, signal processing, data fusion, sensor interfaces, human-computer interactions, sensing systems, and localization and object tracking.

2. Special Issue Papers

The above-mentioned topics have been included in the following thematic sections. The research undertaken by the authors of the following sections can be grouped according to the type of equipment being analyzed, particularly including articles on ship machinery and navigational equipment. Specific topics are related to broadly-understood measurement issues, as well as condition monitoring and predicting.

2.1. Supporting the Ship Navigation

In this section, we present articles on the subject of aiding navigation, maneuvering, and operating ships in various operating conditions.

The first article in this section, "Full-Scale Maneuvering Trials Correction and Motion Modeling Based on Actual Sea and Weather Conditions", was prepared by B. Mei et al. [1]. The authors propose a novel sea trials correction method for ship maneuvering. In the paper, the wind and wave drift forces were calculated according to the measurement data. The authors used a pattern search algorithm to calculate the adjustment parameters for wind, waves, sea surface currents, etc.

Subsequently, the team of Y. Yang et al. presented an article entitled "An Attitude Prediction Method for Autonomous Recovery Operation of Unmanned Surface Vehicle" [2]. The authors presented results of experimental launch and recovery technology for an unmanned surface vehicle (USV). To improve the launch accuracy and reduce the influence of sea waves, the authors proposed a stacking model of a one-dimensional convolutional neural network and long short-term memory neural network to predict the attitude of the USV. The authors tested the efficiency and effectiveness of the launch and recovery system, which were demonstrated by its successful application in actual environments.

The subject of unmanned surface vehicle control is taken up by T. Szelangiewicz et al. in their article, "Application of Measurement Sensors and Navigation Devices in Experimental Research of the Computer System for the Control of an Unmanned Ship Model" [3]. The main objective of the research presented in the paper was to design and build a prototype computer system with the necessary measurement sensors and navigation devices to autonomously control an unmanned ship model. The authors presented this type of system and verified its operation on open water.

2.2. *Optimizing the Operation of Ship Machinery*

The second section presents issues related to condition monitoring and optimizing the operation of a ship's onboard propulsion and power generation.

The first article by D. Kim et al. is entitled, "An Ensemble-Based Approach to Anomaly Detection in Marine Engine Sensor Streams for Efficient Condition Monitoring and Analysis" [4]. The authors proposed an unsupervised anomaly detection method using sensor streams from a marine two-stroke diesel engine to detect anomalous system behaviors that may be a sign of system failure. After detecting an anomaly, clustering analysis was conducted on the anomalous observation to examine anomaly patterns.

In turn, in the article by M. Giernalczyk and P. Kaminski entitled, "Assessment of the Propulsion System Operation of the Ships Equipped with the Air Lubrication System" [5], the authors present the results of the measurements of indicators describing the operational effectiveness of an Air Lubrication System installed on a modern passenger ship. This analysis presents some critical observations regarding the efficiency of the system.

Another article prepared by A. Bogdanowicz and T. Kniaziewicz is entitled, "Marine Diesel Engine Exhaust Emissions Measured in Ship's Dynamic Operating Conditions" [6]. The article deals with the measurement of emissions from marine engines in dynamic states. The authors proposed a measurement methodology using an exhaust gas analyzer with simultaneous recording of the load indicator, engine speed, inclinometer, and global positioning system (GPS) data. A neural network algorithm was used to model the concentrations of ingredients contained in engine exhaust gases during dynamic states. The proposed method enabled the calculation of emissions of the composition of exhaust gases from the marine diesel engine and also the calculation of the route emissions of the tested vessel.

The fourth article in this section by D. Kim et al. is entitled, "Data-Driven Prediction of Vessel Propulsion Power Using Support Vector Regression with Onboard Measurement and Ocean Data" [7]. The authors proposed a data-driven approach to predict the propulsion power of a vessel. In this study, support vector regression (SVR) was used to learn from big data obtained from onboard measurements and the National Oceanic and Atmospheric Administration (NOAA) database.

In turn, the team of S. German-Galkin and D. Tarnapowicz wrote an article entitled, "Energy Optimization of the 'Shore to Ship' System—A Universal Power System for Ships at Berth in a Port" [8]. The authors present an analysis of steady-state electromagnetic and energy processes, allowing the determination of the active and reactive power and losses in a shore-to-ship (STS) system. The presented analytical research enables the development of a control algorithm that optimizes the system's energy efficiency.

2.3. Measurements of Ship Machinery Components

This section includes three articles devoted to geometry measurements and quality assessment of machine driveshaft elements.

The first article by L. Chybowski et al. is entitled "Evaluation of Model-Based Control of Reaction Forces at the Supports of Large-Size Crankshafts" [9]. The article discusses a support control automation system employing force sensors to a large-size crankshaft main journal's flexible support system. The support reaction forces were changed to minimize the crankshaft elastic deflection as a function of the crank angle. The aim of this research was to verify the hypothesis that the mentioned change can be expressed by a monoharmonic model regardless of the crankshaft structure. The authors' investigations confirmed this hypothesis. It was also shown that an algorithmic approach improved the mathematical model mapping with the reaction forces due to faster and more accurate calculations of the phase shift angle. The verification of the model for crankshafts with different structural designs made it possible to assess how well the model fit the coefficients of determination that were calculated with finite element analysis (FEA).

Another article was written by K. Miądlicki et al., "Remanufacturing System with Chatter Suppression for CNC Turning" [10]. The article presents the concept of a support system for the manufacture of machine spare parts. The operation of the system is based on a reverse-engineering module enabling feature recognition based on a 3D parts scan. Then, a CAD geometrical model is generated, on the basis of which a machining strategy using the CAM system is developed. The operation of the described system was presented using the example of machining parts of the shaft class. The result is a replacement part, the accuracy of which was compared using the iterative closest point algorithm to obtain the root mean square error at the level of the scanner accuracy.

The third article in this group was prepared by K. Nozdrzykowski et al. and is entitled "The Effect of Deflections and Elastic Deformations on Geometrical Deviation and Shape Profile Measurements of Large Crankshafts with Uncontrolled Sup-ports" [11]. This article presents a multi-criteria analysis of the errors that may occur while measuring geometric deviations of crankshafts that require multi-point support. The analysis in the paper confirmed that the currently-used conventional support method—in which the journals of large crankshafts rest on a set of fixed, rigid vee-blocks—significantly limits the detectability of their geometric deviations, especially those of the main journal axes' positions. Insights into performing practical measurements, which will improve measurement procedures and increase measurement accuracy, were provided.

2.4. Non-Destructive Testing of Ship Machinery Components

With regard to ship machinery, articles devoted to diagnosing the condition of elements of ship mechanisms using acoustic emission (AE) signals should be mentioned. There are two articles on this subject in this issue.

The first is an article by L. Kyzioł et al. entitled, "Acoustic Emission and K-S Metric Entropy as Methods for Determining Mechanical Properties of Composite Materials" [12]. Composites are now a common material for the construction of elements of modern ships and ship mechanisms. The article concerns the use of AE and Kolmogorov-Sinai (K-S) signals metric entropy to determine the mechanical properties of composites. The authors showed that the application of a modern testing machine and very high-quality instrumentation to record measurement data using the K-S metric entropy method and an AE signal allows the determination of a material's transition from an elastic to a plastic phase.

Another article presenting the application of AE signals is an article by M. Kozak et al. entitled, "Identification of Gate Turn-off Thyristor Switching Patterns Using Acoustic Emission Sensors" [13]. In this case, the material was devoted to assessing the condition of electronic components used in the power supply and control systems of ships. The paper presents an introduction and preliminary tests of a method utilizing an acoustic emission sensor that can be used to detect early-stage damages of the gate turn-off thyristor.

In turn, in the article written by S. Drewing and K. Witkowski, "Spectral Analysis of Torsional Vibrations Measured by Optical Sensors, as a Method for Diagnosing Injector Nozzle Coking in Marine Diesel Engines" [14], the authors discussed the possibility to diagnose the coking of a marine diesel engine injector nozzle by performing a spectral analysis of the crankshaft's torsional vibrations. The authors presented and verified the new method, which enabled the measuring and calculation of torsional vibrations in engine crankshafts.

2.5. Supporting Offshore and Inland Water Operations

In the last section, we present selected issues related to dimensioning of the floating offshore objects and measurements performed from the floating systems.

The first article in this section is a paper prepared by G. Stępień et al. entitled, "Dimensioning Method of Floating Offshore Objects by Means of Quasi-Similarity Transformation with Reduced Tolerance Errors" [15]. The article concerns an improvement in the use of sensors supporting the positioning of floating objects. The accurate measurement of the offsets is vital to establish a mathematical relationship between a sensor and vessel common reference point to achieve sufficient accuracy of the survey data. The authors present the method of transformation by similarity with elements of affine transformation, called Q-ST (Quasi-Similarity Transformation). The method was verified in laboratory conditions, as well as in real conditions.

The second article in this section was prepared by K. Pyrchla et al. and is entitled "Analysis of the Dynamic Height Distribution at the Estuary of the Odra River Based on Gravimetric Measurements Acquired with the Use of a Light Survey Boat—A Case Study" [16]. The authors present possible applications of a dynamic gravity meter for determining the dynamic height along a river. The method described in the article can be applied to measurements in all near-zero-depth areas.

Author Contributions: All guest editors contributed equally to this editorial. All authors have read and agreed to the published version of the manuscript.

Funding: This editorial received no external funding.

Institutional Review Board Statement: Not applicable.

Informed Consent Statement: Not applicable.

Data Availability Statement: Authors of particular papers published in this Special Issue are responsible for the data presented and they will provide additional information if required.

Acknowledgments: We thank the Authors of all papers published in this Special Issue for submitting their wonderful work. We would also like to thank all the Reviewers for their very good work and comments that have improved the quality of this Special Issue. We also thank the MDPI Team for their involvement in the preparation of this Special Issue. We would like to express our special gratitude to our contact editor of Sensors Janetta Li., For her guidance and support throughout the entire publishing process. Without this joint effort, we would not be able to assemble a collection of high-quality research papers.

Conflicts of Interest: The authors declare no conflict of interest.

References

1. Mei, B.; Sun, L.; Shi, G. Full-Scale Maneuvering Trials Correction and Motion Modelling Based on Actual Sea and Weather Conditions. *Sensors* **2020**, *20*, 3963. [CrossRef] [PubMed]
2. Yang, Y.; Pan, P.; Jiang, X.; Zheng, S.; Zhao, Y.; Yang, Y.; Zhong, S.; Peng, Y. An Attitude Prediction Method for Autonomous Recovery Operation of Unmanned Surface Vehicle. *Sensors* **2020**, *20*, 5662. [CrossRef] [PubMed]
3. Szelangiewicz, T.; Żelazny, K.; Antosik, A.; Szelangiewicz, M. Application of Measurement Sensors and Navigation Devices in Experimental Research of the Computer System for the Control of an Unmanned Ship Model. *Sensors* **2021**, *21*, 1312. [CrossRef] [PubMed]
4. Kim, D.; Lee, S.; Lee, J. An Ensemble-Based Approach to Anomaly Detection in Marine Engine Sensor Streams for Efficient Condition Monitoring and Analysis. *Sensors* **2020**, *20*, 7285. [CrossRef] [PubMed]

5. Giernalczyk, M.; Kaminski, P. Assessment of the Propulsion System Operation of the Ships Equipped with the Air Lubrication System. *Sensors* **2021**, *21*, 1357. [CrossRef] [PubMed]
6. Bogdanowicz, A.; Kniaziewicz, T. Marine Diesel Engine Exhaust Emissions Measured in Ship's Dynamic Operating Conditions. *Sensors* **2020**, *20*, 6589. [CrossRef] [PubMed]
7. Kim, D.; Lee, S.; Lee, J. Data-Driven Prediction of Vessel Propulsion Power Using Support Vector Regression with Onboard Measurement and Ocean Data. *Sensors* **2020**, *20*, 1588. [CrossRef] [PubMed]
8. German-Galkin, S.; Tarnapowicz, D. Energy Optimization of the 'Shore to Ship' System—A Universal Power System for Ships at Berth in a Port. *Sensors* **2020**, *20*, 3815. [CrossRef] [PubMed]
9. Chybowski, L.; Nozdrzykowski, K.; Grządziel, Z.; Dorobczyński, L. Evaluation of Model-Based Control of Reaction Forces at the Supports of Large-Size Crankshafts. *Sensors* **2020**, *20*, 2654. [CrossRef] [PubMed]
10. Miądlicki, K.; Jasiewicz, M.; Gołaszewski, M.; Królikowski, M.; Powałka, B. Remanufacturing System with Chatter Suppression for CNC Turning. *Sensors* **2020**, *20*, 5070. [CrossRef] [PubMed]
11. Nozdrzykowski, K.; Adamczak, S.; Grządziel, Z.; Dunaj, P. The Effect of Deflections and Elastic Deformations on Geometrical Deviation and Shape Profile Measurements of Large Crankshafts with Uncontrolled Supports. *Sensors* **2020**, *20*, 5714. [CrossRef] [PubMed]
12. Kyzioł, L.; Panasiuk, K.; Hajdukiewicz, G.; Dudzik, K. Acoustic Emission and K-S Metric Entropy as Methods for Determining Mechanical Properties of Composite Materials. *Sensors* **2020**, *21*, 145. [CrossRef] [PubMed]
13. Kozak, M.; Bejger, A.; Tomczak, A. Identification of Gate Turn-off Thyristor Switching Patterns Using Acoustic Emission Sensors. *Sensors* **2020**, *21*, 70. [CrossRef] [PubMed]
14. Drewing, S.; Witkowski, K. Spectral Analysis of Torsional Vibrations Measured by Optical Sensors, as a Method for Diagnosing Injector Nozzle Coking in Marine Diesel Engines. *Sensors* **2021**, *21*, 775. [CrossRef] [PubMed]
15. Stępień, G.; Tomczak, A.; Loosaar, M.; Ziębka, T. Dimensioning Method of Floating Offshore Objects by Means of Quasi-Similarity Transformation with Reduced Tolerance Errors. *Sensors* **2020**, *20*, 6497. [CrossRef] [PubMed]
16. Pyrchla, K.; Tomczak, A.; Zaniewicz, G.; Pyrchla, J.; Kowalska, P. Analysis of the Dynamic Height Distribution at the Estuary of the Odra River Based on Gravimetric Measurements Acquired with the Use of a Light Survey Boat—A Case Study. *Sensors* **2020**, *20*, 6044. [CrossRef] [PubMed]

Article

Data-Driven Prediction of Vessel Propulsion Power Using Support Vector Regression with Onboard Measurement and Ocean Data

Donghyun Kim [1], Sangbong Lee [2] and Jihwan Lee [3,*]

1 Korea Marine Equipment Research Institute, Busan 49111, Korea; kimdonghyun9942@gmail.com
2 Lab021, Busan 48508, Korea; sblee@lab021.co.kr
3 Division of Systems Management and Engineering, Pukyong National University, Busan 48513, Korea
* Correspondence: jihwan@pknu.ac.kr; Tel.: +82-51-629-6492

Received: 16 February 2020; Accepted: 10 March 2020; Published: 12 March 2020

Abstract: The fluctuation of the oil price and the growing requirement to reduce greenhouse gas emissions have forced ship builders and shipping companies to improve the energy efficiency of the vessels. The accurate prediction of the required propulsion power at various operating condition is essential to evaluate the energy-saving potential of a vessel. Currently, a new ship is expected to use the ISO15016 method in estimating added resistance induced by external environmental factors in power prediction. However, since ISO15016 usually assumes static water conditions, it may result in low accuracy when it is applied to various operating conditions. Moreover, it is time consuming to apply the ISO15016 method because it is computationally expensive and requires many input data. To overcome this limitation, we propose a data-driven approach to predict the propulsion power of a vessel. In this study, support vector regression (SVR) is used to learn from big data obtained from onboard measurement and the National Oceanic and Atmospheric Administration (NOAA) database. As a result, we show that our data-driven approach shows superior performance compared to the ISO15016 method if the big data of the solid line are secured.

Keywords: vessel power prediction; data-driven prediction; support vector regression; ISO15016; onboard measurement data; ocean whether data; predictive analytics

---

1. Introduction

The fluctuation of the oil price and unstable shipping rates have enforced ship builders and shipping companies to improve the energy efficiency of vessels [1]. Since the fuel cost is the largest portion of the operating cost of a vessel, improving energy efficiency can result in huge savings in the total operating cost [2]. Achieving good energy efficiency is also a prerequisite to cope with demanding environmental regulations because more energy-efficient vessels can reduce fuel consumption and greenhouse gas emissions [3]. Therefore, several technological solutions have been proposed to improve the energy efficiency of vessels [4]. Design optimization technology such as 'hull form optimization' [5] or 'propeller configuration' [6] try to design vessels to improve the fuel efficiency. On the other hand, operational optimization technologies aim to improve the operational performance by finding optimal speed and optimized voyage routes of vessels [7,8]. For a more comprehensive list of energy saving technologies, the reader is referred to Tilling et al. [4].

To measure the effectiveness of energy-saving technology, it is necessary to measure the speed/power performance of the vessel [9]. Today, many ship owners leave the responsibility of the delivery trials including speed-power trials with the shipyard [10]. However, the speed/power relationship obtained from a testing environment cannot be generalized to realistic operating conditions and the propulsion performance of the vessel may be affected by several external factors such as wind, tide, wave or hull fouling [11]. Thus, in a dynamic environment, it is difficult to determine exactly whether the increase in energy efficiency is due to newly adopted energy-saving solutions or other factors.

As an internationally recognized standard, ISO15016 is proposed to measure the speed/power relationship from the delivery trials [12]. ISO15016 also introduces several calculation procedures to adjust the impact of external factors on the propulsion performance. Unfortunately, those procedures are computationally expensive because many input parameters are required and their related equations are complex [13]. As an alternative measure, the Energy Efficiency Operation Index (EEOI) [14] was proposed to measure the energy efficiency of the vessel. However, EEOI is just an aggregated index that is obtained by the total amount of fuel usage and cannot be used to predict the relationship between the propulsion performance and other related factors.

To overcome this limitation, this study proposes a data-driven approach to predict the propulsion power by learning data obtained from sensors installed on board the vessel. In this study, support vector regression (SVR) which shows excellent performance in prediction and pattern recognition is used to learn from data. For the illustration, the data obtained from an actual bulk is used to predict its propulsion power at various speeds and dynamic operational conditions. The prediction accuracy was also compared with that of ISO15016. The result shows that if the operational data are obtainable, the proposed model outperforms that of ISO15016.

The remainder of the paper is as follows. Section 2 briefly introduces the traditional methods that have been used to predict vessel propulsion power. Section 3 explains data that are used to train the prediction model. Section 4 explains the model development procedure in detail. Section 5 discusses the prediction result and compare them with other methods. Finally, Section 6 presents the conclusion and future works.

2. Related Works

Predicting the propulsion power of a vessel at various operating conditions is considered a difficult problem because horsepower can be affected by several external factors such as wind, waves, and hull fouling. Those external factors would increase the vessel resistance, the force working against the vessel movement, requiring more propulsion power to maintain a certain vessel speed. Thus, several methodologies have analyzed the added resistance and speed loss to accurately predict the propulsion power.

As previously noted, the conventional method estimates the added resistance of the vessel, and uses this value in prediction of the required propulsion power [15]. The standard model for measuring the total resistance of the vessel is determined by the following equation:

$$R_{total} = R_{friction} + R_{residual} + R_{wind} \tag{1}$$

where $R_{friction}$, $R_{residual}$ and R_{wind} are resistance induced by the friction of the hull, residual resistance, and wind, respectively. $R_{friction}$ can be calculated by the following:

$$R_{friction} = R_{hull} + R_{fouling} + R_{draft} \tag{2}$$

where R_{hull} is the size of the hull's wetted area, $R_{fouling}$ is fouling of the hull and R_{draft} is specific frictional resistance coefficient. The residual resistance $R_{residual}$ is determined by the following:

$$R_{residual} = R_{wave} + R_{eddy} \tag{3}$$

where R_{wave} is the energy loss caused by waves created by the vessel during its propulsion through the water, while R_{eddy} is eddy resistance which refers to the loss caused by flow separation which creates eddies, particularly at the aft end of the ship. In calm weather, R_{wind} is proportional to the square of the ship's speed, and proportional to the cross-sectional area of the ship above the waterline.

Air resistance normally represents about 2% of the total resistance. Finally, the efficient propulsion power (*EHP*) of the vessel at a certain speed is determined by the following:

$$EHP = SPEED \cdot R_{total} \tag{4}$$

where the *SPEED* is a specific vessel speed and R_{total} is total resistance obtained from the above equation.

In ISO15016, several procedures are introduced to correct added resistance [12]. For example, the added resistance due to the effect of short crested irregular wave, R_{aw} can be calculated by the following:

$$R_{aw} = 2\int_0^{2\pi}\int_0^{\infty} \frac{R_{wave}(\omega, \alpha; V_S)}{\zeta_A^2} E(\omega, \alpha) d\omega d\alpha \tag{5}$$

where ζ_A is wave amplitude, ω is circular frequency of regular wave, α is angle between ship heading and incident regular wave, V_S is the vessel speed through the water, and E is directional spectrum. In calculation of R_{aw}, several methods such as the STAWAVE-1, STAWAVE-2, theoretical methods or the Seakeeping model test were proposed in [12]. However, such methods proposed in ISO15016 may not be appropriate to predict the propulsion performance in the actual operating condition because it requires data from the speed trial environment and requires too many input parameters. Even if such big data from operational conditions are available, it would take too much time and cost because the analysis procedure consists of complex equations which are computationally expensive.

Alternatively, Holtrop and Mennen's method [16–18] is also widely used to estimate the resistance and propulsion power requirement during the vessel design phase. It is an empirical method that utilizes data accumulated from many model tests results. The model coefficient and equation are obtained from statistical analysis and regression applied to the data. Although this method requires less computation and numerical analysis, it only provides rough approximation of the resistance and propulsion power which may result in a less accurate result. Moreover, since data is obtained from the model experiments, prediction result cannot be applied to the actual operational condition.

Another stream of works utilizes computational fluid dynamics (CFD) in the analysis of the vessel resistance and the propulsion performance [19–21]. CFD is a branch of fluid mechanics that uses numerical analysis and data structures to analyze and solve problems that involve fluid flows. In the vessel design phase, CFD have been utilized to predict the resistance by simulating the fluid performance of the designed hull form. Although a CFD-based model can accurately predict the resistance, the calculation of the fluid performance is computationally expensive, and usually requires several hours to complete. Thus, it is difficult to use CFD to predict the propulsion power at various operating conditions. To overcome the limitation of the above methodologies, this study proposes using machine-learning technique in the prediction of propulsion power using the data obtained from the actual operating environment.

3. Data Collection

To predict the propulsion power, a 200,000-ton bulk cargo ship was chosen. The general arrangement of our target vessel is shown in Figure 1. Also, the detailed specification is depicted in Table 1. The target ship is dedicated to iron ore transportation and operates only on two regular routes: from South Korea to the US and from South Korea to Australia. Since it operates on a small number of stable routes, the data quality was thought to be reliable. For the propulsion system of the target ship, a fixed-pitch propeller whose pitch angle cannot be changed is adopted.

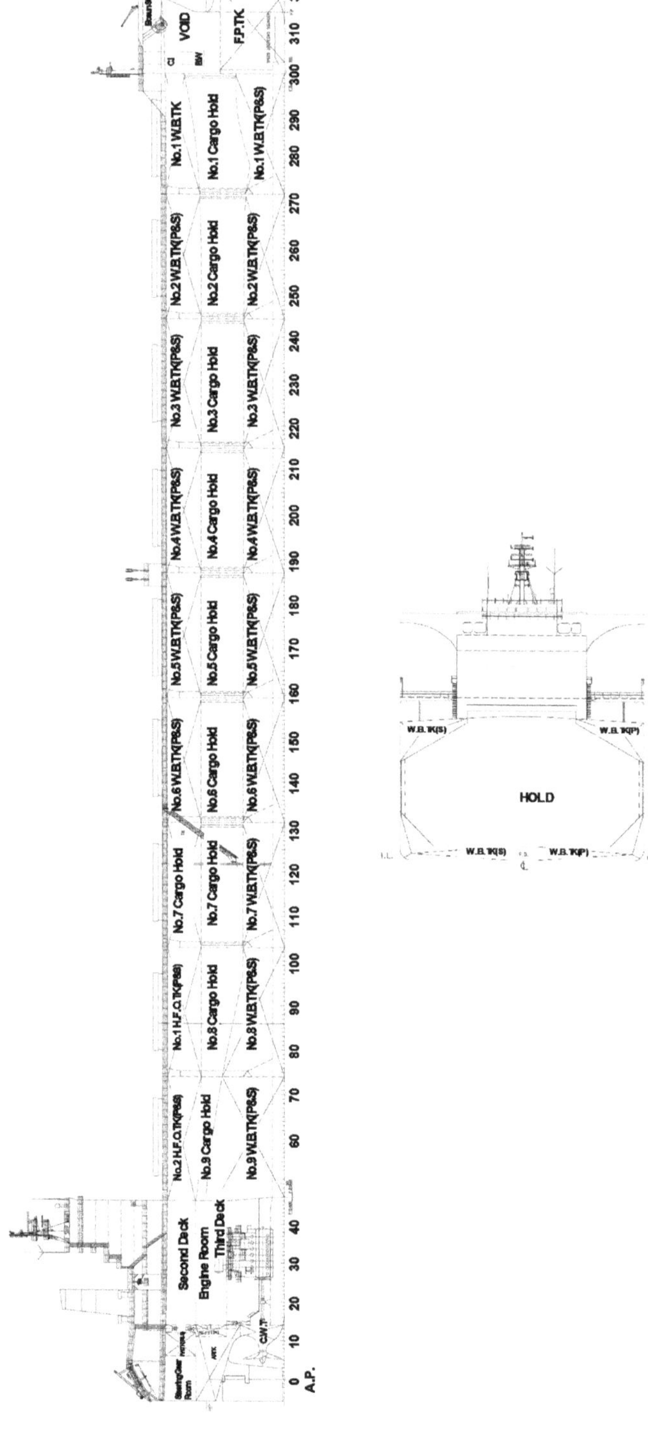

Figure 1. General arrangement of the vessel.

Table 1. Specification of the target vessel.

Specification	
Length Overall	269.36 m
Length betw. perpendiculars	259.00 m
Breadth	43.00 m
Depth	23.80 m
Draught	17.3 m
Deadweight	152.517 metric t

The detailed description of the dataset is shown in Table 2. Most of the data were collected using sensors installed onboard the vessel. The target feature to predict is shaft horsepower which indicates the amount of mechanical power that is delivered by the engine to the propeller shaft. For the input features, velocity, draft, rotation per minute of shaft (RPM), sea depth, tide, wave height, and wind vectors are chosen. Sensor information of each data feature is also illustrated in the same table. The collected data were then processed through the Voyage Data Recorder (VDR) system which digitizes, compress and stores the information in an externally mounted protective unit. Under regulations of the IMO (International Maritime Organization), a cargo ship that weighs more than 20,000 tons should be equipped with VDR. Although the primary purpose of the VDR is to assist marine causality investigation, the recorded data also can be used for performance monitoring of the vessel. The picture of the VDR system installed on the vessel is shown in Figure 2.

Table 2. Data feature description.

Type	Feature Name	Description (unit)	Source	Sensors (Sensing Methods/Protocol)
Input Features	Ship Velocity	Velocity of the ship measured with Differential GPS. (knot)	Onboard Sensors	GPS Sensor (Position, date, time using GPS/NMEA0183)
	Draft	Vertical distance between waterline and the bottom of the hull. Draft is mainly affected by the weight of the cargo on board. (m)	Onboard Sensors	Draft Sensor (using Hydrostatic Level Pressure Transmitters/NMEA0183)
	RPM	Rotation per minute of shaft (RPM)	Onboard Sensors	RPM Indicator
	Sea Depth	Sea depth below the ship. Sea depth is measured with depth log recorded in VDR. (m)	Onboard Sensors	Echo Sounder (sonar wave is used to measure the time interval between emission and return of a pulse/NMEA0183)
	Tide	Tide around ship. Measured with the difference between speed through the water (STW) and speed over the ground (SOG). (m/s)	Onboard Sensors	Doppler log (using ultrasound and applying the Doppler effect to measures the speed of surface ship through water/NMEA0183)
	Wave Height	Height of wave.	NOAA database	Indirectly measured from heave acceleration of the buoys. (m)
	Wind Vector	A vector measure that indicates the speed and direction of the wind around the ship. (m/s)	Onboard Sensors	Anemometer (NMEA0183)
Output Features	Propulsion Power	Shaft horsepower of the ship. (hp)	Onboard Sensors	Shaft Torque Sensor (using strain gauge that converts torque into a change in electrica resistance/MODBUS)

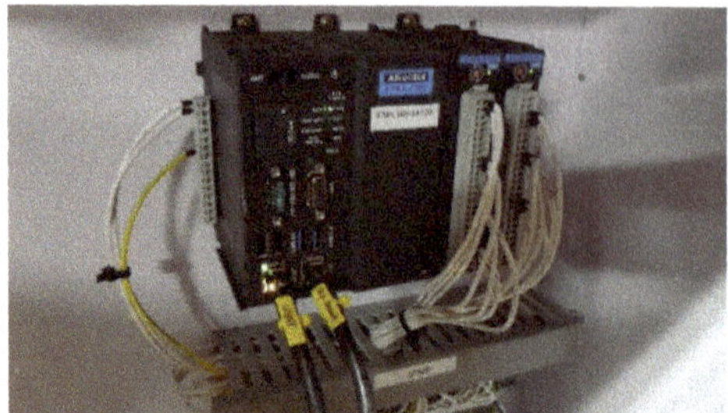

Figure 2. Data collection system in Voyage Data Recorder (VDR).

In processing the raw data obtained from different sensors, many computational steps would be required. For example, the shaft horsepower can be measured with the torque which can be measured with strain. The loading torque T is calculated by the following:

$$T = \frac{\varepsilon \pi G D^3}{8} \tag{6}$$

where ε is shaft strain, G is shaft material shear modulus (pa) and D is shaft diameter (m). To accurately measure the torque, a strain gauge is attached to the propeller shaft of the vessel. The strain gauge measures the change in electrical resistance of the shaft and converts it into the torque. Several commercial system providers [22,23] as well as software vendors [24] are available for shaft power measurement. For our target ship, the SpecsVision-TPM system [25] is used for measuring and processing the shaft power.

In addition to the shaft power measure, the raw data collected from the different sensors were processed by the data acquisition unit (DAU) of the VDR system. DAU transforms each sensor data according to the industrial standard protocol and synthesizes the data from different sensors into a structured data table wherein each datum is recorded at the same time periods. As a result, the synthesized data are transmitted throughout the satellite-based communication system and stored into the cloud database. As a result, each data observation is generated every 10 s, and overall 178,000 observations were collected over seven months (from 2016.01 to 2016.07). The data protocol information used to process the sensor data is also shown in Table 2.

To collect the wave height, which cannot be measured directly on board the vessel, the National Oceanic and Atmospheric Administration (NOAA) database is utilized. NOAA collects spectral wave data using accelerometers on board the buoys which measure the heave acceleration of the buoy. A fast Fourier transform is then applied to obtain the wave height. For more detailed information about wave height measure of the NOAA database, the reader is referred to [26].

4. Development of Propulsion Power Prediction Model

4.1. Support Vector Regression (SVR)

As a machine learning model to predict the propulsion power of the vessel, SVR is applied. SVR, which was introduced by Drucker et al. [27], is an extension of support vector machine (SVM) that considers a regression problem as well. Like SVM, SVR tries to determine the hyperplane that maximizes the margin while ensuring that the error is tolerated.

Suppose that a training data $\{(\mathbf{x}_1, y_1), (\mathbf{x}_2, y_2), \ldots, (\mathbf{x}_n, y_n)\}$ is given, SVR assumes the relationship between $\mathbf{x}$ and y represented by the following:

$$f(\mathbf{x}) = \mathbf{w}^T \varphi(\mathbf{x}) + b \tag{7}$$

where $\varphi(\mathbf{x})$ denotes a kernel function that transforms the data into a higher dimensional space to make it possible to perform the linear separation, w is weight vector associated with vector $\mathbf{x}$ and b is coefficient. To find good estimator, SVR tries to find linear function that is as flat as possible. This can be formulated as a convex optimization problem and formulated as follows:

$$J(\mathbf{w}) = \frac{1}{2}\mathbf{w}^T\mathbf{w} \tag{8}$$

subject to $\forall n : \left|y_n - f(x_n)\right| \leq \varepsilon$.

As shown in the above equation, SVR tries to minimize the norm of w while ensuring all residuals having a value less than ε. However, functions $f(\mathbf{x})$ may not be exist for satisfying the above condition. To deal with such an infeasible constraint, we slightly modify the problem as follows:

$$J(\mathbf{w}) = \frac{1}{2}||\mathbf{w}||^2 + C\sum_{i=1}^{N}(\xi + \xi^*) \tag{9}$$

subject to:

$$\begin{aligned}\forall i &: y_i - J(\mathbf{w}) \leq \varepsilon + \xi_i \\ \forall i &: J(\mathbf{w}) - y_i \leq \varepsilon + \xi_i^* \\ \forall i &: \xi_i \geq 0 \\ \forall i &: \xi_i^* \geq 0.\end{aligned}$$

The slack variables ξ_i and ξ_i^* serves as soft margin that allows the regression error of each data point i to exist up value ξ_i and ξ_i^*. The constraint C controls the overfitting of the model by imposing a penalty on observations that lie outside the margin ε.

The solution of Equation (7) can be optimized by solving a dual problem that is computed as follows:

$$L(a) = \frac{1}{2}\sum_{i=1}^{n}\sum_{j=1}^{n}(a_i - a_i^*)(a_j - a_j^*)K(x_i, x_j) + \varepsilon\sum_{i=1}^{n}(a_i + a_i^*) - \sum_{i=1}^{n}y_i(a_i + a_i^*) \tag{10}$$

subject to:

$$\begin{aligned}\sum_{i=1}^{n}(a_i - a_i^*) &= 0 \\ \forall n &: 0 \leq a_i \leq C \\ \forall n &: 0 \leq a_i \leq C\end{aligned}$$

The function $K(x_i, x_j) = <\varphi(x_i), \varphi(x_j)>$ is called as kernel function where $\varphi(x_i)$ is a transformation that maps x_i to a high-dimensional space. If we assume a linear model, linear kernel function $K(x_i, x_j) = x_i^T x_j$ is utilized. If the non-linear relationship is assumed, radial basis function (RBF) kernel $K(x_i, x_j) = \exp(-||x_i - x_j||^2)$ or sigmoid kernel $K(x_i, x_j) = \tanh(\gamma(x_i^T x_j) + \theta)$ are widely adopted. Now the function that is used to predict new value x_n is equal to Equation (9):

$$f(x) = \sum_{i=1}^{n}(a_n - a_n^*)K(x_n, x) + b \tag{11}$$

The predictive performance of SVR is significantly affected by the hyper-parameters. If we assume RBF as the kernel function, the performance of SVR is affected by hyper-parameters C, ε and σ. Parameters ε and C affect the way of penalizing the training error. The parameter ε controls the amount of error allowed in the model. If the residual is larger than ε, then the training error is penalized by C. Thus, too small C values may result in the overfitting, while too large C values may result in the underfitting of the model. On the other hand, the parameter σ determines the correlation of the kernel function. Thus, an optimal set of hyper-parameters among its possible combinations should be found. For a more detailed description of SVR, the reader is referred to [28].

4.2. Data Preprocessing

Before applying SVR, several preprocessing tasks were performed. In this step, data records that may lower the prediction performance were omitted from the dataset. Firstly, the observations with zero speed were omitted because it means that the vessel does not use propulsion power at all. Moreover, the observations whose vessel speed is less than 6 knots were also omitted because this indicates that the vessel is near a harbor and is likely to undergo frequent course changes. Such a situation may not be adequate for the propulsion power analysis.

Also, some outlier data that has abnormal shaft power value (output feature) is omitted from the dataset. To systematically determine the outlier, Chauvenet's criterion, is utilized [29]. Given N samples of a dataset, an observation is considered as an outlier if $NP(>|z|) < 0.5$ where $P(>|z|)$ is the cumulative probability of the observation is being more than z standard deviation of the mean. An example of outlier detection is shown in Figure 3. As shown in this figure, the outlier that shows a large deviation from the normal shaft power value is omitted from the dataset.

Finally, feature scaling is applied to the dataset. Feature scaling is important for SVR, since it tries to maximize the distance between the separating plane and the support vectors. If one feature has larger magnitudes than others, it will dominate the other features when calculating the distance. In this study, we normalize each feature between the −1 to +1 intervals. All the preprocessing procedure was conducted using Pandas module with a python 3.6 environment.

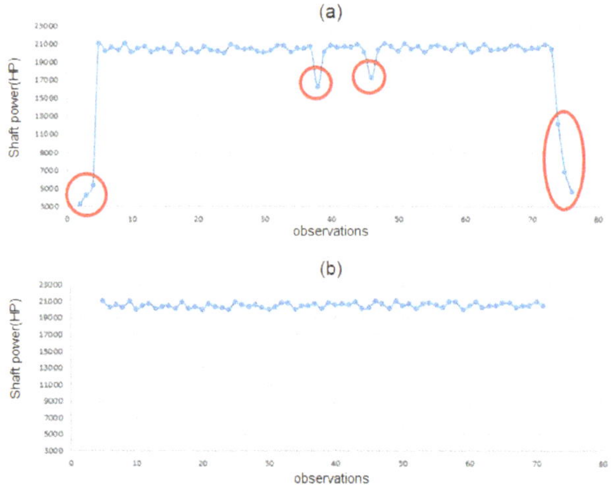

Figure 3. An example of outlier detection using Chauvenent's criterion; 75 observations were randomly chosen from the dataset. (**a**) Before outlier detection. (**b**) After outlier detection.

4.3. Model Learning

The SVR algorithm was applied to the training dataset. In this study, RBF was chosen for the kernel function. The RBF is a widely adopted kernel function because the number of hyper parameters is small compared to that of other models without losing too much prediction performance.

As previously mentioned, finding the optimal parameter is a crucial step for learning the SVR model. For the SVR model with RBF kernel function, three hyper-parameters C, ε and σ are required. In this study, to find the optimal hyper-parameter set, every possible parameter combination was validated by a K-fold cross validation approach. As a result, the optimal hyper parameter for the training dataset was C = 4950, σ= 0.6 and ε= 1.0. The experiment was conducted with Scikit-learn which is a famous open-sourced machine-learning library of python.

5. Result

To validate our model, we divided the data into training and testing datasets. The data collected during first five months was used to train the model while the remaining two months were used to test the model. To measure the predictive performance of the model, RMSE (residual mean squared error) and R2-score were used. The evaluation result is shown in Table 3. The R2 score is 89.78% which indicates a fairly good performance for the regression.

Table 3. Prediction result.

Training dataset	2016.01.01 ~2016.05.31
Testing dataset	2016.06.01~2016.07.30
Kernel Function	RBF kernel
Hyper parameter	C = 4950, r = 0.6, epsilon = 1.0
R2 score	89.78%
RMSE	54 kW
Entry 2	data

In Figure 4, actual propulsion power vs. speed record is compared with the dataset predicted by our SVR model. Each red dot is actual propulsion power in the testing data set given at specific speed of the vessel. As shown in this figure, given the same speed level, the propulsion power shows large deviation. This deviation can be explained by the impact of external factors (wind, wave and tide). The blue dots are the data predicted by the SVR model. As shown in this figure, the SVR model shows almost the same pattern as actual data, which indicates that our SVR is a reliable tool to predict the propulsion power of the vessel.

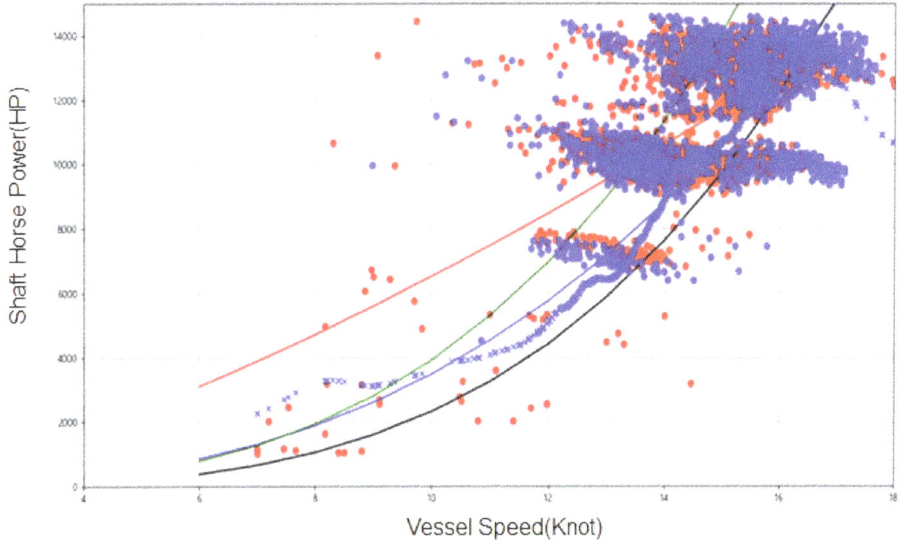

Figure 4. Comparison between predicted propulsion power vs. actual data.

Figure 5 shows the relationship between the speed and the propulsion power of the vessel. The green and black lines are estimated speed vs. power curve obtained from ISO15016 method. The green is obtained when the vessel is loaded with the cargo while the black line is obtained when the vessel is empty. The blue cross dots are speed vs. power relationship predicted by the SVR model while eliminating the impact of external factors. As shown in this figure, the predicted propulsion performance lies between green and black lines, which indicates that SVR model shows generic S/P relationship when it is not affected by the external factors (wind, wave, and tide). The SVR model shows an abnormal pattern when the input is over 18 notes because the data measured with vessel speed being over 18 notes is not available.

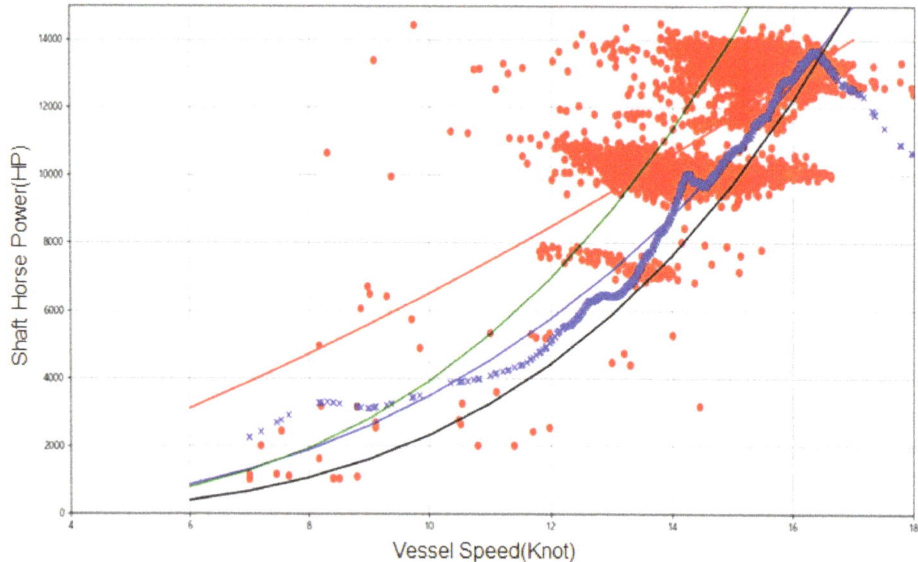

Figure 5. Predicted propulsion power without external effect.

We also have compared the prediction performance of the proposed model with that of an ISO15016 method. In order to do this, we have compared the prediction results of the two methods using the same testing dataset (2016.06.01~2016.08.01). When applying ISO15016 [12], the wind measure is adjusted based on the Fujiwara method (p.41) and the wave feature is adjusted based on the STAWAVE-2 method (p.45). STAWAVE-2 approximates the R_{aw} in Equation (5) by the following:

$$\begin{aligned} R_{wave} &= R_{AWML} + R_{AWRL} \\ R_{AWML} &= 4\rho s g \varsigma_A^2 \frac{B^2}{L_{PP}} \overline{r_{aw}}(\omega) \\ R_{AWRL} &= \frac{1}{2}\rho s g \varsigma_A^2 B \alpha_1(\omega) \end{aligned} \tag{12}$$

where R_{AWML} and R_{AWRL} is mean resistance increase due to wave motion and reflection respectively. $\overline{r_{aw}}(\omega)$ and $\alpha_1(\omega)$ are functions of circular frequency of regular wave ω which are explained in detail in [12]. For a more detailed procedure of STAWAVE-2, the reader is referred to [12].

The comparison result is shown in Table 4. As shown in the table, the SVR-based model outperforms the ISO15016 in both R2 score and RMSE. The ISO15016 method underperforms the SVR-based model especially when the vessel is in bad weather conditions.

Table 4. Comparison with ISO15016 method.

Date	2016.06.01 ~ 2016.07.30
Wind Correction	Fujiwara method
Wave Adjustment	STAWAVE-2
R2 score	30.23%
RMSE	985 kW

6. Conclusions and Future Works

This study proposes a data-driven approach to predict propulsion power. Although several prediction methods are available, most of them are computationally expensive, and suffer from low prediction accuracy. Instead, we propose to use a machine-learning model that utilizes actual operational data of the vessel. In this study, support vector regression (SVR) is used to learn from 178,000 onboard data observations obtained from a bulk carrier that operates on solid lines. Compared to the conventional methods, the proposed model does not require complex equations and showed superior performance if only the big data of the solid line are secured.

There are, however, further issues to explore to improve the model performance. Currently, the data feature related to the vessel status such as the ship damage, roughness of the hull, or engine performance degradation are not addressed. Accommodating such additional features may improve the model performance. Another issue is to compare numerous competing machine-learning algorithms and find the best models.

Author Contributions: Conceptualization, D.K.; Data curation, S.L.; Formal analysis, J.L.; Investigation, J.L.; Methodology, J.L.; Project administration, D.K.; Supervision, J.L.; Validation, J.L.; Visualization, S.L.; Writing—original draft, J.L. All authors have read and agreed to the published version of the manuscript.

Conflicts of Interest: The authors declare no conflict of interest.

References

1. UNCTAD. Review of Maritime Transport 2014. Available online: https://unctad.org/en/PublicationsLibrary/rmt2014_en.pdf (accessed on 3 February 2020).
2. Journee, J.M.; Meijers, J. *Ship Routing for Optimal Performance*; In Transactions IME 1980; Report 0529-P; Delft University of Technology: Delft, The Netherlands, 1980.
3. Chang, Y.T.; Song, Y.; Roh, Y. Assessing Greenhous Gas Emissions from Port Vessel Operations at the Port of Incheon. *Transp. Res. Part D Transp. Environ.* **2013**, *25*, 1–4. [CrossRef]
4. Tilling, F. A generic energy systems model for efficient ship design and operation. *Proc. Inst. Mech. Eng. Part M J. Eng. Marit. Environ.* **2017**, *231*, 649–666.
5. Papanikolaou, A. Holistic ship design optimization. *Comput.-Aided Des.* **2010**, *42*, 1028–1044. [CrossRef]
6. Calcagni, D.; Bernardini, G.; Salvatore, F. Automated marine propeller optimal design combining hydrodynamics models and neural networks. In Proceedings of the 11th International Conference on Computer Applications and Information Technology in the Maritime Industries, Montreal, QC, Canada, 2–5 July 2012.
7. Hellstrom, T. Optimal Pitch, Speed and Fuel Control at Sea. *J. Mar. Sci. Technol.* **2004**, *12*, 71–77.
8. Lin, Y.H.; Fang, M.C.; Yeung, R.W. The optimization of ship weather-routing algorithm based on the composite influence of multi-dynamic elements. *Appl. Ocean Res.* **2013**, *43*, 184–194. [CrossRef]
9. van den Boom, H.J.; Hasselaar, T.W. Ship speed-power performance assessment. In *Transactions—The Society of Naval Architects and Marine Engineers*; Society of Naval Architects and Marine Engineering (SNAME): Alexandria, VA, USA, 2014; p. 122.
10. van den BOOM, H.J.; Huisman, H.; Mennen, F. *New Guidelines for Speed/Power Trials: Level Playing Field Established for IMO EEDI*; SWZ Maritime, Schip en Werf de Zee Foundation, Rotterdam: Wageningen, The Netherlands, 2013; pp. 1–11.
11. Arribas, F.P. Some methods to obtain the added resistance of a ship advancing in waves. *Ocean Eng.* **2007**, *34*, 946–955. [CrossRef]

12. ISO15016. Ships and Marine Technology—Guidelines for the Assessment of Speed and Power Performance by Analysis of Speed Trial Data. Available online: https://www.iso.org/standard/61902.html (accessed on 3 February 2020).
13. Insel, M. Uncertainty in the analysis of speed and powering trials. *Ocean Eng.* **2008**, *35*, 1183–1193. [CrossRef]
14. IMO. Guidelines for Voluntary Use of the Ship Energy Efficiency Operational Indicator (EEOI). Available online: http://www.imo.org/en/OurWork/Environment/PollutionPrevention/AirPollution/Pages/Technical-and-Operational-Measures.aspx (accessed on 3 February 2020).
15. MAN Diesel & Turbo. *Basic Principles of Ship Propulsion*; Man Diesel & Turbo: Copenhagen, Denmark, 2011.
16. Holtrop, J. A statistical re-analysis of resistance and propulsion data. *Int. Shipbuild. Prog.* **1984**, *31*, 272–276.
17. Holtrop, J.; Mennen, G.G.J. A statistical power prediction method. *Int. Shipbuild. Prog.* **1978**, *25*, 290. [CrossRef]
18. Holtrop, J.; Mennen, G.G.J. An approximate power prediction method. *Int. Shipbuild. Prog.* **1982**, *29*, 166–170. [CrossRef]
19. El Moctar, B.; Kaufmann, J.; Ley, J.; Oberhagemann, J.; Shigunov, V.; Zorn, T. Prediction of ship resistance and ship motions using RANSE. In Proceedings of the Gothenburg—A Workshop on Numerical Ship Hydrodynamics, Gothenburg, Sweden, 8–10 December 2010.
20. Simonsen, C.D.; Otzen, J.F.; Joncquez, S.; Stern, F. EFD and CFD for KCS heaving and pitching in regular head waves. *J. Mar. Sci. Technol.* **2013**, *18*, 435–459. [CrossRef]
21. Tezdogan, T.; Demirel, Y.K.; Kellett, P.; Khorasanchi, M.; Incecik, A.; Turan, O. Full-scale unsteady RANS CFD simulations of ship behaviour and performance in head seas due to slow steaming. *Ocean Eng.* **2015**, *97*, 186–206. [CrossRef]
22. Datum electronics. Marine Shaft Power Meter Product Overview. Available online: https://www.datum-electronics.co.uk (accessed on 3 March 2020).
23. Kyma, A.S. Kyma Test Power Meter. Available online: https://kyma.no/wp-content/uploads/2019/03/TPM-Kyma-test-power-meter-brochure.pdf (accessed on 3 March 2020).
24. Keysight Technologies. Data Acquisition Control & Analysis. Available online: https://www.keysight.com/us/en/software/application-sw/benchvue-software/data-acquisition-control---analysis.html (accessed on 3 March 2020).
25. Specs. Specvision. Available online: http://www.specs.co.kr/eng/spmseemspage_specsvision_1.htm. (accessed on 3 March 2020).
26. NOAA (National Oceanic and Atmospheric Administration). How Are Spectral Wave Data Derive from buoy Measurements? Available online: https://www.ndbc.noaa.gov/wave.shtml (accessed on 3 March 2020).
27. Drucker, H.; Burges, C.J.; Kaufman, L.; Smola, A.J.; Vapnik, V. Support vector regression machines. In *Advances in Neural Information Processing Systems*; 10 (NIPS 1997); The MIT Press: Cambridge, CA, USA, 1997; pp. 155–161.
28. Smola, A.J.; Schölkopf, B. A tutorial on support vector regression. *Stat. Comput.* **2004**, *14*, 199–222. [CrossRef]
29. Ross, S.M. Peirce's criterion for the elimination of suspect experimental data. *J. Eng. Technol.* **2013**, *20*, 38–41.

Article

Evaluation of Model-Based Control of Reaction Forces at the Supports of Large-Size Crankshafts

Leszek Chybowski [1,*], Krzysztof Nozdrzykowski [1], Zenon Grządziel [1] and Lech Dorobczyński [2]

[1] Faculty of Marine Engineering, Maritime University of Szczecin, ul. Willowa 2-4, 71-650 Szczecin, Poland; k.nozdrzykowski@am.szczecin.pl (K.N.); z.grzadziel@am.szczecin.pl (Z.G.)

[2] Faculty of Mechatronics and Electrical Engineering, Maritime University of Szczecin, ul. Willowa 2-4, 71-650 Szczecin, Poland; l.dorobczynski@am.szczecin.pl

* Correspondence: l.chybowski@am.szczecin.pl; Tel.: +48-914-809-412

Received: 6 April 2020; Accepted: 3 May 2020; Published: 6 May 2020

Abstract: A support control automation system employing force sensors to a large-size crankshaft main journals' flexible support-system was studied. The current system was intended to evaluate the geometric condition of crankshafts in internal combustion diesel engines. The support reaction forces were changed to minimize the crankshaft elastic deflection as a function of the crank angle. The aim of this research was to verify the hypothesis that the mentioned change can be expressed by a monoharmonic model regardless of a crankshaft structure. The authors' investigations have confirmed this hypothesis. It was also shown that an algorithmic approach improved the mathematical model mapping with the reaction forces due to faster and more accurate calculations of a phase shift angle. The verification of the model for crankshafts with different structural designs made it possible to assess how well the model fits the coefficients of determination that were calculated with the finite element analysis (FEA). For the crankshafts analyzed, the coefficients of determination R^2 were greater than 0.9997, while the maximum relative percentage errors δ_{max} were up to 1.0228%. These values can be considered highly satisfactory for the assessment of the conducted study.

Keywords: geometry measurements; large crankshafts; marine diesel engines; flexible crankshaft supports; forces at supports; force control; mathematical modeling; changing forces; model-based control

---

1. Introduction

The measurement of geometric deviations comprises issues that focus mostly on measuring small-sized components [1–4]. This limitation is due to the use of small structural components in machinery and mechanisms and the availability of comprehensive instrumentation for measurements. It is also arbitrarily assumed [5,6] that the elastic deflections and deformations of such components due to their own weight are negligible and do not affect the results of measurements. Therefore, this paper does not analyze the practical issues of how an object's support affects the object's elastic deformation. In addition, this issue is treated marginally for large components of machinery. This is particularly true for the so-called slender and large-sized components that are of low and variable rigidity with high susceptibility to flexural deformations [7,8]. Important examples of such components are crankshafts of internal combustion diesel engines—the primary units of a main propulsion, ancillary engines and generator sets [9]. More specifically, the crankshafts of piston power machinery used not only in shipbuilding but also in other modes of transport such as railway or automobiles, agriculture, industrial construction and emergency power sources for military facilities and public utilities such as hospitals and offices. These shafts, in addition to their considerable weight and dimensions, have relatively small ratio of cross-section to length. A number of other structural details also make them different from smaller straight shafts used in smaller engines. Another challenge is that these components have different cross-sectional areas along the axis, and the centers of gravity of each section are located at different positions and in different directions relative to the shaft axis.

The known solutions regarding the equipment or systems for the measurement of large-size crankshafts most often involve mounting a shaft on several fixed rigid V-block supports [5,6,10]. For this type of support, it is virtually impossible to obtain reliable measurement results due to initial deflections and geometrical elastic deformations. For these two major reasons, such measurements have elastic deflection errors, whose vectors and magnitude change with the shaft's rigidity varying as it rotates. Considering the impossibility of eliminating the mentioned deflections in a shaft supported by rigid V-blocks, these deflections are tolerated in this research in order to accommodate the spring action of crank webs [11]. Due to these limitations, the results of spring action measurements are currently the basic indicators for evaluating the correct manufacture of crankshafts.

However, deformation values in large crankshafts can be significant [12–15] and jeopardize the results of geometric measurements [6,11,16]. Based on our previous research, we considered large crankshafts to be those where the length-to-diameter ratio (L/d) is greater than 12/15, whereas the shape factor α_k determining the nature of cross-sectional changes may take on significant values $\alpha_k > 1$ (for a straight shaft with a constant diameter, affected by no sudden changes in cross-section $\alpha_k = 1$) [17]. In such crankshafts deformation, under the influence of their dead weight is checked with usage of deflection (springing) measurements as an indirect measure of bearings reactions. To obtain reliable measurements, these deformations must be analyzed and, where possible, eliminated or minimized, which can be achieved by a combination of suitable object support conditions and appropriate tooling that implements the predetermined support conditions.

Proper conditions for measuring a shaft geometry may be ensured by a measuring system fitted with a so-called "flexible shaft support system". A design solution for these flexible supports enables the application of variable and pre-set reaction forces. These are predicted with specialized, finite element analysis (FEA) software to ensure zero deflections at crankshaft main journals being supported. The operation of a flexible support system is assisted by a computer application that monitors the variation of the required reaction forces and works together with the automatic control elements. The computer monitoring application records the variation of forces predicted with FEA software. It uses mathematical solvers described previously by the authors in a study where the monoharmonic model was recommended for the item under testing [18].

According to the experimental results, to accommodate the elastic deflections at the crankshaft, it is necessary to support the main journals during measurements by a set of supports that generate variable reaction forces at the contact of support heads and main journals. To avoid thermal deflections, the temperature of the crankshaft before measurement has to be normalized and must be the same as the ambient temperature. The values of reaction forces should be adjusted to compensate the deflections, both along the shaft axis and at its rotational angle at the supports. A schematic diagram of the main system and essential components of the proposed flexible shaft support system is presented in Figure 1.

The experimental setup is shown in Figure 2. The main system comprises four subsystems: flexible support block, measuring block, turning gear block and the data processing block. The shaft's flexible support block in the current test rig consists of pneumatic supports (4) fitted with V-block heads (5), force sensors (6) and solenoid valves (7). The measuring block consists of a trolley (8), a tripod and a measuring sensor mounted on the trolley. The trolley moves along the shaft in slide bars (11). The turning gear block generates the crankshaft torque during the measurements. This block consists of an electric motor (12) and a belt transmission (13). The last subsystem is the data processing block, which consists of a computer (14) with the software.

The support control algorithm uses a mathematical model that interpolates the values of forces calculated previously with FEA software. The supports are continuously adjusted when the shaft rotates by precision current-controlled valves that operate in feedback with the force sensors measuring the actual force at the contact of support heads and main journals (Figure 3). The whole process is automated using a computer application, developed in-house, to operate the feedback system without any unnecessary delays.

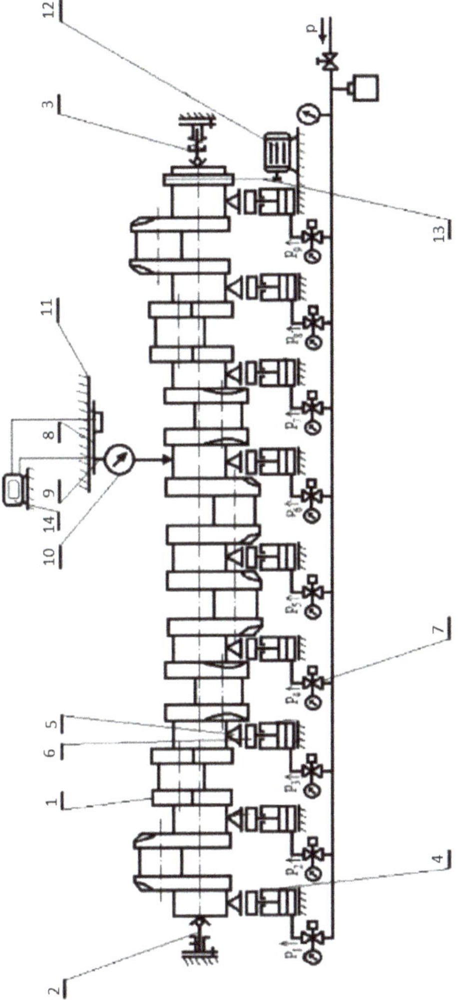

Figure 1. A diagram of the most essential components of the measuring system [6]: (1) crankshaft; (2,3) fixing centers; (4) support with pneumatic actuator; (5) V-block head; (6) force sensor; (7) current-controlled precision valve; (8) trolley; (9) tripod; (10) surface geometry sensor; (11) slide bars; (12) electric motor; (13) belt transmission; (14) PC with software.

This paper is a continuation of the authors' studies on the automation of force-sensor-based flexible support of a shaft by main journal actuators. The base control system was presented in [11], and the possible uses of different models of force variation for a selected crankshaft were analyzed in [18]. The purpose of this article is to prove that the most promising monoharmonic model can be used for crankshafts with different designs. Moreover, the use of an algorithmic approach makes it possible to improve the response time of the reaction forces due to a faster and more accurate determination of a phase shift angle, which is one of the elements of the presented model.

Figure 2. The test rig operating in line with the concept of the measuring system with flexible support of a crankshaft (own source).

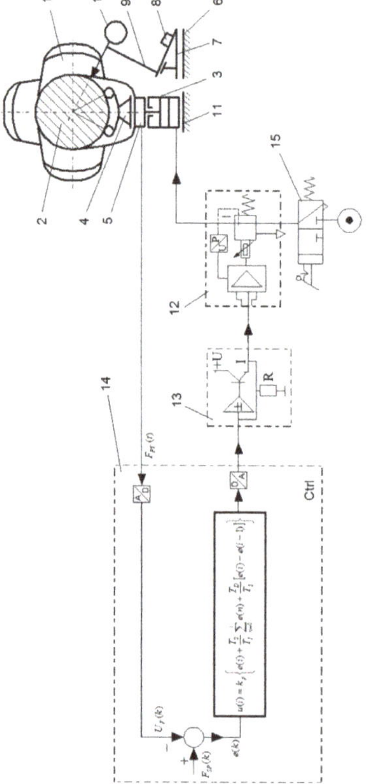

Figure 3. Basic components of the flexible support control unit [11]: (1) crankshaft; (2) shaft's main journal; (3) pneumatic actuator; (4) rolling, articulated, self-adjusting V-block head; (5) force sensor (force transducer); (6) guides; (7) trolley; (8) laser distance meter for measuring the longitudinal coordinate of the measured cross-section; (9) tripod; (10) surface geometry sensor; (11) base; (12) proportional current-controlled reducing valve (controlled proportional regulating valve); (13) current relay; (14) programmable digital controller (control circuit); (15) feed valve; A, analog signal; Ctrl, controller; D, digital signal; $e(k)$, an error signal (an input signal of the PID algorithm); $F_{PV}(t)$, signal of realization force; $F_{SP}(k)$, signal of the set force; I, current signal; *kp(i)*, proportional gain; P, pressure; R, resistance; T_D, differentiation time; T_I, integration time; T_S, sampling time; U, voltage; $U_F(k)$, signal corresponding with $F_{PV}(t)$.

Model verification for crankshafts with different designs will allow analyzing the model's fit and determining the relative percentage errors. In practice, it is assumed that a coefficient of determination (describing how the model fits the given data set) greater than 0.69 renders the fit as substantial [19]. In their earlier papers, the authors assumed that good mapping is provided by the model, for which the coefficient of determination is higher than 0.99 [18].

The authors hypothesize that for a large crankshaft, the change of reaction forces minimizing the elastic deflection of the shaft can be described using a monoharmonic model as a function of the shaft's angular position, regardless of the number and dimensions of the main and crank journals and the shape, dimensions and angular displacements of cranks.

2. Materials and Methods

2.1. Monoharmonic Model

The authors' research to date [18], including the analysis of amplitude spectra of the variation of reaction forces at supports ensuring minimization of elastic deformation of a shaft, has shown that the variation of these forces can be described by the second harmonic. Therefore, the authors suggested using the following monoharmonic model [20]:

$$R(\varphi) = R_0 + C_{R2} \sin(2\varphi + \varphi_{R2}), \tag{1}$$

where C_{R2} is the amplitude of the second harmonic of the reaction force change function; φ_{R2} is the phase shift of the second harmonic of the reaction force change function.

Assuming that the reaction forces calculated with the FEA software at the support of a given journal for successive shaft positions are expressed in the form of a vector:

$$\mathbf{\Phi}_{FEA} = [\varphi_1, \varphi_2, \ldots \varphi_m], \tag{2}$$

$$\mathbf{R}_{FEA} = [R_1, R_2, \ldots R_m], \tag{3}$$

Then, the individual constants of function (1) will be [20]:

$$R_0 = \frac{\sum_{i=1}^{m} R_i}{m}, \tag{4}$$

$$C_{R2} = \frac{\max\limits_{i-1,2,\ldots m} (R_1, R_2, \ldots R_m) - \min\limits_{i-1,2,\ldots m} (R_1, R_2, \ldots R_m)}{2}, \tag{5}$$

$$\varphi_{R2} = \arcsin\left(\frac{R_i - R_0}{C_{R2}}\right) - 2\varphi_i, \tag{6}$$

$$\left(\frac{R_i - R_0}{C_{R2}}\right) \in -1, 1, \tag{7}$$

where ϕ_i is the angle for which the phase shift is determined.

When determining the phase shift, it is important that the variation of the model coincides with the variation of the curve evaluated by the FEA software. For example, both characteristics should be increasing for the shaft position corresponding to the angle φ_i. For the values calculated for the shaft of Buckau Wolf R8VD-136 engine [18], φ_i = 0°CA for even journals and φ_i = 45°CA for odd journals. In general, the analytical determination of the phase shift angle uses the methods given by Mateusz Kowalski [21], modified by the authors.

2.2. The Analytical Method for Determining the Phase Shift

The fit of the model to the FEA calculation results can be described using the quality function of the unfitness $\chi^2(\phi)$ [22], which should be as small as possible (the higher the value of $\chi^2(\phi)$, the worse the fit). This function is given by the formula:

$$\chi^2(\varphi) = \sum\nolimits_{i=1}^{m} \frac{\{R_i - [R_0 + C_{R2}\sin(2\varphi_i - \varphi_{R2}]\}^2}{\sigma_i^2}, \tag{8}$$

The inaccuracy of calculated individual reaction forces is considered constant, hence $\sigma = \sigma_1 = \sigma_2 = \ldots = \sigma_i = \ldots \sigma_m$. Thus, relationship (8) takes the form:

$$\chi^2(\varphi) = \frac{1}{\sigma}\sum\nolimits_{i=1}^{m} \{R_i - [R_0 + C_{R2}\sin(2\varphi_i - \varphi_{R2}]\}^2, \tag{9}$$

The fit of FEA data to a model reaches an extreme for $[\chi^2(\phi)]' = 0$, i.e.,

$$\sum\nolimits_{i=1}^{m} [R_i\cos(2\varphi_i + \varphi_{R2}) - \sin(2\varphi_i + \varphi_{R2})\cos(2\varphi_i + \varphi_{R2})] = 0, \tag{10}$$

Having applied the sum formula for cosine and the sum formula for sine [20]:

$$\cos(2\varphi_i + \varphi_{R2}) = \cos(2\varphi_i)\cos(\varphi_{R2}) - \sin(2\varphi_i)\sin(\varphi_{R2}), \tag{11}$$

$$\begin{aligned} &\sin(2\varphi_i + \varphi_{R2})\cos(2\varphi_i + \varphi_{R2}) = \\ &= \tfrac{1}{2}\cos^2(\varphi_{R2})\sin(4\varphi_i) + \sin(\varphi_{R2})\cos(\varphi_{R2})\cos(4\varphi_i) \\ &-\tfrac{1}{2}sin^2(\varphi_{R2})sin(4\varphi_i), \end{aligned} \tag{12}$$

Relationship (10) takes the form:

$$\begin{aligned} \textstyle\sum_{i=1}^{m}\Big\{&R_i\cos(2\varphi_i)\cos(\varphi_{R2}) - R_i\sin(2\varphi_i)\sin(\varphi_{R2}) - \Big[\tfrac{1}{2}\cos^2(\varphi_{R2})\sin(4\varphi_i) + \\ &\sin(\varphi_{R2})\cos(\varphi_{R2})\cos(4\varphi_i) - \tfrac{1}{2}sin^2(\varphi_{R2})sin(4\varphi_i)\Big]\Big\} = 0, \end{aligned} \tag{13}$$

Relationship (13) can be expressed as the sum of sums [21]:

$$\begin{aligned} &\cos(\varphi_{R2})\textstyle\sum_{i=1}^{m} R_i\cos(2\varphi_i) - \sin(\varphi_{R2})\sum_{i=1}^{m} R_i\sin(2\varphi_i) - \tfrac{1}{2}\cos^2(\varphi_{R2})\sum_{i=1}^{m}\sin(4\varphi_i) - \\ &\sin(\varphi_{R2})\cos(\varphi_{R2})\textstyle\sum_{i=1}^{m}\cos(4\varphi_i) + \tfrac{1}{2}\sin^2(\varphi_{R2})\sum_{i=1}^{m}\sin(4\varphi_i) = 0, \end{aligned} \tag{14}$$

After substituting:

$$A = \sum\nolimits_{i=1}^{m} R_i\cos(2\varphi_i), \tag{15}$$

$$B = \sum\nolimits_{i=1}^{m} R_i\sin(2\varphi_i), \tag{16}$$

$$C = \sum\nolimits_{i=1}^{m} \sin(4\varphi_i), \tag{17}$$

$$D = \sum\nolimits_{i=1}^{m} \cos(4\varphi_i), \tag{18}$$

relationship (14) takes the form [21]:

$$A\cos(\varphi_{R2}) - B\sin(\varphi_{R2}) - \frac{1}{2}C\cos^2(\varphi_{R2}) - D\sin(\varphi_{R2})\cos(\varphi_{R2}) + \frac{1}{2}\sin^2(\varphi_{R2}) = 0, \tag{19}$$

Using the identity $\sin^2(\varphi_{R2}) + \cos^2(\varphi_{R2}) = 1$ and raising both sides of Equation (19) to the second power, we get [21]:

$$\begin{gathered}\sin^4(\varphi_{R2})\left(C^2 + D^2\right) + \sin^3(\varphi_{R2})(-2BC - 2AD) + \sin^2(\varphi_{R2})\left(B^2 - C^2 + A^2 - D^2\right) + \\ \sin(\varphi_{R2})(BC + 2AD) + \left(\tfrac{C^2}{4} - A^2\right) = 0,\end{gathered} \tag{20}$$

The next stage of calculation is to solve Equation (20). This can be done by the Ferro method [23] or numerically, using the roots function in MATLAB 2019b (MathWorks, Natick, MA, USA) or Octave 5.1.0 (John W. Eaton et al., GNU General Public License—GPL) [24]. Then, values $w = \sin(\varphi_{R2})$ are determined, with which, for real roots, the values of potential candidates must be determined for the correct phase shift angle:

$$\varphi_{R2} = \arcsin(w), \tag{21}$$

Finally, it is necessary to verify which extremes are minimums of the unfitness function, which requires calculating expression (8) and incorporating the value for which the unfitness function is the smallest into the model.

2.3. The Recursive Method for Determining the Phase Shift

An alternative to the analytical solution is determining the phase shift angle using a recursive algorithm. This solution simplifies the search for the constants of a model and decreases the evaluation time. The authors used their experience to make an educated guess, assuming that it would be sufficient to determine the crank angle to an accuracy of 1° (CA). The proposed calculation algorithm for a single main journal is presented in Figure 4.

The algorithm was implemented as code for MATLAB 2019b (MathWorks, Natick, MA, USA) and is presented in Appendix A.

To evaluate the usability of the proposed monoharmonic model and the algorithm presented in the paper, the model fit results and FEA calculations were analyzed and compared. The FEA calculations were done using the Midas NFX 2019 R1 (MSC Software Corporation, Newport Beach, CA, USA).

At main journal supports of shafts, the active forces are analyzed, and reaction forces are generated to minimize the elastic deflection of the shaft for individual angular positions of the shaft. These forces are compiled in Appendix B, in Tables A1–A12. The 3D models of analyzed crankshafts are presented in Appendix C, in Figures A1–A12. The graphical presentation of FEA results for the selected crankshaft designs is provided in the paper. The analysis assumes that the initial position of 0°CA corresponds to the top dead center (TDC) of the first journal on the flywheel side (the assumption was made according to the specification of the Buckau Wolf R8VD-136 engine). The direction of shaft rotation was assumed to be clockwise when viewed from the free end of the shaft.

The authors proposed a monoharmonic model with coefficients calculated in a recursive algorithm, which was implemented in the form of a code in MATLAB 2019b (MathWorks, Natick, MA, USA).

In order to consider different design options and assess their impact on the extent of applicability of the monoharmonic model proposed, the authors analyzed eight crankshafts with 10 main journals and three crankshafts with three main journals. To ensure the traceability, shafts were labeled in the following manner to provide information on their selected dimensions: S-MMMMM-XXXX-YY-ZZ-AAA-BBB-D-EE. Here, individual terms are defined as follows: S, shaft; MMMMM, shaft weight in newton (N); XXXX, shaft length in millimeters (mm); YY, number of main journals; ZZ, number of crank journals; AAA, main journal diameter in millimeters (mm); BBB, crank journal diameter in millimeters (mm); D, shape of crank webs (O—oval, C—circular, Z—figure formed by two circles connected by tangents); EE, specific version including crank dimensions, their relative angular offsets and the total shaft weight.

The coefficients of determination R^2 and maximum relative percentage error δ_{max} were determined for each main journal to assess the accuracy of the monoharmonic model's fit for the reaction forces. Finally, the results were obtained, and conclusions were drawn for the suitability of the proposed models.

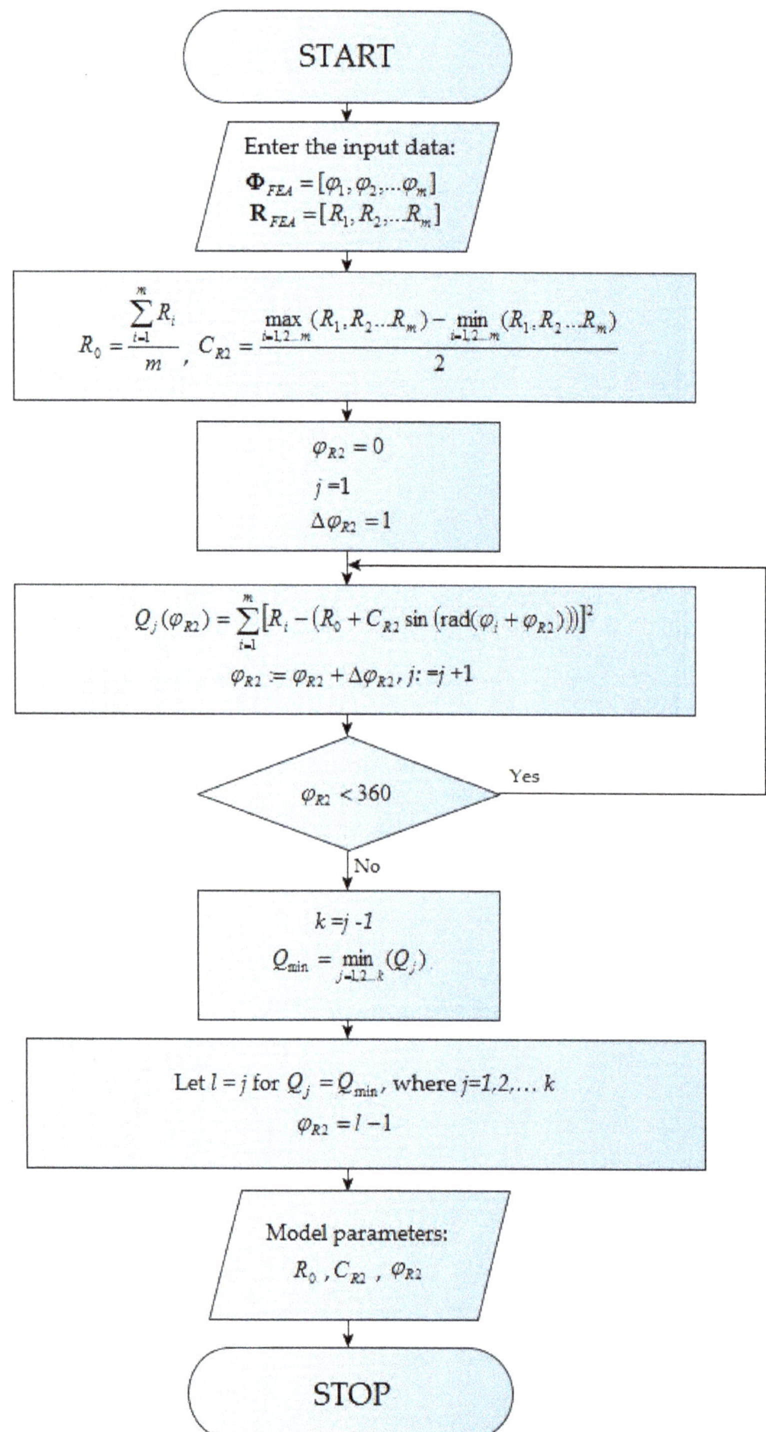

Figure 4. Recursive algorithm for determination of model parameters.

3. Results and Discussion

3.1. Crankshaft S-9285-3600-10-8-149-144-O-01

The first crankshaft analyzed was the shaft of the Buckau Wolf R8DV-136 (manufacturer: VEB SKL—Magdeburg, number of cylinders: 8, nominal effective power: 220 kW, nominal speed: 360 rpm, nominal specific fuel oil consumption: 238 g/kWh, cylinder bore: 240 mm, piston stroke: 360 mm). This was done to validate the accuracy of the model. The basic parameters of the shaft are as follows: weight of 9285 N, length of 3600 mm, ten main journals 149 mm in diameter, eight crank journals 144 mm in diameter, journal length of 100 mm, oval crank webs measuring 252 mm × 358 mm. Reaction force monoharmonic model coefficients that minimize the elastic deflection are listed in Table 1.

Table 1. Calculated coefficients of the monoharmonic model of reaction forces at main journals of crankshaft S-9285-3600-10-8-149-144-O-01.

Journal	1	2	3	4	5	6	7	8	9	10
R_0 (N)	780.17	881.40	997.47	1032.26	984.30	953.88	985.50	1018.80	1070.50	580.07
C_{R2} (N)	52.69	124.04	174.53	220.96	188.99	169.82	188.28	212.82	136.54	34.52
ϕ_{R2} (deg)	246	60	224	37	224	55	224	35	217	42

The quality of the model's fit to the FEA values and relative percentage errors of the reaction forces are presented in Table 2.

Table 2. Model's fit and maximum relative errors for main journals of the crankshaft S-9285-3600-10-8-149-144-O-01.

Journal	1	2	3	4	5	6	7	8	9	10
R^2 (-)	1.0000	1.0000	0.9999	1.0000	1.0000	1.0000	1.0000	1.0000	1.0000	1.0000
δ (%)	0.0579	0.0735	0.7211	0.2029	0.7788	0.1468	0.1513	0.1513	0.1118	0.1382

The smallest coefficient of determination, equal to 0.9999, corresponds to the model of reaction forces at the support of the main journal No. 3, while the largest relative percentage error of the reaction forces, equal to 0.7788%, corresponds to the model of reaction forces at the support of the main journal No. 5.

3.2. Crankshaft S-8658-9285-3600-10-8-149-114-O-02

Subsequently, the shaft that was redesigned by modifying the journal diameter was analyzed. The basic parameters of the shaft are as follows: weight of 8658 N, length of 3600 mm, ten main journals 149 mm in diameter, eight crank journals 114 mm in diameter, journal length of 100 mm, oval crank webs measuring 252 mm × 358 mm. Reaction force monoharmonic model coefficients that minimize the elastic deflection are listed in Table 3.

Table 3. Calculated coefficients of the monoharmonic model of reaction forces at main journals of the crankshaft S-8658-3600-10-8-149-114-O-02.

Journal	1	2	3	4	5	6	7	8	9	10
R_0 (N)	782.93	816.13	935.03	951.08	913.18	870.04	915.33	942.25	975.95	556.39
C_{R2} (N)	71.1	153.5	200.25	248.9	214.25	191.25	213.65	241.35	159.05	42.9
ϕ_{R2} (deg)	243	57	223	38	225	53	225	37	217	40

The quality of the model's fit to the FEA values and relative percentage errors of reaction forces are presented in Table 4.

The smallest coefficient of determination, equal to 0.9999, corresponds to the model of reaction forces at the support of the main journal No. 1, while the largest relative percentage error of the

reaction forces, equal to 1.0481%, corresponds to the model of reaction forces at the support of the main journal No. 5.

Table 4. Model's fit and maximum relative errors for main journals of the crankshaft S-8658-3600-10-8-149-114-O-02.

Journal	1	2	3	4	5	6	7	8	9	10
R^2 (-)	0.9999	1.0000	1.0000	1.0000	1.0000	1.0000	1.0000	1.0000	1.0000	1.0000
δ (%)	0.0759	0.0726	0.7760	0.3426	1.0481	0.2239	1.0351	0.2567	0.1566	0.1253

3.3. Crankshaft S-16942-3600-10-8-149-144-C-03

Next, the shaft that was redesigned by modifying the crank shape was analyzed. The basic parameters of the shaft are as follows: weight of 16,942 N, length of 3600 mm, ten main journals with a diameter of 149 mm, eight crank journals with a diameter of 144 mm, journal length of 100 mm, circular crank webs 450 mm in diameter. Reaction force monoharmonic model coefficients that minimize the elastic deflection are listed in Table 5.

Table 5. Calculated coefficients of the monoharmonic model of reaction forces at main journals of the crankshaft S-16942-3600-10-8-149-144-C-03.

Journal	1	2	3	4	5	6	7	8	9	10
R_0 (N)	636.94	1459.39	1998.68	2058.99	1874.93	1904.53	1875.13	2040.33	2144.96	947.85
C_{R2} (N)	97.65	238.70	346.95	446.75	383.85	333.40	382.85	433.55	274.05	70.95
ϕ_{R2} (deg)	247	59	222	35	223	54	223	33	212	34

The quality of the model's fit to the FEA values and relative percentage errors of reaction forces are presented in Table 6.

Table 6. Model's fit and maximum relative errors for main journals of the crankshaft S-16942-3600-10-8-149-144-C-03.

Journal	1	2	3	4	5	6	7	8	9	10
R^2 (-)	1.0000	1.0000	1.0000	0.9999	1.0000	1.0000	1.0000	1.0000	1.0000	1.0000
δ (%)	0.1748	0.1053	0.4610	0.1922	0.6581	0.1469	0.6564	0.1089	0.0661	0.0413

The smallest coefficient of determination, equal to 0.9999, corresponds to the model of reaction forces at the support of the main journal No. 4, while the largest relative percentage error of the reaction forces, equal to 0.6581%, corresponds to the model of reaction forces at the support of the main journal No. 5.

3.4. Crankshaft S-12075-3600-10-8-149-144-C-04

Subsequently, the analysis was done for a shaft with a design as in the case of the shaft S-12075-3600-10-8-149-144-C-03, but with a reduced diameter of crank webs. The basic parameters of the shaft are as follows: weight of 12,075 N, length of 3600 mm, ten main journals with a diameter of 149 mm, eight crank journals with a diameter of 144 mm, journal length of 100 mm, circular crank webs 358 mm in diameter. Reaction force monoharmonic model coefficients that minimize the elastic deflection are listed in Table 7.

The quality of the model's fit to the FEA values and relative percentage errors of reaction forces are presented in Table 8.

The smallest coefficient of determination, equal to 0.9999, corresponds to models of reaction forces at the support of the main journals Nos. 1, 4, 7 and 10, while the largest relative percentage error of the

reaction forces, equal to 0.5482%, corresponds to the model of reaction forces at the support of the main journal No. 5.

Table 7. Calculated coefficients of the monoharmonic model of reaction forces at main journals of the crankshaft S-12075-3600-10-8-149-144-C-04.

Journal	1	2	3	4	5	6	7	8	9	10
R_0 (N)	725.58	1110.52	1294.29	1445.64	1307.01	1334.94	1255.87	1435.69	1443.20	722.07
C_{R2} (N)	68.80	166.36	264.50	330.56	276.49	238.18	283.67	321.66	204.46	53.14
ϕ_{R2} (deg)	239	51	218	33	221	50	219	31	210	32

Table 8. Model's fit and maximum relative errors for main journals of the crankshaft S-12075-3600-10-8-149-144-C-04.

Journal	1	2	3	4	5	6	7	8	9	10
R^2 (-)	0.9999	1.0000	1.0000	0.9999	1.0000	1.0000	0.9999	1.0000	1.0000	0.9999
δ (%)	0.0706	0.2168	0.2763	0.2052	0.5482	0.3299	0.4426	0.0542	0.0487	0.0588

3.5. Crankshaft S-9283-3600-10-8-149-144-O-05

The redesigned shaft with relative angles between cranks modified by 90° was then analyzed. The basic parameters of the shaft are as follows: weight of 9283 N, length of 3600 mm, ten main journals of 149 mm diameter, eight crank journals 144 mm in diameter, journal length of 100 mm, oval crank webs measuring 252 mm × 358 mm, offset by the angle of 90° relative to each other. Reaction force monoharmonic model coefficients that minimize the elastic deflection are listed in Table 9.

Table 9. Calculated coefficients of the monoharmonic model of reaction forces at main journals of the crankshaft S-9283-3600-10-8-149-144-O-05.

Journal	1	2	3	4	5	6	7	8	9	10
R_0 (N)	747.64	918.19	990.79	1034.98	986.30	947.27	987.34	1032.79	1035.96	601.33
C_{R2} (N)	51.05	113.85	145.53	171.88	133.92	90.08	136.82	168.35	106.89	29.47
ϕ_{R2} (deg)	248	58	210	1	163	0	198	3	166	332

The quality of the model's fit to the FEA values and relative percentage errors of reaction forces are presented in Table 10.

Table 10. Model's fit and maximum relative errors for main journals of the crankshaft S-9283-3600-10-8-149-144-O-05.

Journal	1	2	3	4	5	6	7	8	9	10
R^2 (-)	1.0000	1.0000	0.9999	1.0000	1.0000	0.9999	1.0000	0.9999	1.0000	1.0000
δ (%)	0.0713	0.0086	0.1208	0.0046	0.4026	0.0704	0.3515	0.1262	0.3415	0.0114

The smallest coefficient of determination, equal to 0.9999, corresponds to models of reaction forces at the support of the main journals Nos. 3, 6 and 8, while the largest relative percentage error of the reaction forces, equal to 0.4026%, corresponds to the model of reaction forces at the support of the main journal No. 5.

3.6. Crankshaft S-9283-3600-10-8-149-144-O-06

Subsequently, the analysis was carried out for a shaft redesigned by modifying the relative angles between cranks with a 180° offset in succession. The basic parameters of the shaft are as follows: weight of 9283 N, length of 3600 mm, ten main journals 149 mm in diameter, eight crank journals 144 mm in diameter, journal length of 100 mm, oval crank webs measuring 252 mm × 358 mm, offset by the angle of 180° relative to each other. Reaction force monoharmonic model coefficients that minimize the elastic deflection are listed in Table 11.

Table 11. Calculated coefficients of the monoharmonic model of reaction forces at main journals of the crankshaft S-9283-3600-10-8-149-144-O-06.

Journal	1	2	3	4	5	6	7	8	9	10
R_0 (N)	892.40	751.37	1088.55	882.24	1097.81	878.18	1101.47	872.23	1146.03	572.31
C_{R2} (N)	40.39	90.09	101.45	104.52	103.24	101.03	100.60	91.53	59.37	17.87
ϕ_{R2} (deg)	269	89	268	89	270	88	267	90	273	93

The quality of the model's fit to the FEA values and relative percentage errors of reaction forces are presented in Table 12.

Table 12. Model's fit and maximum relative errors for main journals of the crankshaft S-9283-3600-10-8-149-144-O-06.

Journal	1	2	3	4	5	6	7	8	9	10
R^2 (-)	1.0000	0.9999	1.0000	1.0000	1.0000	1.0000	1.0000	1.0000	1.0000	1.0000
δ (%)	0.0178	0.0887	0.0654	0.0795	0.0246	0.0578	0.0575	0.0189	0.0312	0.0123

The smallest coefficient of determination, equal to 0.9999, corresponds to the model of reaction forces at the support of the main journal No. 2, while the largest relative percentage error of the reaction forces, equal to 0.0887%, corresponds to the model of reaction forces at the support of the main journal No. 2.

3.7. *Crankshaft S-9283-3600-10-8-149-144-O-07*

Subsequently, the analysis was carried out for a shaft redesigned by modifying the relative angles between cranks with a 120° offset in succession. The basic parameters of the shaft are as follows: weight of 9283 N, length of 3600 mm, ten main journals 149 mm in diameter, eight crank journals 144 mm in diameter, journal length of 100 mm, oval crank webs measuring 252 mm × 358 mm, offset by the angle of 120° relative to each other. Reaction force monoharmonic model coefficients that minimize the elastic deflection are listed in Table 13.

Table 13. Calculated coefficients of the monoharmonic model of reaction forces for main journals of the crankshaft S-9283-3600-10-8-149-144-O-07.

Journal	1	2	3	4	5	6	7	8	9	10
R_0 (N)	752.76	895.23	1032.15	1024.88	916.93	1025.31	1023.51	904.80	1129.57	577.39
C_{R2} (N)	47.79	106.74	125.58	96.40	29.89	96.19	95.13	37.92	66.79	30.90
ϕ_{R2} (deg)	257	73	245	50	146	249	51	155	271	83

The quality of the model's fit to the FEA values and relative percentage errors of reaction forces are presented in Table 14.

Table 14. Model's fit and maximum relative errors for main journals of the crankshaft S-9283-3600-10-8-149-144-O-07.

Journal	1	2	3	4	5	6	7	8	9	10
R^2 (-)	1.0000	0.9999	1.0000	1.0000	0.9999	1.0000	1.0000	1.0000	0.9999	1.0000
δ (%)	0.1934	0.3983	0.0982	0.1577	0.0280	0.1276	0.1261	0.0226	0.0478	0.0421

The smallest coefficient of determination, equal to 0.9999, corresponds to models of reaction forces at the support of the main journals Nos. 2, 5 and 9, while the largest relative percentage error of the

reaction forces, equal to 0.3983%, corresponds to the model of reaction forces at the support of the main journal No. 2.

3.8. Crankshaft S-8479-3600-10-8-149-144-O-08

Next, an analysis was carried out for the shaft with a design modified by increasing the so-called drive ratio of main and crank journals, which was done by reducing the distance between the axis of the main and crank journals. The basic parameters of the shaft are as follows: weight of 8479 N, length of 3600 mm, ten main journals 149 mm in diameter, eight crank journals 144 mm in diameter, journal length of 100 mm, oval crank webs measuring 252 mm × 358 mm, reduced distance between main journal axis and crank journal axis. Reaction force monoharmonic model coefficients that minimize the elastic deflection are listed in Table 15.

Table 15. Calculated coefficients of the monoharmonic model of reaction forces for main journals of the crankshaft S-8479-3600-10-8-149-144-O-08.

Journal	1	2	3	4	5	6	7	8	9	10
R_0 (N)	779.92	789.67	912.92	916.33	895.67	863.50	896.17	904.08	972.00	549.00
C_{R2} (N)	36.00	80.50	113.00	143.50	120.00	105.50	119.50	137.00	89.00	23.00
ϕ_{R2} (deg)	246	58	222	37	225	55	225	35	213	35

The quality of the model's fit to the FEA values and relative percentage errors of reaction forces are presented in Table 16.

Table 16. Model's fit and maximum relative errors for main journals of the crankshaft S-8479-3600-10-8-149-144-O-08.

Journal	1	2	3	4	5	6	7	8	9	10
R^2 (-)	0.9999	1.0000	1.0000	1.0000	1.0000	0.9999	0.9999	1.0000	1.0000	0.9999
δ (%)	0.1056	0.0821	0.3048	0.1585	0.4840	0.1391	0.4811	0.1087	0.0715	0.0519

The smallest coefficient of determination, equal to 0.9999, corresponds to models of reaction forces at the support of the main journals Nos. 1, 6, 7 and 10, while the largest relative percentage error of the reaction forces, equal to 0.4840%, corresponds to the model of reaction forces at the support of the main journal No. 5.

3.9. Crankshaft S-7051-3600-10-8-149-144-Z-09

Next, the shaft was redesigned in relation to the shaft S-8658-3600-10-8-149-144-C-03 to form a complex shape consisting of two circles connected by two tangents. The basic parameters of the shaft are as follows: weight of 7051 N, length of 3600 mm, ten main journals 149 mm in diameter, eight crank journals 144 mm in diameter, journal length of 100 mm, crank webs with a complex shape of maximum dimensions 168 mm × 331 mm. Reaction force monoharmonic model coefficients that minimize the elastic deflection are listed in Table 17.

Table 17. Calculated coefficients of the monoharmonic model of reaction forces for main journals of the crankshaft S-7051-3600-10-8-149-114-Z-09.

Journal	1	2	3	4	5	6	7	8	9	10
R_0 (N)	817.63	670.72	723.53	742.93	715.18	683.13	715.45	736.93	755.70	490.03
C_{R2} (N)	58.95	125.40	156.90	190.60	164.35	147.35	163.95	183.95	121.20	32.75
ϕ_{R2} (deg)	242	56	224	39	225	53	225	38	218	41

The quality of the model's fit to the FEA values and relative percentage errors of reaction forces are presented in Table 18.

Table 18. Model's fit and maximum relative errors for main journals of the crankshaft S-7051-3600-10-8-149-114-Z-09.

Journal	1	2	3	4	5	6	7	8	9	10
R^2 (-)	1.0000	0.9999	1.0000	1.0000	1.0000	1.0000	1.0000	1.0000	1.0000	1.0000
δ (%)	0.0477	0.1537	0.8284	0.4294	1.0228	0.2191	1.0130	0.3207	0.1859	0.1279

The smallest coefficient of determination, equal to 0.9999, corresponds to the model of reaction forces at the support of the main journal No. 2, while the largest relative percentage error of the reaction forces, equal to 1.0228%, corresponds to the model of reaction forces at the support of the main journal No. 5.

3.10. Crankshaft S-1977-0740-3-2-149-144-O-10

Subsequently, the analysis was carried out on a shaft with double crank and with crank webs that were offset by 180°. The basic parameters of the shaft are as follows: weight of 1977 N, length of 740 mm, three main journals with a diameter of 149 mm, outermost journal length of 50 mm, other journals' length of 100 mm, two crank journals with a diameter of 144 mm, oval crank webs measuring 252 mm × 358 mm located on the opposite sides of one plane. Reaction force monoharmonic model coefficients that minimize the elastic deflection are listed in Table 19.

Table 19. Calculated coefficients of the monoharmonic model of reaction forces for main journals of the crankshaft S-1977-0740-3-2-149-144-O-10.

Journal	1	2	3
R_0 (N)	388.82	1199.63	388.88
C_{R2} (N)	19.55	39.10	19.55
φ_{R2} (deg)	271	91	271

The quality of the model's fit to the FEA values and relative percentage errors of reaction forces are presented in Table 20.

Table 20. Model's fit and maximum relative errors for main journals of the crankshaft S-1977-0740-3-2-149-144-O-10.

Journal	1	2	3
R^2 (-)	1.0000	1.0000	1.0000
δ (%)	0.0159	0.0096	0.0149

The coefficients of determination equal 1.0000 for all journals, being accurate to 4 decimal places, while the largest relative percentage error of the reaction forces equals 0.0159%, which corresponds to the model of reaction forces at the support of the main journal No. 1.

3.11. Crankshaft S-1977-0740-3-2-149-144-O-11

Next, the analysis was carried out on a shaft with a double crank and crank journal positions allocated at the same axis. The basic parameters of the shaft are as follows: weight of 1977 N, length of 740 mm, three main journals with a diameter of 149 mm, outermost journal length of 50 mm, other journals' length of 100 mm, two crank journals with a diameter of 144 mm, oval crank webs measuring 252 mm × 358 mm located in one plane, on the same side. Reaction force monoharmonic model coefficients that minimize the elastic deflection are listed in Table 21.

Table 21. Calculated coefficients of the monoharmonic model of reaction forces for main journals of the crankshaft S-1977-0740-3-2-149-144-O-11.

Journal	1	2	3
R_0 (N)	401.47	1173.83	401.48
C_{R2} (N)	36.27	72.53	36.27
ϕ_{R2} (deg)	90	270	90

The quality of the model's fit to the FEA values and relative percentage errors of reaction forces are presented in Table 22.

Table 22. Model's fit and maximum relative errors for main journals of the crankshaft S-1977-0740-3-2-149-144-O-11.

Journal	1	2	3
R^2 (-)	0.9997	0.9997	0.9997
δ (%)	0.2077	0.1416	0.2077

The coefficients of determination equal 0.9997 for all journals, while the largest relative percentage error of the reaction forces equals 0.2077%, which corresponds to models of reaction forces at the support of main journals Nos. 1 and 3.

3.12. Crankshaft S-1977-0740-3-2-149-144-O-12

The last analysis was carried out on a shaft with a double crank and crank webs offset by 90°. The basic parameters of the shaft are as follows: weight of 1977 N, length of 740 mm, three main journals with a diameter of 149 mm, outermost journal length of 50 mm, other journals' length of 100 mm, two crank journals with a diameter of 144 mm, oval crank webs sized 252 mm × 358 mm, mutually perpendicular. Reaction force monoharmonic model coefficients that minimize the elastic deflection are listed in Table 23.

Table 23. Calculated coefficients of the monoharmonic model of reaction forces for main journals of the crankshaft S-1977-0740-3-2-149-144-O-12.

Journal	1	2	3
R_0 (N)	420.35	1136.60	420.40
C_{R2} (N)	42.76	85.50	42.78
ϕ_{R2} (degree)	133	313	133

The quality of the model's fit to the FEA values and relative percentage errors of reaction forces are presented in Table 24.

Table 24. Model's fit and maximum relative errors for main journals of the crankshaft S-1977-0740-3-2-149-144-O-12.

Journal	1	2	3
R^2 (-)	1.0000	1.0000	1.0000
δ (%)	0.3099	0.2700	0.3143

The coefficients of determination equal 1.0000 for all journals, being accurate to 4 decimal places, similar to the previous case, while the largest relative percentage error of the reaction forces equals 0.3143%, which corresponds to the model of reaction forces at the support of the main journal No. 3.

3.13. Comparative Analysis

The results obtained for all 12 crankshafts under analysis are compared in Table 25; for each model, the maximum relative error δ_{max} and the minimum coefficient of determination $R^2{}_{min}$ are listed. The calculated results for the monoharmonic model based on the second harmonic for all 12 crankshafts with different designs are found to be in complete agreement the FEA data. For all crankshafts analyzed, the coefficients of determination R^2 are greater than 0.9997, and this value is valid for all main journals of the crankshaft S-1977-0740-3-2-149-144-O-11. The maximum relative percentage error does not exceed 1.0228%, which is valid for the main journal No. 5 of the crankshaft S-7051-3600-10-8-149-144-Z-09.

Table 25. The comparison of model fit values and maximum relative errors for the shafts analyzed.

Crankshaft	$R^2{}_{min}$ (-)	δ_{max} (%)
S-9283-3600-10-8-149-144-O-01	0.9999	0.7788
S-8658-3600-10-8-149-114-O-02	0.9999	1.0481
S-16942-3600-10-8-149-144-C-03	0.9999	0.6581
S-12075-3600-10-8-149-144-C-04	0.9999	0.5482
S-9283-3600-10-8-149-144-O-05	0.9999	0.4026
S-9283-3600-10-8-149-144-O-06	0.9999	0.0887
S-9283-3600-10-8-149-144-O-07	0.9999	0.3983
S-8479-3600-10-8-149-144-O-08	0.9999	0.4840
S-7051-3600-10-8-149-144-Z-09	0.9999	1.0228
S-1977-0740-3-2-149-144-O-10	1.0000	0.0159
S-1977-0740-3-2-149-144-O-11	0.9997	0.2077
S-1977-0740-3-2-149-144-O-12	1.0000	0.3143

The analysis has shown that the monoharmonic model maps the reaction forces at supports of main journals, regardless of how the shaft is designed. In particular, the designs were changed by modifying the following features:

- the number of crankshaft's main journals,
- the number of crankshaft's crank journals,
- the length of main journals,
- the diameter of main journals,
- the diameter of crank journals,
- dimensions of the crank webs,
- the shape of crank webs,
- the relative angular offset between subsequent crank webs.

As the phase shift angle was determined using model algorithms, there was a significant improvement in the quality of the model's fit to the FEA data. This is true for an angular increment of 15°. With the algorithm, it is possible to increase the number of interpolation nodes, as shown in this article for an angular step of 1°. The authors' previous studies [18] have shown calculations for every 15° and the phase shift given by the formula (6), the shaft S-9283-3600-10-8-149-144-O-01 was modeled with the fit resulting in the coefficient of determination $R^2{}_{min}$ = 0.9959 and the maximum relative percentage error δ_{max} = 1.52899. Thus, according to the data presented in Table 25, the refinement of interpolation to 360 nodes per 1 revolution allowed reducing the error by less than 51%.

4. Model Validation

The operational tests for verification of the proposed system were conducted using a test rig built in the Szczecin AM, equipped with the flexible shaft support system presented in the Section 1.

The testing included measurements of the geometry deviations in main journals of the crankshaft of a Buckau Wolf R8DV-136 engine (crankshaft designated S-9285-3600-10-8-149-144-O-01). Measurements were carried out in which the crankshaft was supported by a set of rigid V-block supports and also with the shaft being supported by a set of flexible supports. The outer faces of the shaft were fixed at the centers. For this variant, at the shaft support points, we calculated the reaction forces beforehand to ensure zero deflection at the main journals, in 15 angular degree increments. Then, these values were replaced with the model given in Section 2.3, and the measurement results were compared with the assumed reference results. The reference measurements were performed using an operative measuring system with a MUK 25-600 measuring head equipped with SAJD software that was previously tested under industrial conditions.

The reference measurement system was designed to measure the roundness profiles of cylindrical surfaces using a reference-based method. The MUK 25-600 head was seated directly on the surface of the tested journal and assessed the shape profile independent of the measured item's support conditions. This system was selected for reference measurements because we could compare the measurements using similar mathematical tools, including harmonic analysis of profiles. The measurement procedure was based on polar coordinates. In individual cross-sections, consecutive changes in the radius r_{ji} of a specific angle of shaft rotation φ_{ji} were measured.

The roundness profiles, reference $r_1(\varphi)$ and corresponding tested one $r_2(\varphi)$, were pre-filtered for harmonics in the range of $n = 2$–15 and then comparatively evaluated using a standardized intercorrelation function given by:

$$\rho(\gamma_\phi) = \frac{2\int_0^{2\pi} r_1(\phi) r_2(\phi + \gamma_\phi) d\phi}{\int_0^{2\pi} r_1(\phi)^2 d\phi + \int_0^{2\pi} r_2(\phi)^2 d\phi} \tag{22}$$

where: $r_1(\varphi)$ is the roundness profile measured with the reference method, $r_2(\varphi)$ is the roundness profile measured by the evaluated method and γ_φ is the phase shift between the graphs being compared.

For a shaft supported by a set of rigid V-blocks, the compared profiles showed moderate overlap. The coefficient of intercorrelation between the tested journals had values from 0.7665 to 0.8132. This is shown qualitatively in Figure 5 using superimposed measured and reference roundness profiles mapped in polar and Cartesian coordinates for the selected main journal (No. 4). The intercorrelation determined for this journal was $\rho(\gamma_\varphi) = 0.7987$.

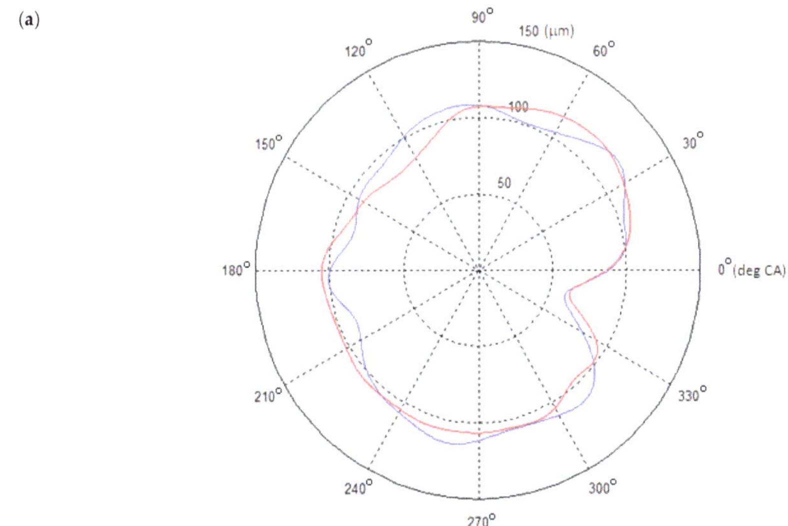

Figure 5. *Cont.*

(b)

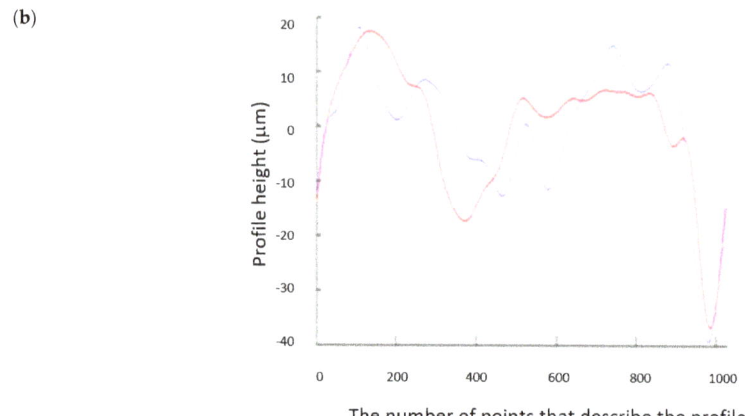

Figure 5. The measured (blue) and reference (red) profiles of journal No. 4, obtained for a shaft supported by a set of rigid V-blocks: (**a**) in the polar coordinate system; (**b**) in the Cartesian coordinate system.

The measurements on a flexible shaft support, assisted by the algorithm described in Section 2.3, were compared with the reference measurements (Figure 6). As in the preceding case, the comparison is made in both polar and Cartesian coordinates.

When we applied the proposed algorithm to a shaft support with controlled reaction forces, the results showed a high correlation between the compared profiles. The intercorrelation coefficient between the tested journal profiles ranged from 0.9113 to 0.9399. The intercorrelation coefficient of main journal No. 4 was determined to be $\rho(\gamma_\varphi) = 0.9266$.

The experimental studies have confirmed the suitability of the presented models. This is a major step in the development of a geometry measurement system with flexible shaft support for large-size crankshafts.

(a)

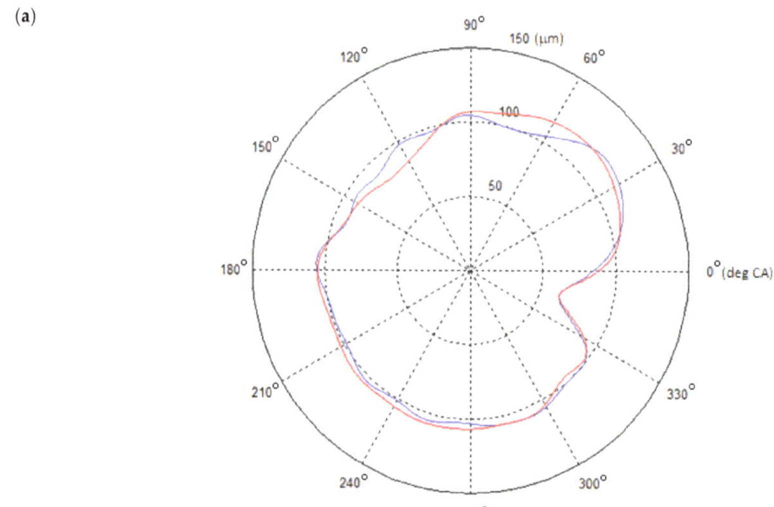

Figure 6. *Cont.*

(b)

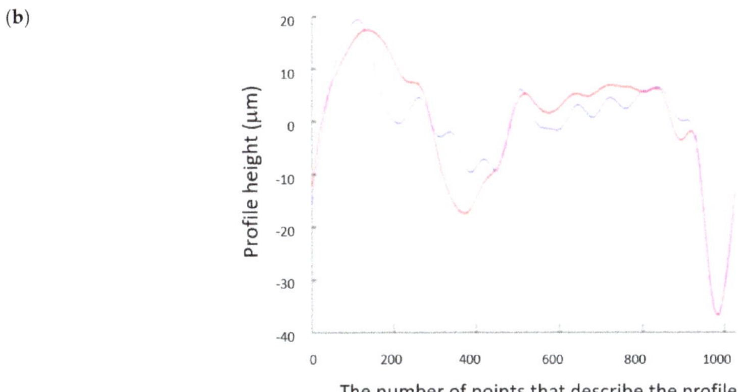

Figure 6. The measured (blue) and reference (red) profiles of journal No. 4 obtained for the controlled-reaction-force shaft support: (**a**) in polar coordinates; (**b**) in Cartesian coordinates.

5. Conclusions

The monoharmonic model provides results which are in agreement with the calculated data. These values can be used in the algorithmic control of reaction forces at the supports of a crankshaft geometry measuring system. The model has higher degree of accuracy and simplicity compared to the other potential options such as the basic polyharmonic model and spline-based polyharmonic model. The current model is not dependent on spectral analysis of the input data to describe the individual harmonics for the function of reaction force variation calculated with the FEA software.

It should be noted that for all modifications in the design of shafts analyzed, the common feature is the symmetry of the crank webs relative to the plane that contains the longitudinal axis of the main journal and crank journal, which are located in the immediate vicinity of the crank in question. Therefore, it can be said that it is possible to maintain the regularity for virtually all large-size crankshafts used in modern internal combustion high-power engines.

The proposed models make it possible to automate the control of flexible supports of crankshaft's main journals, thereby increasing the accuracy of the measurement process by minimizing the elastic deflection of the measured shaft.

6. Patents

1. Chybowski L., Kazienko D., *Universal resistance system supporting the calibration of resistance sensors in energetic machines monitoring systems, preferably internal combustion engines.* Polish Patent Office, P.429574.
2. Chybowski L., Kazienko D., *Housing of the detector of diagnostic signals values deviations.* Polish Patent Office, Wp.27118, Registred design nr 25673.
3. Nozdrzykowski, K. *Device for measuring positional deviation of axis of crankshaft pivot set.* Polish Patent Office, PL393829-A1; PL218653-B1.

Author Contributions: Conceptualization, L.C. and K.N.; methodology, L.C., K.N., Z.G. and L.D.; software, L.C., Z.G. and L.D.; validation, L.C., K.N. and L.D., formal analysis, L.C. and K.N.; investigation, L.C., K.N., Z.G., and L.D.; resources, L.C., K.N., Z.G. and L.D.; data curation, L.C., K.N., Z.G. and L.D.; writing—original draft preparation, L.C. and K.N.; writing—review and editing, L.C., K.N., and L.D.; visualization, L.C., K.N., Z.G. and L.D.; supervision, L.C. and K.N.; project administration, L.C. and K.N.; funding acquisition, L.C. and K.N. All authors have read and agreed to the published version of the manuscript.

Funding: This research was co-funded by the Ministry of Science and Higher Education of Poland from Grant 1/S/KPBMiM/20.

Conflicts of Interest: The authors declare no conflict of interest.

Appendix A

```
% Recursive algorithm
clear,clc,format compact
FiFEA=[0:15:345];%whole turn of the shaft 360 deg, every 15 deg -> reaction forces 0–345 deg
  %example shaft: S-12075-3600-10-8-149-144-C-04, example main journal: #10, FEA reaction forces:
RFEA=[749.87 768.80 775.20 767.37 747.40 720.64 694.26 675.33 668.93 676.77 696.74 723.50 749.87 768.80 775.20
767.37 747.40 720.64 694.26 675.33 668.93 676.77 696.74 723.50];

DeltaFi=1;% calculation step (deg)
R0=mean(RFEA);
disp(['Mean value of the reaction forces: ' num2str(R0) ' N'])
CR2=0.5*(max(RFEA)-min(RFEA));
disp(['Reaction forces amplitude of the 2nd harmonic: ' num2str(CR2) ' N'])
for j=1:360 % calculation loop
  FiR2=(j-1)*DeltaFi;
  P1=RFEA-(R0+CR2*sind(FiFEA*2+FiR2));
  Q(j)=P1*P1'; % sum of squares of deviations
end
k=length(Q);
[Qmin,L]=min(Q);
[Qmax,H]=max(Q);
FiR2a=L-1;
FiR2b=H+1;

disp(['Reaction forces phase shift of the 2nd harmonic: ' num2str(FiR2a) ' deg'])
```

Appendix B

Table A1. Reaction forces based on FEA calculations for the crankshaft S-9285-3600-10-8-149-144-O-01.

Angular Position (°CA)	Main Journal Number (-)									
	1	2	3	4	5	6	7	8	9	10
	Reaction Forces in Main Journals (N)									
0	731.62	988.50	871.12	1166.33	847.89	1093.01	852.04	1142.22	988.15	603.47
15	727.48	1005.43	823.76	1237.24	796.42	1123.70	799.74	1212.93	944.12	613.51
30	737.46	989.14	822.94	1253.23	795.30	1108.89	797.22	1231.62	933.95	614.59
45	758.88	943.98	868.88	1210.01	844.82	1052.54	845.15	1193.29	960.37	606.42
60	786.00	882.05	949.28	1119.16	931.72	969.75	930.69	1108.20	1016.30	591.19
75	811.57	819.94	1042.60	1005.03	1032.70	882.71	1030.91	999.16	1086.75	572.98
90	828.72	774.30	1123.82	898.19	1120.71	814.74	1118.97	895.38	1152.84	556.67
105	832.87	757.35	1171.18	827.29	1172.17	784.05	1171.26	824.67	1196.87	546.62
120	822.89	773.65	1172.00	811.30	1173.29	798.87	1173.78	805.98	1207.04	545.54
135	801.47	818.82	1126.05	854.53	1123.77	855.22	1125.85	844.31	1180.62	553.71
150	774.33	880.76	1045.65	945.37	1036.88	938.00	1040.31	929.40	1124.69	568.94
165	748.76	942.87	952.34	1059.50	935.90	1025.04	940.09	1038.44	1054.24	587.16
180	731.62	988.50	871.12	1166.33	847.89	1093.01	852.04	1142.22	988.15	603.47
195	727.48	1005.43	823.76	1237.24	796.42	1123.70	799.74	1212.93	944.12	613.51
210	737.46	989.14	822.94	1253.23	795.30	1108.89	797.22	1231.62	933.95	614.59
225	758.88	943.98	868.88	1210.01	844.82	1052.54	845.15	1193.29	960.37	606.42
240	786.00	882.05	949.28	1119.16	931.72	969.75	930.69	1108.20	1016.30	591.19
255	811.57	819.94	1042.60	1005.03	1032.70	882.71	1030.91	999.16	1086.75	572.98
270	828.72	774.30	1123.82	898.19	1120.71	814.74	1118.97	895.38	1152.84	556.67
285	832.87	757.35	1171.19	827.29	1172.17	784.05	1171.26	824.67	1196.87	546.62
300	822.89	773.65	1172.00	811.30	1173.29	798.87	1173.78	805.98	1207.04	545.54
315	801.47	818.82	1126.05	854.53	1123.77	855.22	1125.85	844.31	1180.62	553.71
330	774.33	880.76	1045.65	945.37	1036.88	938.00	1040.31	929.40	1124.69	568.94
345	748.76	942.86	952.34	1059.50	935.90	1025.04	940.09	1038.44	1054.24	587.16
360	731.62	988.50	871.12	1166.33	847.89	1093.01	852.04	1142.22	988.15	603.47

Table A2. Reaction forces based on FEA calculations for the crankshaft S-8658-3600-10-8-149-114-O-02.

Angular Position (°CA)	Main Journal Number (-)									
	1	2	3	4	5	6	7	8	9	10
	Reaction Forces in Main Journals (N)									
0	719.80	944.80	793.60	1105.90	757.00	1024.40	760.30	1088.20	880.10	584.20
15	711.80	969.60	737.80	1184.20	699.30	1061.30	702.50	1165.80	828.80	597.20
30	723.00	953.30	734.80	1200.00	698.90	1046.90	701.70	1183.60	816.90	599.30
45	750.20	900.30	785.40	1149.10	756.00	985.20	758.10	1136.60	847.60	589.90
60	786.10	824.70	876.10	1045.10	855.10	892.60	856.70	1037.60	912.70	571.50
75	821.20	746.80	982.60	916.00	969.80	793.90	971.00	913.00	994.80	549.00
90	846.10	687.50	1076.40	796.30	1069.30	715.70	1070.40	796.30	1071.80	528.60
105	854.00	662.60	1132.30	718.00	1127.00	678.80	1128.20	718.70	1123.10	515.60
120	842.90	678.90	1135.30	702.20	1127.40	693.20	1129.00	700.90	1135.00	513.50
135	815.70	732.00	1084.70	753.10	1070.40	754.90	1072.50	747.90	1104.30	522.90
150	779.80	807.60	993.90	857.00	971.30	847.50	973.90	846.90	1039.20	541.30
165	744.60	885.50	887.40	986.10	856.60	946.10	859.70	971.50	957.10	563.70
180	719.80	944.80	793.60	1105.90	757.00	1024.40	760.30	1088.20	880.10	584.20
195	711.80	969.60	737.80	1184.20	699.30	1061.30	702.50	1165.80	828.80	597.20
210	723.00	953.30	734.80	1200.00	698.90	1046.90	701.70	1183.60	816.90	599.30
225	750.10	900.30	785.40	1149.10	756.00	985.20	758.10	1136.60	847.60	589.90
240	786.10	824.70	876.10	1045.10	855.10	892.60	856.70	1037.60	912.70	571.50
255	821.20	746.80	982.60	916.00	969.80	793.90	971.00	913.00	994.80	549.00
270	846.10	687.50	1076.40	796.30	1069.30	715.70	1070.40	796.30	1071.80	528.60
285	854.00	662.60	1132.30	718.00	1127.00	678.80	1128.20	718.70	1123.10	515.60
300	842.90	678.90	1135.30	702.20	1127.40	693.20	1129.00	700.90	1135.00	513.50
315	815.70	732.00	1084.70	753.10	1070.40	754.90	1072.50	747.90	1104.30	522.90
330	779.80	807.60	993.90	857.00	971.30	847.50	973.90	846.90	1039.20	541.30
345	744.60	885.50	887.40	986.10	856.60	946.10	859.70	971.50	957.10	563.70
360	719.80	944.80	793.60	1105.90	757.00	1024.40	760.30	1088.20	880.10	584.20

Table A3. Reaction forces based on FEA calculations for the crankshaft S-16942-3600-10-8-149-144-C-03.

Angular Position (°CA)	Main Journal Number (-)									
	1	2	3	4	5	6	7	8	9	10
	Reaction Forces in Main Journals (N)									
0	546.10	1663.20	1762.90	2313.40	1607.30	2176.70	1608.20	2275.10	2000.90	987.90
15	539.30	1698.10	1662.20	2463.80	1498.80	2237.90	1500.00	2426.20	1903.50	1011.90
30	558.60	1669.00	1651.70	2505.70	1491.10	2209.80	1492.30	2473.90	1870.90	1018.80
45	598.90	1583.70	1734.20	2428.00	1586.20	2099.90	1587.10	2405.40	1911.70	1006.60
60	649.40	1465.10	1887.50	2251.40	1758.60	1937.60	1759.20	2239.10	2015.00	978.70
75	696.60	1345.00	2070.70	2023.20	1962.30	1766.50	1962.30	2019.60	2153.10	942.60
90	727.80	1255.50	2234.50	1804.60	2142.50	1632.40	2142.00	1805.60	2289.10	907.80
105	734.60	1220.70	2335.20	1654.20	2251.00	1571.10	2250.30	1654.50	2386.40	883.80
120	715.30	1249.80	2345.60	1612.20	2258.80	1599.20	2258.00	1606.80	2419.00	876.90
135	675.00	1335.10	2263.20	1690.00	2163.70	1709.20	2163.10	1675.20	2378.20	889.10
150	624.40	1453.70	2109.80	1866.60	1991.20	1871.40	1991.10	1841.50	2274.90	917.00
165	577.30	1573.80	1926.70	2094.80	1787.60	2042.60	1788.00	2061.10	2136.80	953.10
180	546.10	1663.20	1762.90	2313.40	1607.30	2176.70	1608.20	2275.10	2000.90	987.90
195	539.30	1698.10	1662.20	2463.80	1498.80	2237.90	1500.00	2426.20	1903.50	1011.90
210	558.60	1669.00	1651.70	2505.70	1491.10	2209.80	1492.30	2473.90	1870.90	1018.80
225	598.90	1583.70	1734.20	2428.00	1586.20	2099.90	1587.10	2405.40	1911.70	1006.60
240	649.40	1465.10	1887.50	2251.40	1758.60	1937.60	1759.20	2239.10	2015.00	978.70
255	696.60	1345.00	2070.70	2023.20	1962.30	1766.50	1962.30	2019.60	2153.10	942.60
270	727.80	1255.50	2234.50	1804.60	2142.50	1632.40	2142.00	1805.60	2289.10	907.80
285	734.60	1220.70	2335.20	1654.20	2251.00	1571.10	2250.30	1654.50	2386.40	883.80
300	715.30	1249.80	2345.60	1612.20	2258.80	1599.20	2258.00	1606.80	2419.00	876.90
315	675.00	1335.10	2263.20	1690.00	2163.70	1709.20	2163.10	1675.20	2378.20	889.10
330	624.40	1453.70	2109.80	1866.60	1991.20	1871.40	1991.10	1841.50	2274.90	917.00
345	577.30	1573.80	1926.70	2094.80	1787.60	2042.60	1788.00	2061.10	2136.80	953.10
360	546.10	1663.20	1762.90	2313.40	1607.30	2176.70	1608.20	2275.10	2000.90	987.90

Table A4. Reaction forces based on FEA calculations for the crankshaft S-12075-3600-10-8-149-144-C-04.

Angular Position (°CA)	Main Journal Number (-)									
	1	2	3	4	5	6	7	8	9	10
	Reaction Forces in Main Journals (N)									
0	666.85	1241.62	1131.48	1628.25	1120.83	1520.32	1073.28	1600.71	1341.58	749.87
15	656.78	1276.88	1047.58	1741.92	1039.89	1573.12	986.67	1716.67	1266.49	768.80
30	665.14	1267.57	1029.79	1776.19	1030.53	1562.10	972.19	1757.34	1238.75	775.20
45	689.71	1216.17	1082.87	1721.90	1095.24	1490.21	1033.73	1711.83	1265.79	767.37
60	723.88	1136.46	1192.60	1593.57	1216.71	1376.72	1154.78	1592.32	1340.37	747.40
75	758.51	1049.80	1329.59	1425.61	1362.37	1252.03	1302.93	1430.84	1442.50	720.64
90	784.31	979.42	1457.11	1263.01	1493.19	1149.56	1438.46	1270.66	1544.83	694.26
105	794.38	944.17	1541.00	1149.35	1574.13	1096.76	1525.07	1154.70	1619.92	675.33
120	786.02	953.49	1558.78	1115.08	1583.50	1107.78	1539.54	1114.03	1647.66	668.93
135	761.47	1004.88	1505.69	1169.38	1518.78	1179.67	1478.01	1159.54	1620.61	676.77
150	727.30	1084.58	1395.96	1297.70	1397.32	1293.16	1356.95	1279.05	1546.03	696.74
165	692.67	1171.23	1258.99	1465.66	1251.66	1417.85	1208.81	1440.53	1443.90	723.50
180	666.85	1241.62	1131.48	1628.25	1120.83	1520.32	1073.28	1600.71	1341.58	749.87
195	656.78	1276.88	1047.58	1741.92	1039.89	1573.12	986.67	1716.67	1266.49	768.80
210	665.14	1267.57	1029.79	1776.19	1030.53	1562.10	972.19	1757.34	1238.75	775.20
225	689.71	1216.17	1082.87	1721.90	1095.24	1490.21	1033.73	1711.83	1265.79	767.37
240	723.88	1136.46	1192.60	1593.57	1216.71	1376.72	1154.78	1592.32	1340.37	747.40
255	758.51	1049.80	1329.59	1425.61	1362.37	1252.03	1302.93	1430.84	1442.50	720.64
270	784.31	979.42	1457.11	1263.01	1493.19	1149.56	1438.46	1270.66	1544.83	694.26
285	794.38	944.17	1541.00	1149.35	1574.13	1096.76	1525.07	1154.70	1619.92	675.33
300	786.02	953.49	1558.78	1115.08	1583.50	1107.78	1539.54	1114.03	1647.66	668.93
315	761.47	1004.88	1505.69	1169.38	1518.78	1179.67	1478.01	1159.54	1620.61	676.77
330	727.30	1084.58	1395.96	1297.70	1397.32	1293.16	1356.95	1279.05	1546.03	696.74
345	692.67	1171.23	1258.99	1465.66	1251.66	1417.85	1208.81	1440.53	1443.90	723.50
360	666.85	1241.62	1131.48	1628.25	1120.83	1520.32	1073.28	1600.71	1341.58	749.87

Table A5. Reaction forces based on FEA calculations for the crankshaft S-9283-3600-10-8-149-144-O-05.

Angular Position (°CA)	Main Journal Number (-)									
	1	2	3	4	5	6	7	8	9	10
	Reaction Forces in Main Journals (N)									
0	699.84	1014.79	919.05	1037.94	1026.79	947.93	943.38	1042.84	1062.58	587.43
15	696.59	1032.04	865.35	1123.49	955.73	992.88	882.97	1125.67	1005.57	602.29
30	707.02	1018.78	845.25	1185.31	892.87	1025.61	850.51	1183.61	956.70	616.90
45	728.34	978.57	864.15	1206.86	855.04	1037.35	854.72	1201.14	929.07	627.34
60	754.82	922.18	916.99	1182.35	852.38	1024.95	894.47	1173.56	930.08	630.80
75	779.39	864.72	989.59	1118.35	885.60	991.73	959.09	1108.26	959.46	626.37
90	795.44	821.59	1062.52	1032.02	945.81	946.60	1031.29	1022.74	1009.34	615.23
105	798.69	804.34	1116.22	946.47	1016.86	901.65	1091.71	939.91	1066.35	600.37
120	788.26	817.60	1136.32	884.65	1079.73	868.92	1124.16	881.97	1115.22	585.76
135	766.95	857.81	1117.42	863.10	1117.56	857.18	1119.95	864.44	1142.85	575.33
150	740.46	914.20	1064.60	887.61	1120.20	869.58	1080.20	892.02	1141.80	571.86
165	715.90	971.66	991.98	951.61	1087.00	902.80	1015.58	957.32	1112.46	576.29
180	699.84	1014.79	919.05	1037.94	1026.79	947.93	943.38	1042.84	1062.58	587.43
195	696.59	1032.04	865.35	1123.49	955.73	992.88	882.97	1125.67	1005.57	602.29
210	707.02	1018.78	845.25	1185.31	892.87	1025.61	850.51	1183.61	956.70	616.90
225	728.34	978.57	864.15	1206.86	855.04	1037.35	854.72	1201.14	929.07	627.34
240	754.82	922.18	916.99	1182.35	852.38	1024.95	894.47	1173.56	930.08	630.80
255	779.39	864.72	989.59	1118.35	885.60	991.73	959.09	1108.26	959.46	626.37
270	795.44	821.59	1062.52	1032.02	945.81	946.60	1031.29	1022.74	1009.34	615.23
285	798.69	804.34	1116.22	946.47	1016.86	901.65	1091.71	939.91	1066.35	600.37
300	788.26	817.60	1136.32	884.65	1079.73	868.92	1124.16	881.97	1115.22	585.76
315	766.95	857.81	1117.42	863.10	1117.56	857.18	1119.95	864.44	1142.85	575.33
330	740.46	914.20	1064.59	887.61	1120.22	869.58	1080.21	892.02	1141.84	571.86
345	715.90	971.66	991.98	951.61	1087.00	902.80	1015.58	957.32	1112.46	576.29
360	699.84	1014.79	919.05	1037.94	1026.79	947.93	943.38	1042.84	1062.58	587.43

Table A6. Reaction forces based on FEA calculations for the crankshaft S-9283-3600-10-8-149-144-O-06.

Angular Position (°CA)	Main Journal Number (-)									
	1	2	3	4	5	6	7	8	9	10
	Reaction Forces in Main Journals (N)									
0	852.00	841.46	987.10	986.76	994.57	979.21	1000.87	963.76	1086.66	590.18
15	857.14	830.51	999.28	973.32	1008.27	967.19	1012.02	951.58	1095.99	587.28
30	871.73	798.35	1035.37	935.47	1045.96	931.31	1047.15	918.14	1118.73	580.37
45	891.85	753.61	1085.72	883.36	1097.54	881.20	1096.82	872.40	1148.78	571.31
60	912.12	708.26	1136.82	830.95	1149.20	830.28	1147.74	826.61	1178.09	562.51
75	927.11	674.47	1174.99	792.28	1187.08	792.20	1186.26	793.04	1198.82	556.33
90	932.79	661.28	1190.00	777.71	1201.05	777.15	1202.06	780.70	1205.40	554.44
105	927.65	672.23	1177.83	791.16	1187.35	789.18	1190.91	792.88	1196.07	557.33
120	913.07	704.39	1141.73	829.00	1149.66	825.05	1155.79	826.32	1173.33	564.24
135	892.94	749.13	1091.38	881.11	1098.08	875.16	1106.11	872.07	1143.28	573.31
150	872.67	794.48	1040.28	933.53	1046.42	926.08	1055.19	917.86	1113.97	582.11
165	857.69	828.27	1002.11	972.19	1008.54	964.17	1016.67	951.42	1093.24	588.28
180	852.00	841.46	987.10	986.76	994.57	979.21	1000.87	963.76	1086.66	590.18
195	857.14	830.51	999.28	973.32	1008.27	967.19	1012.02	951.58	1095.99	587.28
210	871.73	798.35	1035.37	935.47	1045.96	931.31	1047.15	918.14	1118.73	580.37
225	891.85	753.61	1085.72	883.36	1097.54	881.20	1096.82	872.40	1148.78	571.31
240	912.12	708.26	1136.82	830.95	1149.20	830.28	1147.74	826.61	1178.09	562.51
255	927.11	674.47	1174.99	792.28	1187.08	792.20	1186.26	793.04	1198.82	556.33
270	932.79	661.28	1190.00	777.71	1201.05	777.15	1202.06	780.70	1205.40	554.44
285	927.65	672.23	1177.83	791.16	1187.35	789.18	1190.91	792.88	1196.07	557.33
300	913.07	704.39	1141.73	829.00	1149.66	825.05	1155.79	826.32	1173.33	564.24
315	892.94	749.13	1091.38	881.11	1098.08	875.16	1106.11	872.07	1143.28	573.31
330	872.67	794.48	1040.28	933.53	1046.42	926.08	1055.19	917.86	1113.97	582.11
345	857.69	828.27	1002.11	972.19	1008.54	964.17	1016.67	951.42	1093.24	588.28
360	852.00	841.46	987.10	986.76	994.57	979.21	1000.87	963.76	1086.66	590.18

Table A7. Reaction forces based on FEA calculations for the crankshaft S-9283-3600-10-8-149-144-O-07.

Angular Position (°CA)	Main Journal Number (-)									
	1	2	3	4	5	6	7	8	9	10
	Reaction Forces in Main Journals (N)									
0	704.97	1000.45	918.33	1100.18	933.89	934.39	1098.46	920.77	1062.78	608.29
15	705.70	1001.97	906.56	1121.28	919.26	929.13	1118.64	901.35	1072.58	606.06
30	719.03	974.88	928.44	1116.55	904.00	949.64	1113.33	882.86	1097.65	596.14
45	741.40	926.46	978.11	1087.25	892.21	990.43	1083.95	870.24	1131.27	581.20
60	766.81	869.66	1042.26	1041.24	887.04	1040.57	1038.38	866.88	1164.44	565.24
75	788.46	819.71	1103.70	990.85	889.88	1086.62	988.83	873.69	1188.26	552.54
90	800.54	790.00	1145.96	949.57	899.96	1116.24	948.57	888.83	1196.35	546.50
105	799.82	788.48	1157.73	928.48	914.59	1121.50	928.39	908.25	1186.55	548.73
120	786.49	815.57	1135.85	933.21	929.85	1100.99	933.70	926.74	1161.48	558.65
135	764.12	864.00	1086.18	962.51	941.64	1060.20	963.07	939.36	1127.86	573.59
150	738.70	920.79	1022.03	1008.51	946.82	1010.06	1008.64	942.71	1094.69	589.55
165	717.05	970.74	960.60	1058.91	943.98	964.01	1058.20	935.91	1070.87	602.25
180	704.97	1000.45	918.33	1100.18	933.89	934.39	1098.46	920.77	1062.78	608.29
195	705.70	1001.97	906.56	1121.28	919.26	929.13	1118.64	901.35	1072.58	606.06
210	719.03	974.88	928.44	1116.55	904.00	949.64	1113.33	882.86	1097.65	596.14
225	741.40	926.46	978.11	1087.25	892.21	990.43	1083.95	870.24	1131.27	581.20
240	766.81	869.66	1042.26	1041.24	887.04	1040.57	1038.38	866.88	1164.44	565.24
255	788.46	819.71	1103.70	990.85	889.88	1086.62	988.83	873.69	1188.26	552.54
270	800.54	790.00	1145.96	949.57	899.96	1116.24	948.57	888.83	1196.35	546.50
285	799.82	788.48	1157.73	928.48	914.59	1121.50	928.39	908.25	1186.55	548.73
300	786.49	815.57	1135.85	933.21	929.85	1100.99	933.70	926.74	1161.48	558.65
315	764.12	864.00	1086.18	962.51	941.64	1060.20	963.07	939.36	1127.86	573.59
330	738.70	920.79	1022.03	1008.51	946.82	1010.06	1008.64	942.71	1094.69	589.55
345	717.05	970.74	960.60	1058.91	943.98	964.01	1058.20	935.91	1070.87	602.25
360	704.97	1000.45	918.33	1100.18	933.89	934.39	1098.46	920.77	1062.78	608.29

Table A8. Reaction forces based on FEA calculations for the crankshaft S-8479-3600-10-8-149-144-O-08.

Angular Position (°CA)	Main Journal Number (-)									
	1	2	3	4	5	6	7	8	9	10
	Reaction Forces in Main Journals (N)									
0	747.00	858.00	835.00	1003.00	808.00	951.00	810.00	983.00	923.00	562.00
15	744.00	870.00	803.00	1049.00	776.00	969.00	777.00	1029.00	893.00	570.00
30	750.00	861.00	800.00	1060.00	776.00	959.00	777.00	1041.00	883.00	572.00
45	765.00	833.00	827.00	1032.00	808.00	924.00	808.00	1017.00	898.00	568.00
60	784.00	793.00	877.00	973.00	864.00	872.00	863.00	962.00	932.00	559.00
75	801.00	752.00	937.00	899.00	928.00	818.00	927.00	892.00	977.00	547.00
90	813.00	721.00	990.00	830.00	984.00	776.00	983.00	825.00	1021.00	536.00
105	816.00	709.00	1023.00	783.00	1016.00	758.00	1015.00	780.00	1051.00	528.00
120	809.00	718.00	1026.00	773.00	1015.00	768.00	1016.00	767.00	1061.00	526.00
135	795.00	747.00	999.00	801.00	983.00	803.00	984.00	791.00	1046.00	530.00
150	776.00	787.00	949.00	859.00	927.00	855.00	929.00	846.00	1012.00	539.00
165	759.00	827.00	889.00	934.00	863.00	909.00	865.00	916.00	967.00	551.00
180	747.00	858.00	835.00	1003.00	808.00	951.00	810.00	983.00	923.00	562.00
195	744.00	870.00	803.00	1049.00	776.00	969.00	777.00	1029.00	893.00	570.00
210	750.00	861.00	800.00	1060.00	776.00	959.00	777.00	1041.00	883.00	572.00
225	765.00	833.00	827.00	1032.00	808.00	924.00	808.00	1017.00	898.00	568.00
240	784.00	793.00	877.00	973.00	864.00	872.00	863.00	962.00	932.00	559.00
255	801.00	752.00	937.00	899.00	928.00	818.00	927.00	892.00	977.00	547.00
270	813.00	721.00	990.00	830.00	984.00	776.00	983.00	825.00	1021.00	536.00
285	816.00	709.00	1023.00	783.00	1016.00	758.00	1015.00	780.00	1051.00	528.00
300	809.00	718.00	1026.00	773.00	1015.00	768.00	1016.00	767.00	1061.00	526.00
315	795.00	747.00	999.00	801.00	983.00	803.00	984.00	791.00	1046.00	530.00
330	776.00	787.00	949.00	859.00	927.00	855.00	929.00	846.00	1012.00	539.00
345	759.00	827.00	889.00	934.00	863.00	909.00	865.00	916.00	967.00	551.00
360	747.00	858.00	835.00	1003.00	808.00	951.00	810.00	983.00	923.00	562.00

Table A9. Reaction forces based on FEA calculations for the crankshaft S-7051-3600-10-8-149-144-Z-09.

Angular Position (°CA)	Main Journal Number (-)									
	1	2	3	4	5	6	7	8	9	10
	Reaction Forces in Main Journals (N)									
0	765.70	775.40	611.20	863.50	594.90	801.60	595.70	850.70	680.70	511.90
15	758.70	796.10	568.10	922.60	550.90	830.50	551.70	908.80	642.40	521.50
30	767.40	783.20	566.60	933.50	550.80	819.90	551.50	920.90	634.50	522.80
45	789.60	740.20	607.20	893.30	594.80	772.70	595.30	883.70	659.10	515.30
60	819.30	678.60	679.00	812.90	671.10	701.40	671.30	807.10	709.60	501.00
75	848.60	614.80	762.70	713.70	759.20	625.30	759.10	711.80	772.40	483.70
90	869.60	566.00	835.90	622.30	835.50	564.70	835.20	623.20	830.70	468.20
105	876.60	545.30	879.00	563.30	879.50	535.80	879.20	565.00	869.00	458.50
120	867.90	558.20	880.40	552.30	879.50	546.30	879.40	553.00	876.90	457.30
135	845.60	601.30	839.80	592.50	835.50	593.60	835.60	590.20	852.30	464.80
150	815.90	662.90	768.10	673.00	759.30	664.80	759.60	666.70	801.80	479.10
165	786.70	726.60	684.40	772.20	671.20	741.00	671.80	762.10	739.00	496.30
180	765.70	775.40	611.20	863.50	594.90	801.60	595.70	850.70	680.70	511.90
195	758.70	796.10	568.10	922.60	550.90	830.50	551.70	908.80	642.40	521.50
210	767.40	783.20	566.60	933.50	550.80	819.90	551.50	920.90	634.50	522.80
225	789.60	740.20	607.20	893.30	594.80	772.70	595.30	883.70	659.10	515.30
240	819.30	678.60	679.00	812.90	671.10	701.40	671.30	807.10	709.60	501.00
255	848.60	614.80	762.70	713.70	759.20	625.30	759.10	711.80	772.40	483.70
270	869.60	566.00	835.90	622.30	835.50	564.70	835.20	623.20	830.70	468.20
285	876.60	545.30	879.00	563.30	879.50	535.80	879.20	565.00	869.00	458.50
300	867.90	558.20	880.40	552.30	879.50	546.30	879.40	553.00	876.90	457.30
315	845.60	601.30	839.80	592.50	835.50	593.60	835.60	590.20	852.30	464.80
330	815.90	662.90	768.10	673.00	759.30	664.80	759.60	666.70	801.80	479.10
345	786.70	726.60	684.40	772.20	671.20	741.00	671.80	762.10	739.00	496.30
360	765.70	775.40	611.20	863.50	594.90	801.60	595.70	850.70	680.70	511.90

Table A10. Reaction forces based on FEA calculations for the crankshaft S-1977-0740-3-2-149-144-O-0-10.

Angular Position (°CA)	Main Journal Number (-)		
	1	2	3
	Reaction Forces in Main Journals (N)		
0	369.30	1238.70	369.30
15	372.00	1233.20	372.10
30	379.30	1218.70	379.40
45	389.10	1199.00	389.20
60	398.90	1179.60	398.90
75	405.90	1165.50	406.00
90	408.40	1160.50	408.40
105	405.60	1166.10	405.70
120	398.30	1180.60	398.40
135	388.50	1200.20	388.60
150	378.80	1219.70	378.80
165	371.70	1233.80	371.80
180	369.30	1238.70	369.30
195	372.00	1233.20	372.10
210	379.30	1218.70	379.40
225	389.10	1199.00	389.20
240	398.90	1179.60	398.90
255	405.90	1165.50	406.00
270	408.40	1160.50	408.40
285	405.60	1166.10	405.70
300	398.30	1180.60	398.40
315	388.50	1200.20	388.60
330	378.80	1219.70	378.80
345	371.70	1233.80	371.80
360	369.30	1238.70	369.30

Table A11. Reaction forces based on FEA calculations for the crankshaft S-1977-0740-3-2-149-144-O-11.

Angular Position (°CA)	Main Journal Number (-)		
	1	2	3
	Reaction Forces in Main Journals (N)		
0	438.42	1099.93	438.43
15	433.11	1110.56	433.12
30	419.10	1138.59	419.10
45	400.64	1175.50	400.65
60	382.89	1211.00	382.90
75	370.35	1236.09	370.35
90	365.89	1245.00	365.90
105	370.35	1236.09	370.35
120	383.14	1210.49	383.15
135	401.00	1174.78	401.00
150	419.46	1137.86	419.47
165	433.33	1110.11	433.34
180	438.42	1099.93	438.43
195	433.11	1110.56	433.12
210	419.10	1138.59	419.10
225	400.64	1175.50	400.65
240	382.89	1211.00	382.90
255	370.35	1236.09	370.35
270	365.89	1245.00	365.90
285	370.35	1236.09	370.35
300	383.14	1210.49	383.15
315	401.00	1174.78	401.00
330	419.46	1137.86	419.47
345	433.33	1110.11	433.34
360	438.42	1099.93	438.43

Table A12. Reaction forces based on FEA calculations for the crankshaft S-1977-0740-3-2-149-144-O-12.

Angular Position (°CA)	Main Journal Number (-)		
	1	2	3
	Reaction Forces in Main Journals (N)		
0	452.34	1072.60	452.39
15	433.01	1111.30	433.06
30	410.28	1156.70	410.33
45	390.25	1196.80	390.30
60	378.30	1221.00	378.30
75	377.60	1222.00	377.60
90	388.40	1201.00	388.40
105	407.69	1161.90	407.74
120	430.41	1116.50	430.46
135	450.44	1076.40	450.49
150	462.41	1052.50	462.46
165	463.10	1051.10	463.15
180	452.34	1072.60	452.39
195	433.01	1111.30	433.06
210	410.28	1156.70	410.33
225	390.25	1196.80	390.30
240	378.29	1220.70	378.34
255	377.59	1222.10	377.64
270	388.35	1200.60	388.40
285	407.69	1161.90	407.74
300	430.41	1116.50	430.46
315	450.44	1076.40	450.49
330	462.40	1052.00	462.50
345	463.10	1051.10	463.15
360	452.34	1072.60	452.39

Appendix C

Figure A1. A 3D model of the crankshaft S-9285-3600-10-8-149-144-O-01.

Figure A2. A 3D model of the crankshaft S-8658-3600-10-8-149-114-O-02.

Figure A3. A 3D model of the crankshaft S-16942-3600-10-8-149-144-C-03.

Figure A4. A 3D model of the crankshaft S-12075-3600-10-8-149-144-C-04.

Figure A5. A 3D model of the crankshaft S-9283-3600-10-8-149-144-O-05.

Figure A6. A 3D model of the crankshaft S-9283-3600-10-8-149-144-O-06.

Figure A7. A 3D model of the crankshaft S-9283-3600-10-8-149-144-O-07.

Figure A8. A 3D model of the crankshaft S-8479-3600-10-8-149-144-O-08.

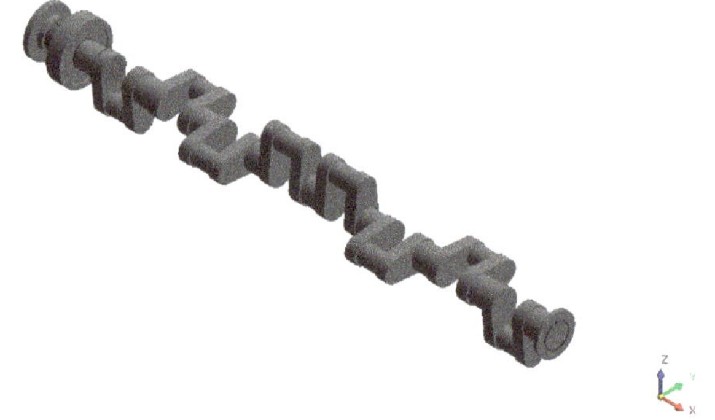

Figure A9. A 3D model of the crankshaft S-7051-3600-10-8-149-114-Z-09.

Figure A10. A 3D model of the crankshaft S-1977-0740-3-2-149-144-O-10.

Figure A11. A 3D model of the crankshaft S-1977-0740-3-2-149-144-O-11.

Figure A12. A 3D model of the crankshaft S-1977-0740-3-2-149-144-O-12.

References

1. Dai, K.; Gao, S.P.; Gao, L. The Applications of Measurement System for Crankshaft. In Proceedings of the 1st International Symposium on Digital Manufacture, Wuhan, China, 15–17 October 2006; pp. 1–3, 833–836.
2. Kim, J.-K.; Lee, M.-C. Development of wireless device for inspecting crankshaft deflection using linear encoder. *J. Korean Soc. Precis. Eng.* **2009**, *26*, 104–111.
3. Kim, M.C. Reliability block diagram with general gates and its application to system reliability analysis. *Ann. Nucl. Energy* **2011**, *38*, 2456–2461. [CrossRef]
4. Quan, X.; Wei, W.; Zhao, J.; Sun, Y.W.; Bianl, L.H. Development of an Automatic Optical Measurement System for Engine Crankshaft. In Proceedings of the International Conference on Computational Science and Applications, Santander, Spain, 20–23 June 2011; pp. 68–75.
5. Adamczak, S.; Janecki, D.; Stepien, K. Problems of mathematical modelling of rotary elements. In Proceedings of the International Symposium for Production Research 2018; Springer Nature Switzerland AG: Cham, Switzerland, 2019; pp. 747–752.
6. Nozdrzykowski, K. Metodyka pomiarów geometrycznych odchyłek powierzchni walcowych wielkogabarytowych elementów maszyn na przykładzie wałów korbowych silników okrętowych. In *Methodology of Geometric Measurements of Deviations of Cylindrical Surfaces of Large-Size Machine Eleme, I*; Wydawnictwo Naukowe Akademii Morskiej w Szczecinie: Szczecin, Poland, 2013.
7. Muhammad, A.; Ali, M.A.H.; Shanono, I.H. Fatigue and Harmonic Analysis of a Diesel Engine Crankshaft Using ANSYS. In Proceedings of the XXIV AIMETA Conference 2019, Rome, Italy, 15–19 September 2019; pp. 371–376.
8. Al-Azirjawi, B.S.K. New Design Solution for Crankshaft. In Proceedings of the ASME 2017 International Mechanical Engineering Congress and Exposition, Tampa, FL, USA, 3–9 November 2017.
9. Bejger, A.; Drzewieniecki, J. The use of acoustic emission to diagnosis of fuel injection pumps of marine diesel engines. *Energies* **2019**, *12*, 4661. [CrossRef]
10. Adamczak, S.; Janusiewicz, A.; Makieła, W.; Stepien, K. Statistical validation of the method for measuring radius variations of components on the machine tool. *Metrol. Meas. Syst.* **2011**, *18*, 35–46. [CrossRef]
11. Nozdrzykowski, K.; Chybowski, L. A force-sensor-based method to eliminate deformation of large crankshafts during measurements of their geometric condition. *Sensors* **2019**, *19*, 3507. [CrossRef] [PubMed]
12. Fonte, M.; Duarte, P.; Anes, V.; Freitas, M.; Reis, L. On the assessment of fatigue life of marine diesel engine crankshafts. *Eng. Fail. Anal.* **2015**, *56*, 51–57. [CrossRef]

13. Sun, J.; Wang, J.; Gui, C. Whole crankshaft beam-element finite-element method for calculating crankshaft deformation and bearing load of an engine. *Proc. Inst. Mech. Eng. Part J J. Eng. Tribol.* **2009**, *224*, 299–303. [CrossRef]
14. Zhao, Y.; Cao, S.Q. Modal analysis of large marine crankshaft. *Appl. Mech. Mater.* **2012**, *271*, 1022–1026. [CrossRef]
15. Król, K.; Wikło, M.; Olejarczyk, K.; Kołodziejczyk, K.; Siemiątkowski, Z.; Żurowski, W.; Rucki, M. Residual stresses assessment in the marine diesel engine crankshaft 12V38 type. *J. KONES Powertrain Transp.* **2017**, *24*, 117–123.
16. Metkar, R.; Sunnapwar, V.K.; Hiwase, S.D. A fatigue analysis and life estimation of crankshaft—A review. *Int. J. Mech. Mater. Eng.* **2011**, *6*, 425–430.
17. Niezgodziński, M.; Niezgodziński, T. *Wytrzymałość Materiałów [Resistance of Materials]*; PWN: Warsaw, Poland, 1979.
18. Nozdrzykowski, K.; Chybowski, L.; Dorobczyński, L. Model-based estimation of the reaction forces in an elastic system supporting large-size crankshafts during measurements of their geometric quantities. *Measurement* **2020**, *155*, 107543. [CrossRef]
19. Henseler, J.; Ringle, C.M.; Sinkovics, R.R. *The Use of Partial Least Squares Path Modeling in International Marketing*; Emerald: Bingley, UK, 2009; pp. 277–319.
20. Bronshtein, I.; Semendyayev, K.; Musiol, G.; Mühlig, H. *Handbook of Mathematics*; Springer-Verlag Berlin and Heidelberg GmbH &, Co. KG: Berlin, Germany, 2015.
21. Kowalski, M. Jak Dopasować Fazę Sinusoidy do Pomiarów? [How to Fit Phase of a Sinusoidal to Measurements?]. TIWM. 2016. Available online: http://www.kowalskimateusz.pl/jak-dopasowac-faze-sinusoidy-do-pomiarow/ (accessed on 13 December 2019).
22. Taylor, J.R. *An Introduction to Error Analysis: The Study of Uncertainties in Physical Measurements*; University Science Books: Sausalito, CA, USA, 1997.
23. O'Connor, J.J.; Robertson, E.F. Quadratic, Cubic and Quartic Equations. Available online: http://mathshistory.st-andrews.ac.uk/HistTopics/Quadratic_etc_equations.html (accessed on 23 December 2019).
24. Kowalski, M. Octave Calculating Script. 2016. Available online: http://www.kowalskimateusz.pl/wp-content/uploads/2016/08/obliczenia.zip (accessed on 13 January 2019).

Article

Energy Optimization of the 'Shore to Ship' System—A Universal Power System for Ships at Berth in a Port

Sergey German-Galkin and Dariusz Tarnapowicz *

Faculty of Mechatronics and Electrical Engineering, Maritime University of Szczecin, 70-500 Szczecin, Poland; s.german-galkin@am.szczecin.pl

* Correspondence: d.tarnapowicz@am.szczecin.pl; Tel.: +48-9148-09-955

Received: 2 June 2020; Accepted: 6 July 2020; Published: 8 July 2020

Abstract: One of the most effective methods of limiting air pollution emissions by ships at a berth in a port is the power connection of ships to the on-shore system. "Shore to Ship" (STS)—A universal system for the connection of the ship's electrical power network with the on-shore network—ensures the adoption of the voltage and frequency of the on-shore network for the exploitation of various types of ships in the port. The realization of such a system is possible due to the use of semiconductor technologies during the construction of mechatronic systems (i.e., systems that ensure the maintenance of electricity parameters). The STS system ensures energy efficiency for high-power ship systems through the use of an active semiconductor converter. This article presents an analysis of steady state electromagnetic and energy processes, allowing the determination of the active and reactive power and losses in the STS system. The presented analytical research enables the development of a control algorithm that optimizes the system energy efficiency. In the article, the control methods allowing the optimization of the energy characteristics of the system are considered and investigated. On the basis of theoretical studies, a model was developed in the Matlab-Simulink environment, which allowed us to study steady and transient processes in the STS system in order to reduce losses in power lines and semiconductor converters.

Keywords: shore connection; "Shore to Ship" system; active and reactive power control; energy optimization; control of the active converter

---

1. Introduction

When mooring ships in a port, a source of electricity is required (i.e., for the operation of all systems). For this purpose, autonomous diesel generator (D-G) sets are commonly used. Depending on the power demand, from one to several D-G units can operate in a ship mooring at a quay. Each of these units can burn up to several tons of fuel every single day. D-G generating sets emit toxic compounds into the atmosphere. When analyzing various sources of air pollution in the port, it was stated that seagoing vessels are the main source [1–6].

Low-emission ports are an essential condition for global sustainable development. One of the most effective methods to reduce the emission of air pollution by ships is to use a "Shore to Ship" (STS) system; i.e., a universal system for the connection of the ship's electricity network with the on-shore network [1,6–9].

When the ship's power supply is provided from the onshore power grid, the ship's generating sets, which are the main source of air pollution and noise in the port, are turned off. When the generator sets are turned off, ship crews and port workers will benefit due to noise reduction (for example: the operation of D-G sets of 1 MW generates noise level of 140 dB) [1]. Apart from the environmental aspect related to the exclusion of autonomous ship generating sets, the crew has the possibility to carry out their repair and inspection. Another important aspect of shore-side deliveries is the reduction in electricity costs, as shore-side electricity is cheaper than energy from ships' generating sets [1].

The versatility of the STS system for the connection of various types of vessels to the on-shore power network requires global standardization of the system.

The main problem in the technical implementation of the STS system is connected with matching the frequency and voltage level of the ship's power network and the on-shore network [1,7–9]. In 2006, the European Union (EU) recommended the construction and development of systems for the collection of electricity from the on-shore network by ships moored in ports [10]. In accordance with the EU recommendations, member states (during the building of systems) should follow the technical solutions presented in the annex to these recommendations (chapter: Technical requirements—typical configuration). The chapter concerning technical requirements defines the main elements of the STS system. Figure 1 shows the configuration of the STS system based on the annex to these recommendations.

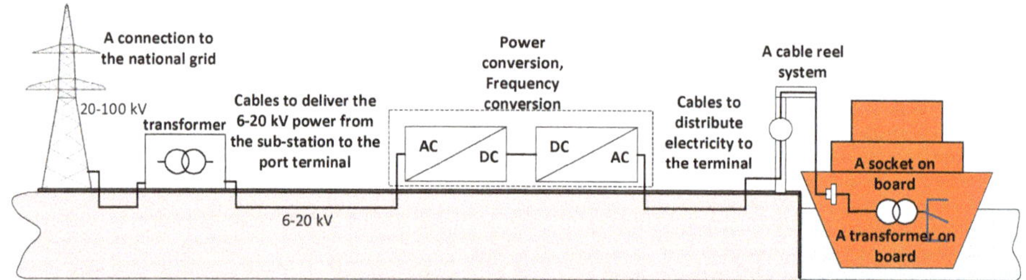

Figure 1. Configuration of the "Shore to Ship" (STS) system in accordance with the European Union (EU) recommendations [10].

The presented configuration of the STS system does not introduce technical details for the construction of the system but only presents a very general solution. In 2012, the IEC (International Electrotechnical Commission), ISO (International Organization for Standardization) and IEEE (Institute of Electrical and Electronics Engineers) developed a global standard that enabled the standardization of the design and construction of STS systems [11,12]. This standardization was intended for ships with a power demand above 1 MW and it was based on high-voltage (HV) installations. In 2014, the standardization of STS systems for ships with a power demand below 1 MW (based on low-voltage (LV) installations) was developed [12,13]. There may be many topologies of the STS system that meet the described standards [13,14]. Figure 2 presents an exemplary topology of the system, which is in accordance with the IEC/ISO/IEEE standardization.

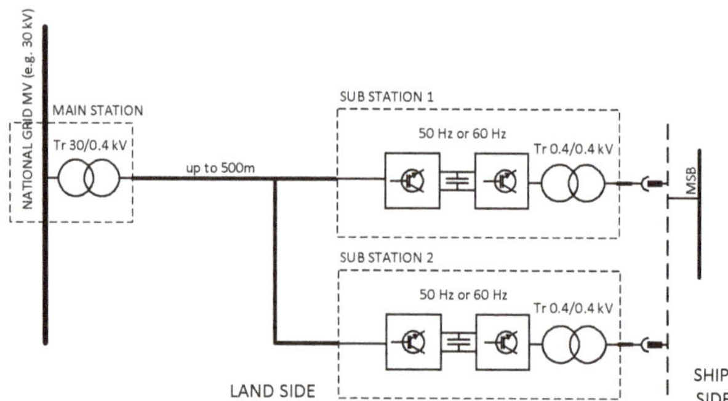

Figure 2. Selected topology of the STS system in accordance with the International Electrotechnical Commission (IEC)/International Organization for Standardization (ISO)/Institute of Electrical and Electronics Engineers (IEEE) standard.

In addition to the main feature of universality, the STS system should ensure uninterrupted switching from the ship's electricity network to the on-shore network and vice versa. The system is switched with the use of synchronization systems [15–18]. The switching of systems (national power grid to ship power grid) should be flexible; i.e., without dynamic load changes. When switching from the ship network to the on-shore network, power should be gently transferred from the ship's generating sets; therefore, the STS system must be able to convert power in both directions.

The article considers and examines the methods for the control of an active semiconductor converter that is able to optimize the energy performance of the system. A similar solution can be used for on-shore power networks.

2. Functional Diagram of the 'Shore to Ship' (STS) Mechatronic System

The main elements of the analyzed STS system are two active converters connected with each other with the use of a direct current (DC) circuit (with a capacitive filter). The general functionality of the system is presented in Figure 3.

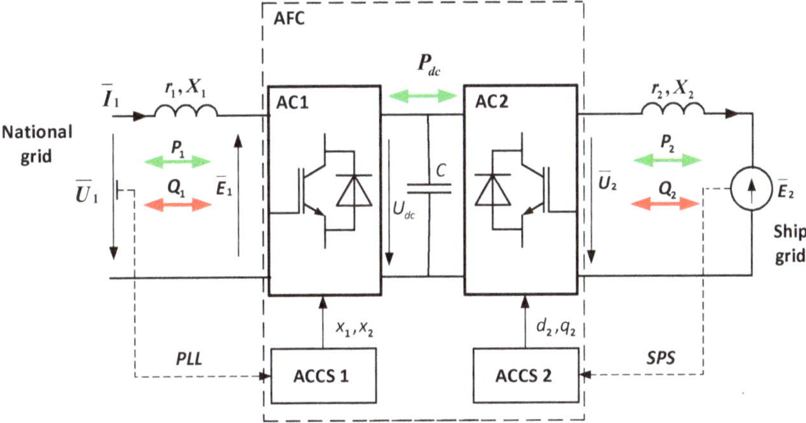

Figure 3. Functional diagram of the STS mechatronic system. AFC: Active frequency converter; AC: Active converter; ACCS: Control system of active converters.

The elements of the system (Figure 3) include the following:

- AFC—Active frequency converter;
- AC_1, AC_2—Active converters;
- ACCS 1, ACCS 2—Control system of active converters;
- $\overline{E}_1, \overline{U}_1, \overline{I}_1$—Electromotive force, voltage and current vectors in the AC_1;
- $\overline{E}_2, \overline{U}_2, \overline{I}_2$—Electromotive force, voltage and current vectors in the AC_2;
- $r_1,\ X_1 = \omega_1 L_1$—Resistance and reactance in the alternating current circuit of the AC_1 converter
- $r_2,\ X_2 = \omega_2 L_2$—Resistance and reactance of the ship load (or synchronous generator of the ship load)
- U_{dc}, C—Voltage and capacity in the direct current circuit.

The operation of AC_1 is synchronized with the network. The angular (fundamental $\omega_1 = 2\pi f_1$) frequency of this network is determined by the automatic phase frequency regulation system (PLL). The operation of AC_2 is synchronized with the generating set on the ship. For example, in systems with synchronous generators, a rotor position sensor (SPS) is used for this purpose. In such a case, the angular frequency is dependent on the mechanical speed of the shaft ($\omega_2 = p\omega_m$). AC_2 can also be connected to a passive load. In such a case, the electromotive force (E2) is replaced by the voltage drop on the load.

If the STS mechatronic system transfers active power from the on-shore network to the ship's network, AC_1 acts as an active rectifier, and AC_2 acts as an inverter. When the STS mechatronic system transfers active power from the ship's network to the on-shore network (when switching the ship's electrical network to the on-shore network), AC_1 acts as an inverter, and AC_2 is an active rectifier. This two-way energy conversion capability (B2B—Back to back) is the main advantage of this system; an additional advantage is its ability to reduce losses to provide a sinusoidal form of current in the electrical network.

In the common DC circuit, only active power (P_{dc}) is transmitted. In AC circuits, active power (P_1, P_2) is transmitted, while reactive power (Q_1, Q_2) is exchanged in the AC circuits of AC_1 and AC_2 converters. The active powers P_1, P_2 are rigidly connected with each other (equal in the case of the negligence of losses in AFC), and the reactive power depends on the method of controlling the AC_1 and AC_2 converters. These powers are not dependent in any way and can be controlled independently of each other (decoupled control).

In [19–21], methods of controlling active converters in an electric drive with a synchronous machine ensuring maintenance and zero reactive power were considered and analyzed.

In this article, the optimal control methods proposed in [19–21] were extended with the control of the AC_1 active converter connected to the on-shore network in the STS system.

Analytical and modeling tests were conducted on the basis of the analysis of electromagnetic processes in quasi-steady modes, described in the classical works of scientists in the field of electrical engineering, electromechanics and automatic control theory [22–29].

3. Analysis of the STS System with Independent Control of AC_1 Active Converter

The mathematical description of electromagnetic processes in the STS system is realized with the use of the spatial vector method [23,24] and method of fundamental components [25,26].

In relation to the mains, the active converter AC_1 (Figure 3) can be represented by electromotive force. Its fundamental component is equal to [27]

$$\begin{aligned} \overline{E}_1 &= \tfrac{m}{2} U_{dc} e^{-j\varphi_m} = E_{1y} - jE_{1x}, \\ E_{1x} &= \tfrac{m}{2} U_{dc} \sin \varphi_m, \\ E_{1y} &= \tfrac{m}{2} U_{dc} \cos \varphi_m. \end{aligned} \quad (1)$$

where:

m—The modulation factor;

φ_m—The shift angle between the network voltage vector $\overline{U}_1$, and electromotive force vector of AC_1 $\overline{E}_1$ (modulation angle);

E_{1x}, E_{1y}—The value of electromotive force in relation to the x and y axis.

The angle φ_m depends on the control signal and the load (current in the circuit). If the modulation factor m and angle φ_m are considered as control signals, they can be set in the control system to a certain extent and used to examine the electromagnetic and energy characteristics of the system. This method of controlling the inverter is called independent control.

Testing the STS system with an independent control method enables the determination of the limits for status variables and properties.

The equation for electromagnetic processes in a determined operating mode based on Kirchhoff's second law for the fundamental component of the STS system (Figure 3) can be presented in the following way:

$$\overline{U}_1 = \overline{E}_1 + r_1 \overline{I}_1 + jX_1 \overline{I}_1 \quad (2)$$

In order to explain the energy properties of the considered system, vector analysis was introduced. The STS vector graph in the synchronously rotating coordinate system (x, y), synchronized with the network, is presented in Figure 4. In Figure 4a, the vector graph corresponds to the case in which AC_1 operates in the mode of an active rectifier. For the case in which AC_1 operates in an inverter mode, the vector diagram of the system is shown in Figure 4b.

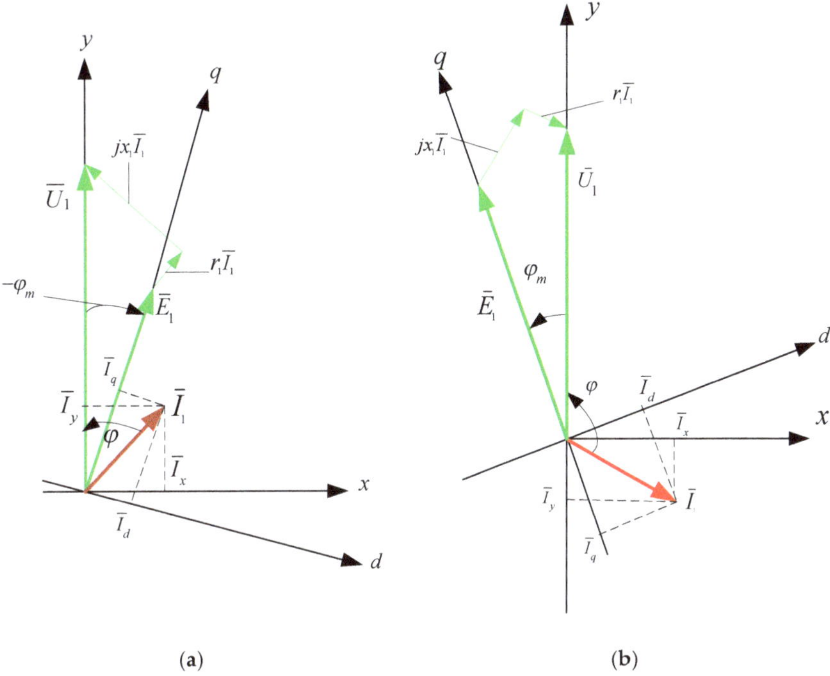

Figure 4. Vector diagram of the STS system, when AC_1 operates in a mode of an active rectifier (**a**) and inverter (**b**).

In addition to the x, y coordinates connected with the network voltage $\overline{U}_1$, so that $\overline{U}_1 = U_y, U_x = 0$, in Figure 4, the rotating axes d, q (connected with the electromotive force of the active converter CA_1) were marked in order to ensure that $\overline{E}_1 = E_q, E_d = 0$. In the steady state, both coordinate systems are fixed to each other. The AC_1 converter is controlled by the ACCS_1 control system in x, y axes.

In relation to the mains, the STS system consumes active power when $\varphi_m < 0^0$ (i.e., AC_1 acts as the active rectifier) or transfers active energy to the network $\varphi_m > 0^0$ (i.e., AC_1 acts as the inverter). The system with independent control is described by a system of non-linear equations. This system has problems connected with the stability of operation in a certain range of changes of control signals. Converter current control should be used for the reliable and stable operation of the STS system.

4. Analysis of the STS System with Current Control

In the active converter connected to the network via a parametric choke (r_1, X_1), current control is usually realized when the current in CA_1 is generated with the use of negative feedback from the transmitter [28,29]. In such a case, the set (control) signals are network currents (I_x, I_y) in the x and y axes. Electromagnetic and energy characteristics in the STS system with current control are calculated with the use of Equations (3) and (4) prepared on the basis of the geometrical relations of the vector graph presented in Figure 4.

$$\begin{aligned} \varphi_m &= -\mathrm{arct}g\frac{X_1 I_y - r_1 I_x}{U_1 - (X_1 I_x + r_1 I_y)}, \\ E_1 &= \sqrt{(U_1 - r_1 I_y - X_1 I_x)^2 + (X_1 I_y - r_1 I_x)^2}, \\ E_x &= E_1 \sin\varphi_m, \\ E_y &= E_1 \cos\varphi_m, \end{aligned} \tag{3}$$

$$\begin{aligned} P_{1(1)} &= 1.5 U_1 I_y, \\ Q_{1(1)} &= 1.5 U_1 I_x, \\ \Delta P_{1(1)} &= 1.5 r_1 (I_x^2 + I_y^2), \\ P_{AC(1)} &= 1.5 (E_x I_x + E_y I_y), \\ Q_{AC(1)} &= 1.5 (E_y I_x - E_x I_y), \\ \Delta P_{AC(1)} &= 1.5 r_{AC} (I_x^2 + I_y^2). \end{aligned} \tag{4}$$

where:

$P_{1(1)}$, $Q_{1(1)}$—the active and reactive power in the power network;
$P_{AC(1)}$, $Q_{AC(1)}$—the active and reactive power in the AC_1 active converter;
$\Delta P_{1(1)}$, $\Delta P_{AC(1)}$—the power losses in the network and in the AC_1 converter;
r_1—the resistance of the whole STS system;
r_{AC}—the equivalent resistance of the AC_1 converter.

The results of analytical tests presenting the energy characteristics for the STS system calculated on the basis of Equations (3) and (4) are presented in Figures 5–7 as dependences of active power $P_{1(1)}$ and reactive power $Q_{1(1)}$ in the load network (Figure 5), as dependences of active power $P_{AC(1)}$ and reactive power $Q_{AC(1)}$ in the active converter AC_1 (Figure 6) and as power losses in the whole STS system and the converter (Figure 7). The calculated energy characteristics calculated and constructed during the change of currents are I_x and I_y.

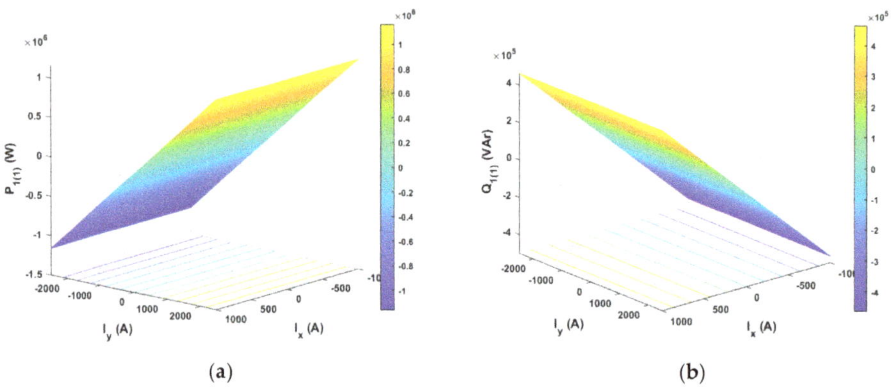

Figure 5. Active (**a**) and reactive (**b**) power of the fundamental component in the STS power supply network with current control.

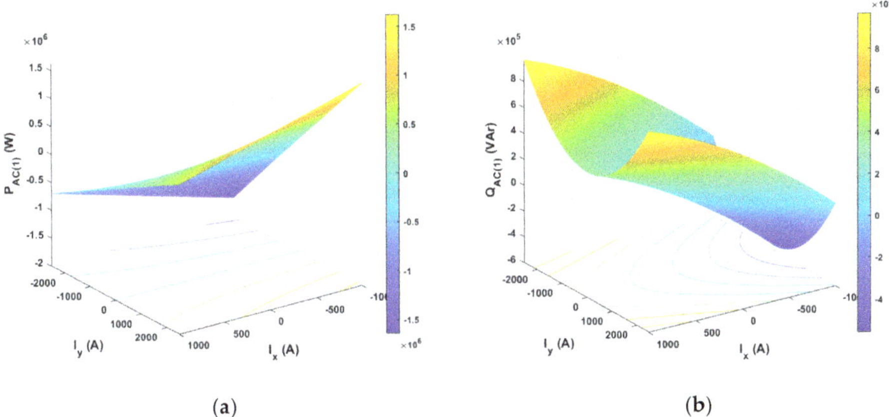

Figure 6. Active (**a**) and reactive (**b**) power of the fundamental component in the AC_1 converter with current control.

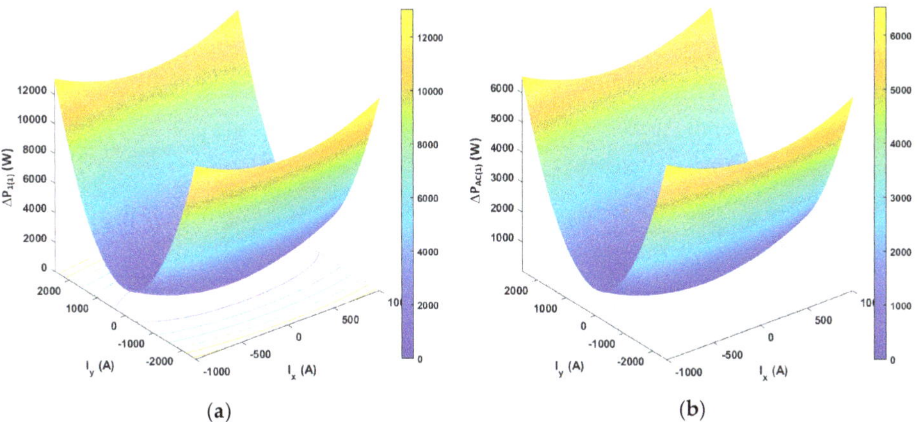

Figure 7. Power losses in the fundamental component (STS system) in the whole STS system (**a**) and in the AC_1 converter (**b**).

When evaluating the energy properties of the system with current control, the following conclusions can be drawn:

- Active power in the network only depends (linearly) on the current I_y, while reactive power in the network only depends (linearly) on the current I_x. Active and reactive power can have positive and negative values (Figure 5).
- Active and reactive power in the active converter depends (non-linearly) on the currents I_x and I_y. Powers can have positive and negative values (Figure 6).
- Power losses in the STS system and in the AC_1 active converter (parabolically) depend on currents I_y and I_x.

5. Analysis of the STS System with Optimized Current Control

The analysis of the energy performance of the STS system with current control AC_1 [19,20] proves that, with a certain dependency between control signals (I_x, I_y), it is possible to maintain reactive power in the supply network at a level of zero, regardless of the value and direction of active power flow transmitted by the STS system. This operating mode is determined as the optimized mode. It should be stressed that the optimized operating mode of the network does not coincide with the optimized mode of the AC_1 active converter. This difference results from the presence of a choke in the common circuit with parameters r_1, X_1.

For the analysis of the STS system with optimized power consumption from the network, Figure 8 presents vector charts with a phase shift between current and voltage in the network of 0° (AC_1 is an active rectifier) or 180° (AC_1 is a network inverter).

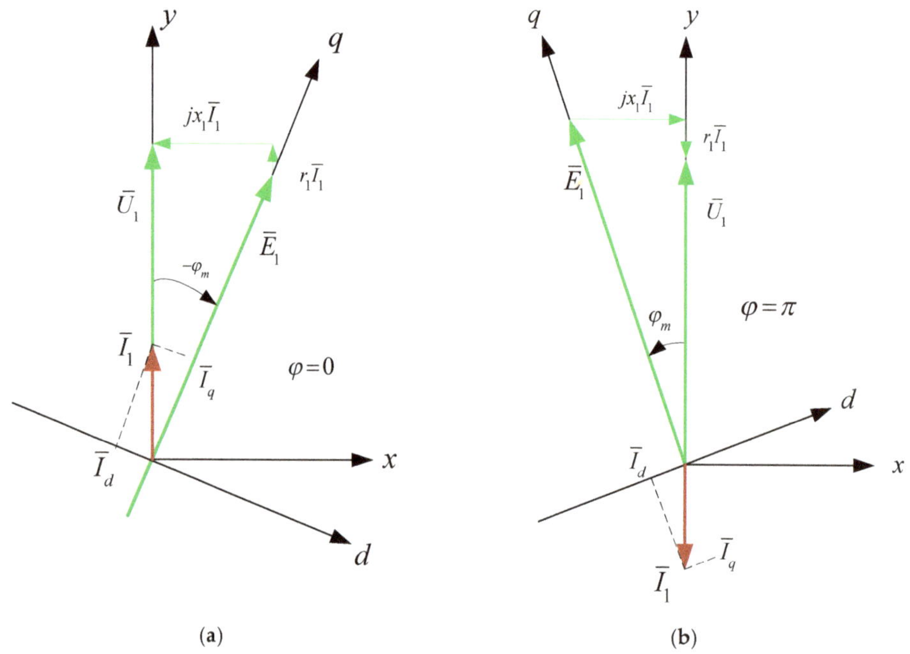

Figure 8. Vector charts of the optimized STS system when AC_1 operates (**a**) in the mode of an active rectifier and (**b**) in the mode of a network inverter.

The control of the active converter (AC_1) is realized with the use of a PLL (phase-locked loop) in x and y axes. In this case, the condition of optimality is met when $\bar{I}_1 = I_1 = I_y, I_x = 0$. Electromagnetic and energy characteristics are calculated on the basis of Equations (5) and (6) received from the geometrical dependencies of the vector diagram (Figure 8). They have the following form:

$$\begin{aligned} &\varphi_m = \text{arctg}\tfrac{X_1 I_1}{U_1 - r_1 I_1}, \\ &I_d = I_1 \sin(-\varphi_m), \\ &I_q = I_1 \cos(-\varphi_m), \\ &E_1 = \sqrt{(U_1 - r_1 I_1)^2 + (X_1 I_1)^2}, \end{aligned} \tag{5}$$

$$\begin{aligned} &P_{1(1)} = 1.5 U_1 I_1, \\ &Q_{1(1)} = 0, \\ &\Delta P_{1(1)} = r_1 I_1^2. \\ &P_{AC(1)} = 1.5 E_1 I_q, \\ &Q_{AC(1)} = 1.5 E_1 I_d, \\ &\Delta P_{AC(1)} = 1.5 r_{AC} (I_d^2 + I_q^2). \end{aligned} \tag{6}$$

The energy characteristics of the STS system for the optimization of the power consumption from the network, calculated with the use of Equations (5) and (6), are presented in Figure 9.

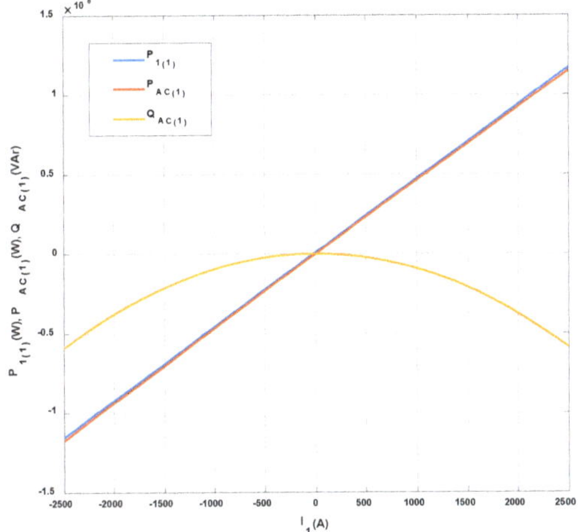

Figure 9. Energy characteristics of the optimized STS system.

During the optimization of power consumption from the network, the reactive power in the AC_1 converter is negative, both in the mode of an active rectifier and in the mode of a network inverter. This means that the network for the AC_1 converter is an active–capacitive load. In this case, the electromotive force (E_1) is greater than the voltage U_1 (Figure 8).

Current control in the AC_1 converter takes place within the limits determined by the ratio of the U_1 network voltage and electromotive force at the AC terminals of the E_1 converter. In the time interval in which the instantaneous valve e_1 exceeds the instantaneous valve u_1, the relay current regulator (in the control system) goes into the state of saturation and loses the ability to switch the converter's transistors. During this time, there is no switching to the converter's branches. This enables the relative time during which the converter's transistors do not switch to be calculated, which can be presented (approximately) in the following way:

$$\omega t \simeq \pi - 2\arcsin\frac{U_1}{E_1} \simeq \pi - 2\arcsin\frac{U_1}{\sqrt{(U_1 - r_1 I_1)^2 + (X_1 I_1)^2}} \tag{7}$$

The presence of such an interval leads to a reduction in losses of ΔP_{AC} in semiconductor elements of the converter; this enables the calculation of the relative coefficient for the loss reduction:

$$\lambda \simeq 1 - \frac{2}{\pi}\arcsin\frac{U_1}{\sqrt{(U_1 - r_1 I_1)^2 + (X_1 I_1)^2}} \tag{8}$$

The dependency of the relative loss reduction coefficient in relation to the current for the STS-optimized system is presented in Figure 10. In abscissae, the current is shown in relative units (pu) defined as the ratio of real current I_1 to short-circuit current $I_{sc} = \frac{U_1}{z_1}$:

$$I_{pu} = \frac{I_1}{I_{sc}} = \frac{z_1 I_1}{U_1} \tag{9}$$

Power losses in the AC_1 converter (taking into account the relative loss reduction coefficient, λ) are determined in accordance with the following formula:

$$\Delta P_{AC(\lambda,1)} = (1 - \lambda)\Delta P_{AC(1)} \tag{10}$$

where:

$\Delta P_{AC(1)}$—Power losses in the AC_1 converter determined on the basis of Equation (6).

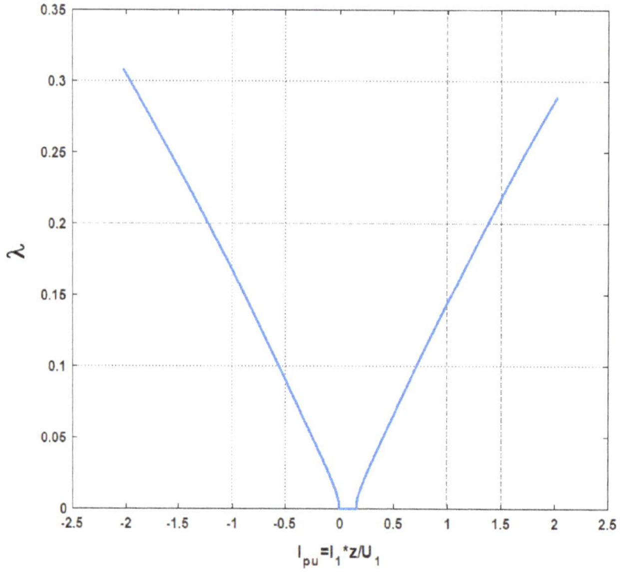

Figure 10. Dependence of the relative loss reduction coefficient λ in the function of the current of the active converter.

The analysis of the STS energy characteristics can be conducted taking into account the losses in the AC_1 converter to the fundamental component, $\Delta P_{AC(1)}$, and the reduction in switching losses λ in the converter, $\Delta P_{AC(\lambda,1)}$.

For the final conclusion regarding losses in the AC_1 converter, it is necessary to calculate not only the fundamental component (taking into account the reduction in switching losses, λ), but also the power losses at all other fundamental components that occur during the switching of transistors to the carrier frequency. These losses will be called switching losses.

When calculating switching losses in the active converter, a distinction should be made between currents in the converter phase and currents in semiconductor elements of the same converter phase. Currents in phases are responsible for power transfer; these currents are sinusoidal, because the transmission of power takes place at the fundamental frequency (modulation frequency). Currents in semiconductor elements are responsible for energy exchange. The exchange is carried out with the carrier frequency. A significant distortion of currents has an impact on switching losses, which can be taken into account with the use of the total harmonic distortion (THD) according to the following formula:

$$\Delta P_{AC(\lambda,n)} = \left[1 + (THD)^2\right]\Delta P_{AC(1)} \tag{11}$$

where:

$\Delta P_{AC(\lambda,n)}$—Switching losses in the AC_1 converter, taking into account all fundamental components, including the loss reduction coefficient, λ.

In the current control, the switching frequency (carrier frequency) in the AC_1 converter depends on the choke time constant ($\tau_1 = \frac{L_1}{r_1}$), the width of the hysteresis loop of the relay current regulator in the control system and the instantaneous value of the current I_1. The switching frequency value for the systems (medium and high power) is in the range of 1–10 kHz.

The calculation of all energy indicators cannot be realized in an analytical way. This is still implemented with the use of imitation (virtual) models in the Matlab-Simulink environment.

6. STS System Studies

The block diagram of the STS system is shown in Figure 11.

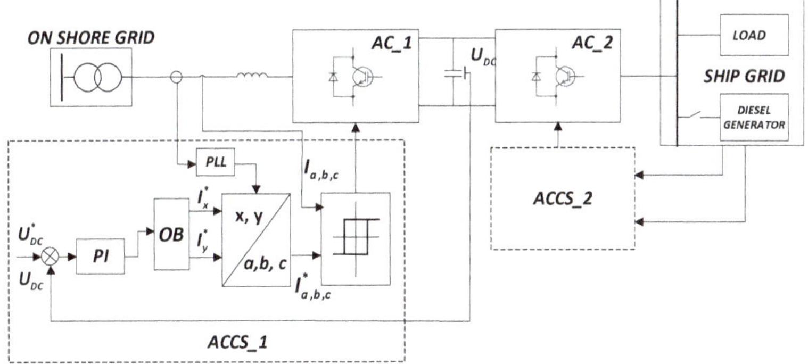

Figure 11. Block diagram of the STS system.

The main task of the STS system is electric energy transfer from the onshore grid to the ship's grid. In this mode of operation of the STS system the AC_1 converter works as an active rectifier. The optimization block (OB) ensures the energy optimization of the system. As a result of the transformation of set currents I^*_x, I^*_y to the three-phase a, b, c coordinate system the preset currents $I^*_{a,b,c}$ were obtained, which in the tracking system (hysteresis) will provide a constant voltage value U_{dc} in the DC circuit. An analogous control algorithm is realized in the ACCS_2 system. Switching between on shore grid and ship grid systems in transition states ensures controlled (soft) transfer of active power between both systems.

The model of the examined system is presented in Figure 12.

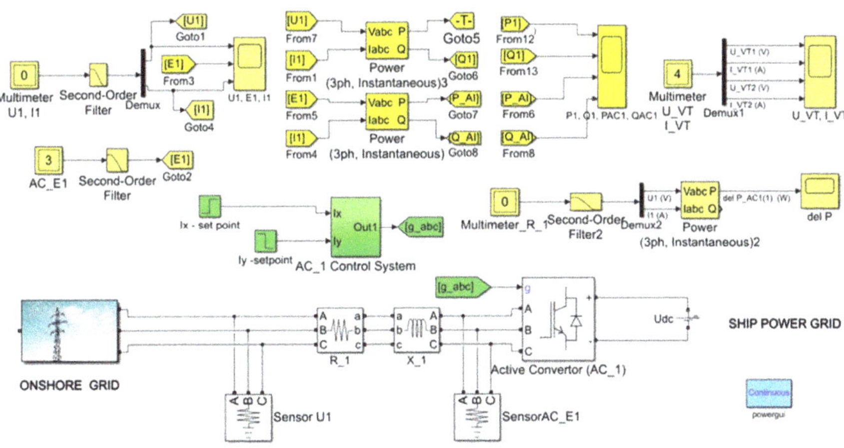

Figure 12. Simulation model of STS system.

The control of transmission in both directions of active power P_1 and reactive power Q_1 in the network is realized by current I_x and I_y (Figure 12; Ix-set point, Iy-set point). The study compared a system with and without optimized power transmission control. The AC_1 converter is built as a secondary voltage stabilization circuit on the capacitor in the DC circuit (Figure 11) [20,27]. In order to minimize the calculations in the simulation model (Figure 12), a DC voltage source was used in the DC circuit, which allowed the possibility of control of active power flow in both directions to be checked, with a possibility to check the idea of energy optimization of the STS system. The use of a DC voltage source in a simulation model provided the control concept shown in the block diagram of Figure 11.

The results of tests for the optimized STS system are indicated in Figures 13–16.

The system is tested in two steady state conditions:

1. The state of power transmission from the mains to the ship's network. In this case, active power is transferred from the on-shore network to the ship's network. The AC_1 converter operates in the mode corresponding to the active rectifier.
2. The state of power transmission from the ship's network to the network. In this case, active power is transferred from the ship's network to the on-shore network. The AC_1 converter operates in the mode corresponding to the network inverter.

The system is tested in a transient status (with changes in the direction of active power's transfer).

The energy processes in the system ($P_{1(1)}$, $Q_{1(1)}$, $P_{AC1(1)}$, $Q_{AC1(1)}$) are presented in Figure 13. In the determined operating modes of the system, the reactive power in the network $Q_{1(1)}$ is zero. A change in reactive power is observed during the transition of AC_1 from the active rectifier mode to the network inverter mode (Figure 13). The reactive power in the converter (in steady states) is always negative and changes only in the transient state.

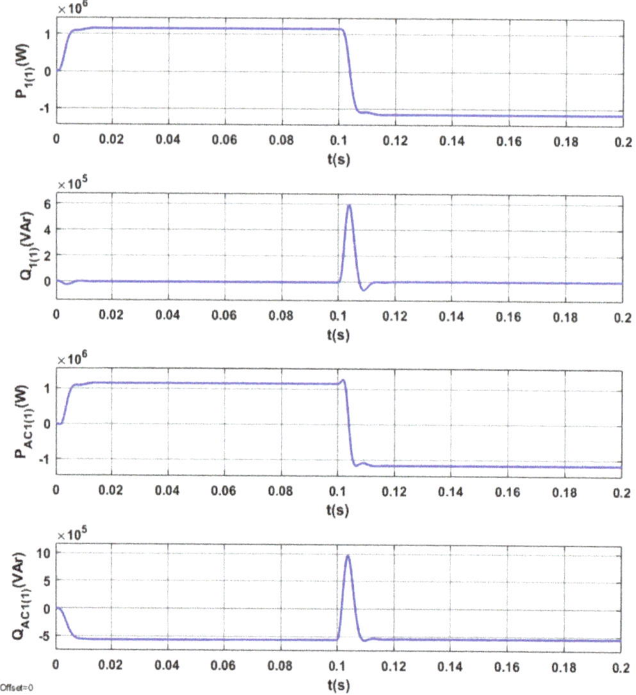

Figure 13. Energy processes $P_{1(1)}$, $Q_{1(1)}$, $P_{AC1(1)}$, $Q_{AC1(1)}$ in the optimized STS system, when switching AC_1 from the operating mode corresponding to the active rectifier to the operating mode with the network inverter.

The power losses in the optimized STS system in steady and transient states in the case of changes in the direction of active power transmission are shown in Figure 14.

The electromagnetic processes in the U_1, E_1, I_1 system in steady and transient states when switching AC_1 from the operating mode corresponding to the active rectifier to the operating mode corresponding to the inverter are presented in Figure 15. In steady states, electromagnetic variables are sinusoidal.

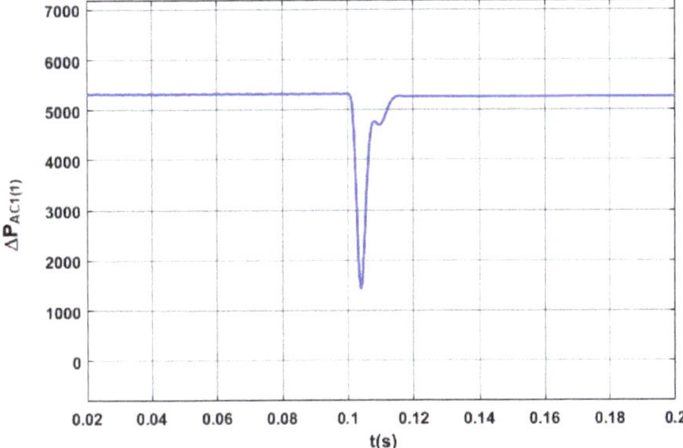

Figure 14. Power losses in the AC_1 converter ($\Delta P_{AC(1)}$) in the optimized STS system for the fundamental component, when switching the AC_1 from the operating mode corresponding to the active rectifier to the operating mode with the network inverter.

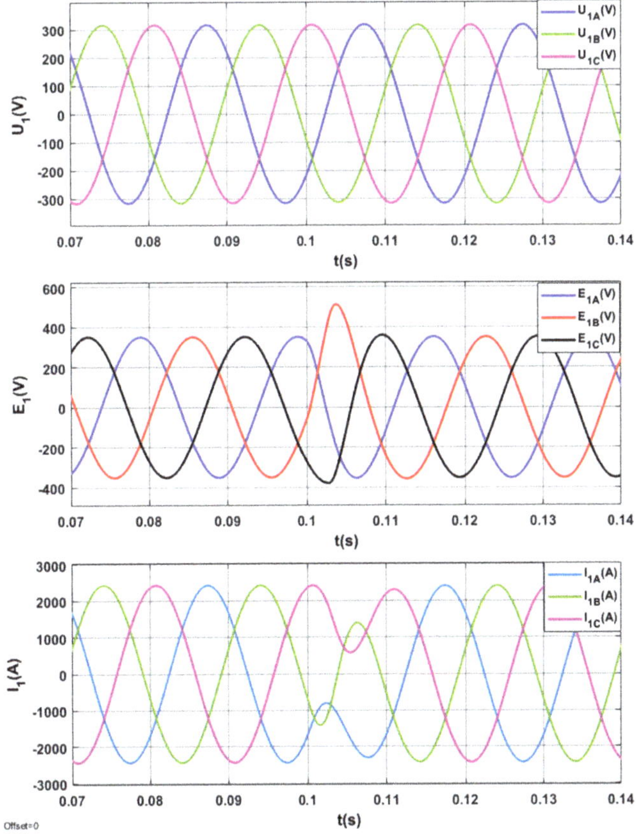

Figure 15. Electromagnetic processes in the optimized STS system, when switching the AC_1 from the operating mode corresponding to the active rectifier to the operating mode with the network inverter.

Figure 16 presents the waveforms of voltage and current for the semiconductor elements in one branch of the AC_1 converter in the optimized STS system for two operating modes corresponding to different directions of active power in the STS system.

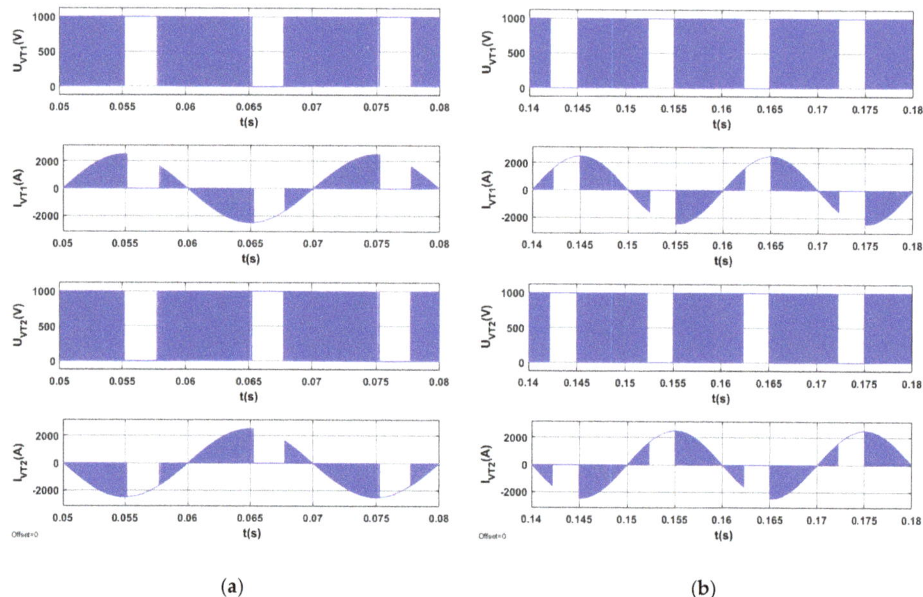

Figure 16. Voltages and currents on semiconductor elements of AC_1's branch in the optimized STS system in the operating mode of an active rectifier (**a**) and in the operating mode of a network inverter (**b**).

The results of tests for the non-optimized STS system are presented in Figures 17–20.

The system is tested in the state of active power transfer from the on-shore network to the ship's network. The system is tested in a steady state and in a transient state during the change of reactive power.

Energy processes in the system ($P_{1(1)}$, $Q_{1(1)}$, $P_{AC1(1)}$, $Q_{AC1(1)}$) when switching the AC_1 from the operating mode corresponding to positive (inductive) reactive power to the operating mode corresponding to negative (capacitive) reactive power are presented in Figure 17.

Power losses in the STS system in the discussed operating modes are presented in Figure 18. Compared with Figure 15, where $Q_{1(1)} = 0$, power losses in steady states increased by more than 10%.

The electromagnetic processes in the system (U_1, E_1, I_1) when switching AC_1 from the operating mode corresponding to positive (inductive) reactive power ($Q_{1(1)} > 0$) to the operating mode corresponding to negative (capacitive) reactive power ($Q_{1(1)} < 0$) are presented in Figure 19.

Voltages and currents in semiconductor elements (U_{VT1}, I_{VT1}, U_{VT2}, I_{VT2}) for AC_1's branch are presented in Figure 20. Figure 20 shows that, in the case of $Q_{1(1)} > 0$, there is no gap in the switching processes of semiconductor elements. In the case in which $Q_{1(1)} < 0$, the switching processes of semiconductor elements do not occur in a time interval lasting approximately 0.4 of the period's duration.

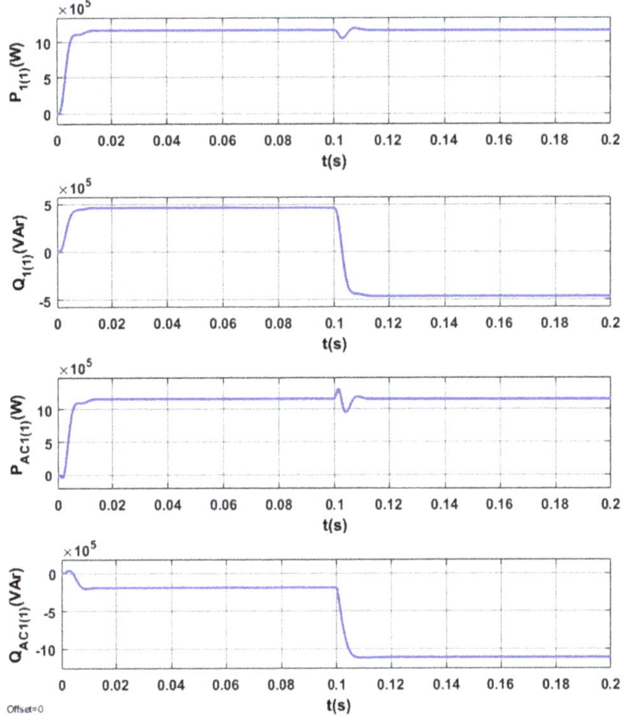

Figure 17. Energy processes $P_{1(1)}$, $Q_{1(1)}$, $P_{AC1(1)}$, $Q_{AC1(1)}$ in the STS system, taking into account the operating mode of the AC_1 corresponding to the active rectifier (I_y = 2500 A) and I_x changing from 1000 A to −1000 A.

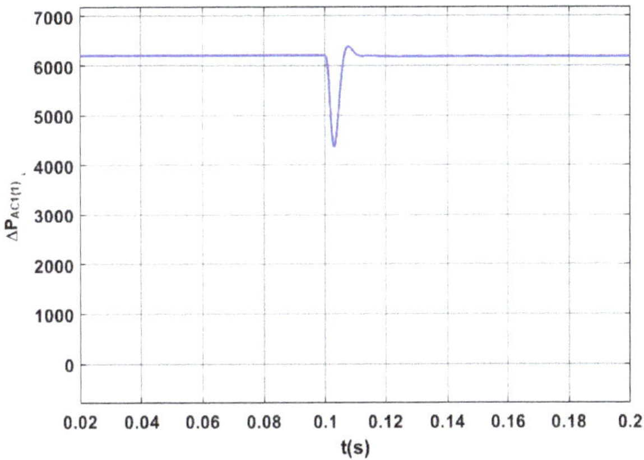

Figure 18. Power losses for the fundamental component in the STS system, taking into account the operating mode of the AC_1 corresponding to the active rectifier (I_y = 2500 A) and I_x changing from 1000 A to −1000 A.

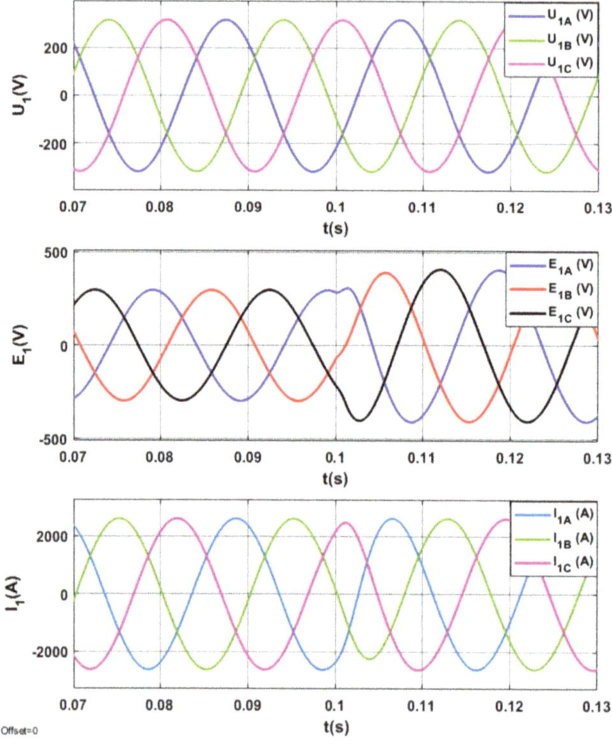

Figure 19. Electromagnetic processes (U_1, E_1, I_1) in the STS system corresponding to the active rectifier (I_y = 2500 A) and I_x changing from 1000 A to −1000 A.

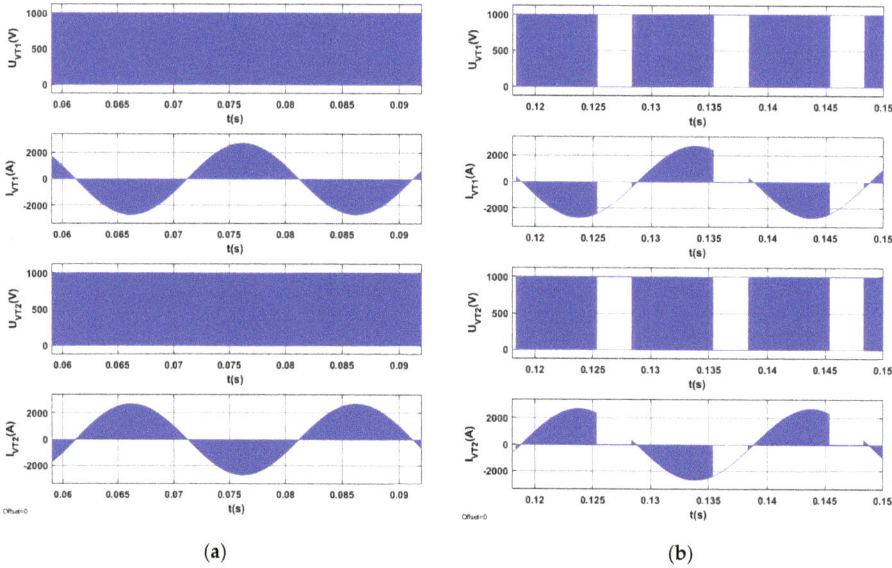

Figure 20. Voltages and currents on semiconductor elements of AC_1's branch, which operates in the active rectifier mode in the STS system, (**a**) Iy = 2500 A, Ix = 1000 A, (**b**) Iy = −2500 A, Ix = −1000 A.

The current spectrum of one semiconductor element in the optimized STS system (where operation of AC_1 corresponds to the active rectifier) is presented in Figure 21a.

The current spectrum of one semiconductor element in the STS system (at $Q_{1(1)} > 0$ corresponding to the active rectifier) is presented in Figure 21b.

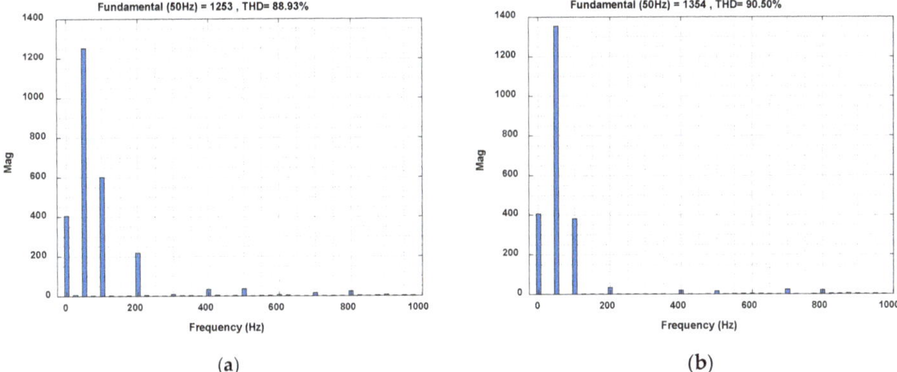

Figure 21. Current spectrum in the semiconductor element of the AC_1 converter operating in the active rectifier mode with the optimized STS system (**a**) and I_y = 2500 A, I_x = 1000 A (**b**) for two of the 10 cycles of the selected current signal I_{VT1}.

Table 1 presents a comparison of the STS system with and without energy optimization in terms of power losses in the transmission line ($\Delta P_{1(1)}$), power loss (fundamental component) in the AC_1 converter ($\Delta P_{AC(1)}$) and switching losses in AC_1 ($\Delta P_{1AC(\lambda,n)}$).

The formulas (4), (11) and the results of the tests presented in Figures 14, 18 and 21 were used in the calculations. The calculations were made for the resistance of the transmission lines $r_1 = 6 \times 10^{-4}$ Ohm and constant active power $P_{1(1)}$ = 1.2 MW. The results are presented for the STS system without optimization $Q_{1(1)} > 0$ (I_y = 2500 A, I_x = 1000 A) and with optimization $Q_{1(1)} = 0$, (I_y = 2500 A, I_x = 0 A).

Table 1. Comparison of power losses in STS system with and without energy optimization.

Power Losses	Power Losses in the STS System without Optimization [W]	Power Losses in the STS System with Optimization [W]	Reduction of Losses through the Use of Optimization [%]
$\Delta P_{1(1)}$	6525	5625	13.8
$\Delta P_{AC(1)}$	6210	5320	14.3
$\Delta P_{1AC(\lambda,n)}$	11296	9527	15.7

It should be noted that for STS systems of higher power (e.g., for STS systems supporting passenger ships, the transferred power is from a dozen to several dozen MW) and with a lower power factor, the value of the proposed system's energy optimization will be more significant.

7. Conclusions

The basic problem in the design and implementation of the STS network is connected with the increase in energy efficiency. Currently, the energy efficiency, normally understood as the ratio of active output power to active input power (P_2/P_1), is relatively large, and it is difficult to expect significant improvement. However, enterprises (companies and production plants) incur costs that are not only connected with active power consumption but also with reactive power consumption.

As mentioned above, it is difficult to minimize costs related to active power consumption. The issue of reactive power compensation is still being developed. Apart from reactive power compensators (constructed from capacitors—passive compensators (PC)), it is also possible to build active compensators (AC), which use active converters (energy electronic converters).

Energy optimization in the STS system (which reduces the reactive power in the supply network to zero) reduces the loss in the entire system for the fundamental components and decreases switching losses on the carrier frequency in the semiconductor elements of the active converter by approximately 15–30%.

In the active converter (when switching semiconductor elements), there is an increase in switching losses proportional to THD. These losses should be taken into account during the calculation of energy indicators.

In addition to the optimization of the above quantities, active converters are also able to change the direction of power flow. The discussed conclusions were confirmed by the results of analytical and model tests; our contributions in this paper are, therefore, as follows:

- A method of analysis of steady state electromagnetic and energetic processes in the "port power grid–active converter" circuit has been developed.
- An analytical method has been developed to determine the active and reactive power in the port power supply network and active converter, as well as to determine power losses in this circuit.
- The optimal control of the active converter was synthesized when the reactive power in the port power supply network was equal to zero.
- For the developed optimal control algorithm, the power in the port's power supply network and in the active converter was calculated.
- The developed method of calculating switching losses was based on the spectral analysis of currents in semiconductor elements of the active converter.
- A simulation (virtual) model of the "port power network–active converter" system has been developed.

The comparison of the simulation results with theoretical results allows us to recommend the developed methods of analysis and calculation for the design of an STS in a wide power range, as well as for use in medium-voltage STS systems. Transmission of reactive energy through power grids has many negative effects, such as voltage drops, heating of cables, and reduction of capacity of infrastructure used for energy transmission. The importance of the proposed increase in the energy efficiency of the STS system (the study examined an STS system with a power of about 1 MW) is growing for systems with higher power due to the economic effects.

It should also be emphasized that the proposed strategy to control the converter with energy optimization has reduced transistor power loss during switching by reducing the number of switches over the period. This increase in the converter's efficiency results in the improvement of the efficiency of the whole STS system, which is particularly important for high-power systems.

Author Contributions: Conceptualization, S.G.-G. and D.T.; methodology, S.G.-G. and D.T.; software, S.G.-G. and D.T.; validation, S.G.-G. and D.T.; formal analysis, S.G.-G. and D.T.; investigation, S.G.-G. and D.T.; resources, S.G.-G. and D.T.; data curation, S.G.-G. and D.T.; writing—original draft preparation, S.G.-G. and D.T.; writing—review and editing, S.G.-G. and D.T.; visualization, S.G.-G. and D.T.; supervision, S.G.-G. and D.T.; project administration, S.G.-G. and D.T.; funding acquisition, S.G.-G. and D.T. All authors have read and agreed to the published version of the manuscript.

Funding: This research received no external funding.

Conflicts of Interest: The authors declare no conflict of interest.

References

1. Tarnapowicz, D.; Borkowski, T. *Shore to Ship System: Alternative Power Supply of Ships in Ports*; Scientific Publishing House of the Maritime University: Szczecin, Poland, 2014; ISBN 9788364434037 8364434039.
2. Borkowski, T.; Nicewicz, G.; Tarnapowicz, D. *Ships Mooring in the Port as a Threat to our Natural Environment*; Management Systems in Production Engineering: Szczecin, Poland, 2012; pp. 22–27.

3. Yuan, J.; Wang, H.; Ng, S.H.; Nian, V. Ship Emission Mitigation Strategies Choice Under Uncertainty. *Energies* **2020**, *13*, 2213. [CrossRef]
4. Peterson, K.; Chavdarian, P.; Islam, M.; Cayanan, C. Tackling ship pollution from the shore. *IEEE Ind. Appl. Mag.* **2009**, *15*, 56–60. [CrossRef]
5. Bailey, D.; Solomon, G. Pollution prevention at ports: Clearing the air. *Environ. Impact Assess. Rev.* **2004**, *24*, 749–774. [CrossRef]
6. Prousalidis, J.; Antonopoulos, G.; Patsios, C.; Greig, A.; Bucknall, R. Green shipping in emission controlled areas: Combining smart grids and cold ironing. In Proceedings of the 2014 International Conference on Electrical Machines (ICEM), Berlin, Germany, 2–5 September 2014; pp. 2299–2305. [CrossRef]
7. Sulligoi, G.; Bosich, D.; Pelaschiar, R.; Lipardi, G.; Tosato, F. Shore-to-Ship Power. *Proc. IEEE* **2015**, *103*, 749–774. [CrossRef]
8. Vlachokostas, A.; Prousalidis, J.; Spathis, D.; Nikitas, M.; Kourmpelis, T.; Dallas, S.; Soghomonian, Z.; Georgiou, V. Ship-to-grid integration: Environmental mitigation and critical infrastructure resilience. In Proceedings of the 2019 IEEE Electric Ship Technologies Symposium (ESTS), Arlington, VA, USA, 14–16 August 2019. [CrossRef]
9. Khersonsky, Y.; Islam, M.; Peterson, K.L. Challenges of Connecting Shipboard Marine Systems to Medium Voltage Shoreside Electrical Power. *IEEE Trans. Ind. Appl.* **2007**, *43*, 838–844. [CrossRef]
10. Commission Recommendation of 8 May 2006 on the Promotion of Shore-Side Electricity for Use by Ships at Berth in Community Ports (Text with EEA Relevance) (2006/339/EC). Available online: https://eur-lex.europa.eu/legal-content/EN/TXT/?uri=CELEX% (accessed on 3 March 2020).
11. IEC/IEEE International Standard—Utility Connections in Port—Part 1: High Voltage Shore Connection (HVSC) Systems—General Requirements. Edition 2.0 March 2019. Available online: https://ieeexplore.ieee.org/document/8666180 (accessed on 3 March 2020).
12. Tarnapowicz, D.; German-Galkin, S. International Standardization in the Design of "Shore to Ship"—Power supply systems of ships in port. *Manag. Syst. Prod. Eng.* **2018**, *26*, 9–13. [CrossRef]
13. IEC/ISO/IEEE 80005-3 Utility Connections in Port—Part 3: Low Voltage Shore Connection (LVSC) Systems—General Requirements, Edition 1.0, 2014. Available online: https://ieeexplore.ieee.org/document/7822875 (accessed on 3 March 2020).
14. Tarnapowicz, D.; German-Galkin, S. Analysis of the Topology of "Shore to Ship" Systems—Power Electronic Connection of Ships with Land. *New Trends Prod. Eng.* **2018**, *1*, 325–333. [CrossRef]
15. Strzelecki, R.; Mysiak, P.; Sak, T.; Ryszard, S. Solutions of inverter systems in Shore-to-Ship Power supply systems. In Proceedings of the 9th International Conference on Compatibility and Power Electronics (CPE), Costa da Caparica, Portugal, 24–26 June 2015. [CrossRef]
16. Smolenski, R.; Benysek, G.; Malinowski, M.; Sedlak, M.; Stynski, S.; Jasinski, M. Ship-to-Shore Versus Shore-to-Ship Synchronization Strategy. *IEEE Trans. Energy Convers.* **2018**, *33*, 1787–1796. [CrossRef]
17. Tarnapowicz, D. Synchronization of national grid network with the electricity ships network in the "shore to ship" system. *Manag. Syst. Prod. Eng.* **2013**, *12*, 9–13.
18. Sedlak, M.; Stynski, S.; Malinowski, M.; Jasinski, M.; Benysek, G.; Smolenski, R. Stress free shore to ship (S2SP) electrical power networks synchronization. In Proceedings of the 10th International Conference on Compatibility, Power Electronics and Power Engineering (CPE-POWERENG), Bydgoszcz, Poland, 29 June–1 July 2016. [CrossRef]
19. German-Galkin, S. MATLAB School: Lesson 26. Optimization of energy properties of a mechatronic system with a electric machine. *Power Electron.* **2018**, *3*, 56–63.
20. German-Galkin, S. MATLAB School: Lesson 28. Synthesis of optimal control of a generating set with a electric machine. *Power Electron.* **2019**, *1*, 56–63.
21. Tarnapowicz, D.; German-Galkin, S. Energy optimization of mechatronic systems with PMSG. *E3S Web Conf.* **2018**, *46*, 16. [CrossRef]
22. Park, R.H. Two-reaction theory of synchronous machines-II. *Trans. Am. Inst. Electr. Eng.* **1933**, *52*, 352–354. [CrossRef]
23. Kovacs, K.P.; Racz, I. *Transient Processes in Alternating Current Machines; Transiente Vorgänge in Wechselstrommaschinen*; Verlag der Ungar: Budapest, Hungary, 1959.
24. Asanbayev, V. *Alternating Current Multi-Circuit Electric Machines*; Springer: Cham, Switzerland, 2015; ISBN -978-3-319-10108-8.

25. Bulgakov, A.A. *New Theory of Controlled Rectifiers*; Science: Moscow, Russia, 1970.
26. Wang, H.; Zhang, S.; Zhang, J. Application of a novel soft phase-locked loop in three-phase PWM rectifier based on coordinate transformation. In Proceedings of the IEEE Conference and Expo Transportation Electrification Asia-Pacific (ITEC Asia-Pacific), Beijing, China, 31 August–3 September 2014. [CrossRef]
27. Rozanov, Y.K.; Ryabchitsky, M.V.; Kvasnyuk, A.A. *Power Electronics*; Publishing House MPEI: Moscow, Russia, 2007; ISBN 978-5-383-00169-1.
28. Brodsky, V.N.; Ivanov, E.S. *Frequency-Controlled Drives*; Energy: Moscow, Russia, 1974; p. 168.
29. German-Galkin, S.; Tarnapowicz, D.; Matuszak, Z.; Jaskiewicz, M. Optimization to Limit the Effects of Underloaded Generator Sets in Stand-Alone Hybrid Ship Grids. *Energies* **2020**, *13*, 708. [CrossRef]

Article

Full-Scale Maneuvering Trials Correction and Motion Modelling Based on Actual Sea and Weather Conditions

Bin Mei [1,*], Licheng Sun [1] and Guoyou Shi [2]

1 Navigation College, Dalian Maritime University, Dalian 116026, China; navi_captain@foxmail.com
2 Collaborative Innovation Research Institute of Autonomous Ship, Dalian Maritime University, Dalian 116026, China; allenimitsg@163.com
* Correspondence: meibindmu@dlmu.edu.cn

Received: 21 June 2020; Accepted: 12 July 2020; Published: 16 July 2020

Abstract: Aiming at the poor accuracy and difficult verification of maneuver modeling induced by the wind, waves and sea surface currents in the actual sea, a novel sea trials correction method for ship maneuvering is proposed. The wind and wave drift forces are calculated according to the measurement data. Based on the steady turning hypothesis and pattern search algorithm, the adjustment parameters of wind, wave and sea surface currents were solved, the drift distances and drift velocities of wind, waves and sea surface currents were calculated and the track and velocity data of the experiment were corrected. The hydrodynamic coefficients were identified by the test data and the ship maneuvering motion model was established. The results show that the corrected data were more accurate than log data, the hydrodynamic coefficients can be completely identified, the prediction accuracy of the advance and tactical diameters were 93% and 97% and the prediction of the maneuvering model was accurate. Numerical cases verify the correction method and full-scale maneuvering model. The turning circle advance and tactical diameter satisfy the standards of the ship maneuverability of International Maritime Organization (IMO).

Keywords: full-scale maneuvering; trials correction; motion modeling; actual sea and weather conditions; reference model and support vector machine (RM-SVM); standards for ship maneuverability

---

1. Introduction

During sea trial, ship motions include maneuvering and drifting. Drift motion is caused by the wind and waves at sea, and the ship shows slow, long periods of movement and even steady movement [1]. In order to obtain accurate trial data, the correction of this is an important step for ship maneuvering modeling. Dating back to 1978, Abkowitz utilized Esso Osaka for sea trials, identified the ship maneuvering mathematical model and verified the feasibility of the identification modeling method [2]. Recently, Zhang et al. [3], Bai et al. [4] and Kim et al. [5] also used full-scale ship data for identification modeling. In the literature [2–4], it should be noted that the log has also been installed underwater, on the ship hull, which is prone to suffering from cross flow, in addition to the ship being affected by the drift forces of wind and wave. Kim et al. [5] employed the method seen in the literature [6–8] to correct the sea trial data and identified the ship maneuvering model, but did not consider the influences of wind and waves. Using trial data to establish a model, one should choose the small-influence trials; otherwise, the influences of wind and waves need to be eliminated.

The International Maritime Organization(IMO) explanations for maneuvering standards [6], Society of Naval Architects and Marine Engineers(SNAME) guidelines [7] and International Towing Tank Conference(ITTC) instructions [8] proposed correction methods for the turning circle test;

otherwise, sea trials should be implemented in deep, calm and non-restricted waters. The IMO, SNAME and ITTC rules solved the current direction, based on the uniform current hypothesis and the steady turning hypothesis [6–8]. However, the influence of wind load on ballast ships, container ships and ro-ro ships is greater than that on full-load oil tankers and bulk carriers. Moreover, the wind coefficient changes with wind direction, indicating that the wind load of the ship motion in turning circle is not constant; therefore, it cannot be regarded as the influence of uniform current. Thus, the installation of instruments and trial environments on sea trial vessels increase the difficulty of identification modeling, and thus the measurement data need to be corrected.

Compared with the ship model test, the sea trial has certain disadvantages which require improvement. Currently, the naval surface warfare center of America has a maneuvering and seakeeping tank to study ship motion in various sea conditions [9]. The National Maritime Research Institute (NMRI) established an actual sea model basin, using wind load and wave load simulation instruments [10,11], to research the performance of a full-scale ship in an actual sea. The indoor model test is organized, operated and validated by a professional organization and equipped with sophisticated towing devices and Charge Coupled Device(CCD) cameras; meanwhile, outdoor trials use high-precision satellite positioning instruments and shore-based wireless positioning devices on the sea. Due to the standardization and diversification of the test, the basin model test data quality is better than the real ship trial. Therefore, compared with basin model test, it is necessary to further process the data of the full-scale ship in order to improve the quality of its data.

In terms of wind force, Isherwood, Blendermann et al. and Fujiwara et al. used wind tunnel test data to fit the wind coefficients [12–14]: firstly, Isherwood proposed the estimation method for the calculation of wind force by formula, then Blendermann and Fujiwara updated the wind force formula structure and coefficients with a new wind tunnel experiment. Currently, the shipping industry focuses on the wind coefficient of container ships with dynamic stowage [15]. Aiming to calculate added mass, Motora proposed a simple method [16] and Zhou reproduced the formula for easy application [17]. For wave disturbance, Daidola used the second-order wave drift force and moment coefficient [18], Li used Daidola's method for ship motion simulation [19]. Yasukawa studied the numerical prediction of second-order wave drift force [20], and Zhang et al. [21] and Hong et al. [22] studied second-order wave force and wave added resistance. Mei et al. [23] established a ship maneuvering model for a basin test; this paper will further explore actual sea ship maneuver modeling.

The paper is organized as follows: Section 2 briefly introduces the traditional methods that have been used to correct the full-scale sea trial. Section 3 proposes the novel correction method by wind, wave and sea surface current calculation. Section 4 explains the reference model and support vector machine (RM-SVM) for maneuvering modeling. Section 5 presents the case of trial correction. Section 6 presents maneuver modeling. Section 7 discusses the results of trial correction and motion modeling, and presents possible options for future works. Finally, Section 8 concludes this paper.

2. Traditional Correction Method

As shown in Figure 1, the literature [6–8] proposed a fast and convenient correction method called traditional correction method. In Figure 1, the blue line represents the turning circle track in a calm environment, the red line represents the turning circle track with disturbances and the green arrow represents the drift vector. The calm water track is a corrected −35° turning circle of a ship called Yukun, while the disturbed track is reproduced by one uniform surface current. The uniform surface current consisted of an east current, 0.5 m/s, and a north current, 0.5 m/s.

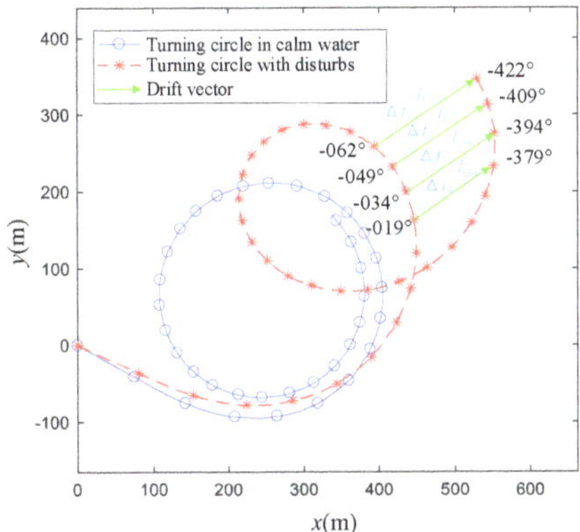

Figure 1. Traditional correction method of turning circle test.

The SNAME [7] requires that the course changing of the turning circle is greater than 540°. It is assumed that the ship reaches a steady turning stage after 360°, and a steady drift velocity can be obtained by using the position data of the steady turning. As shown in the Figure 1, according to the last point of the track, the ship takes 250 s to drift 250 m in the east direction and drift 250 m in the north direction.

The correction process of Figure 1 is shown as the following: Suppose ship position as (x_i, y_i) and ship heading angle as ψ_i at time t_i, and $i \in \{1, 2, \ldots, n\}$. Suppose ship position as $\left(x_i', y_i'\right)$ and ship heading angle as ψ_i' at time t_i', and $\psi_i - \psi_i'$ equals 360° or −360 °. In Figure 1, n is 4. Then, the drift distance between (x_i, y_i) and $\left(x_i', y_i'\right)$ is $l_{\mathrm{D}i}$. The average drift velocity $\Delta l_{\mathrm{D}i}$ between (x_i, y_i) and $\left(x_i', y_i'\right)$ can be estimated by following:

$$\Delta l_{\mathrm{D}i} = \frac{\sum\limits_{i=1}^{n} l_{\mathrm{D}i}}{\sum\limits_{i=1}^{n}\left(t_i - t_i'\right)} = \frac{\sum\limits_{i=1}^{n} \left\|(x_i, y_i) - \left(x_i', y_i'\right)\right\|_2}{\sum\limits_{i=1}^{n}\left(t_i - t_i'\right)} \tag{1}$$

In the recommended process from IMO, SNAME and ITTC [6–8], the corrections were completed based on the assumption of uniform current and steady turning. The influence of a uniform current on a ship track increases linearly and is time-constant. However, the ship drifts induced by wind and waves are related to wind and wave direction angle, and the ship drifts are nonlinear and time-varying. Therefore, it is assumed that wind and wave disturbances are treated as linear; the nonlinear components are ignored. In this paper, the influences of wind, waves and currents are calculated separately on the basis of the hypothesis, and the improved method is proposed.

As this paper focus on maneuvering motion, the rolling, pitching and heaving of the ship are ignored by following explanations. Firstly, the drifts induced by wind and waves are treated as long term motion; meanwhile this manuscript focus on maneuvering motion, that is, only surge, sway and turning will be concerned. The rolling, pitching and heaving of the ship are therefore ignored. Secondly, the maneuvering is simplified as three degrees of freedom motion and is independent of seakeeping. The periodic motion of seakeeping has little effect on maneuvering motion with a large rudder angle. Thirdly, in the measurements, the rolling, pitching and heaving of the ship are periodic; thus, the motion, being periodic, can be filtered. Therefore, the maneuvering data can be used for correction and modelling.

3. Improved Correction Method

The improved correction method is mainly divided into three parts: first, calculate the wind force; then, calculate the wave drift force; finally, calculate the wind and wave drift distance. In this section, the surge and sway are corrected. Yaw is considered for the following reasons: Firstly, the ship hull underwater and the ship superstructure overwater, together, is close to being a box-shape. Thus, the yawing induced by wind and waves is negligible. Secondly, the yaw should be corrected for high precision; however, this will be much more complicated; this is because the yaw makes the heading angle change, and the heading angle changes the surge distance and sway distance.

3.1. Wind Load Calculation

Suppose that ship begins turning at time t_0. At time t, the ship velocity is $V(t)$, heading angle is $\psi(t)$, true wind velocity is $V_T(t)$, true wind direction is $\psi_T(t)$, the frontal wind load is $X_w(t)$ and lateral wind load is $Y_w(t)$. According to the reference [24], the wind force and its components, the force of the earth-centered earth-fixed (ECEF) east and ECEF north, change along with the ship heading. Thus, from t_0 to t, the wind induced drift distance in the earth-centered earth-fixed(ECEF)reference frame are $\Delta x_w(t)$ and $\Delta y_w(t)$, respectively, and are calculated as following:

$$\begin{cases} \frac{\Delta x_w(t)}{\rho_a U_R^2(t)} = \int_{t_0}^{t}\int_{t_0}^{t}\left[\frac{A_{fw}C_{wx}(\alpha_{wR}(t))}{2m+2m_x}dt\cos\psi(t) - \frac{A_{lw}C_{wy}(\alpha_{wR}(t))}{2m+2m_y}dt\sin\psi(t)\right]dt \\ \frac{\Delta y_w(t)}{\rho_a U_R^2(t)} = \int_{t_0}^{t}\int_{t_0}^{t}\left[\frac{A_{fw}C_{wx}(\alpha_{wR}(t))}{2m+2m_x}dt\sin\psi(t) + \frac{A_{lw}C_{wy}(\alpha_{wR}(t))}{2m+2m_y}dt\cos\psi(t)\right]dt \end{cases} \quad (2)$$

where m is ship mass, m_x and m_y are added mass, A_{fw} and A_{lw} are the ship front projected area and lateral projected area, respectively, and C_{wx} and C_{wy} are wind coefficients of the ship front and lateral projected area, respectively. U_R and α_{wR} are relative wind velocity and direction, and can be calculated by ψ_T,V_T,V and ψ. Currently, the wind tunnel test is still the best means to determine the wind coefficient. Due to limited test facilities and high cost, the empirical formula of Blendermann [13] is applied in this paper. The added mass is calculated by the formulas from reference [16,17].

3.2. Wave Drift Force Calculation

Suppose that the wave drift force of ship longitude is $X_d(t)$ and wave drift force of ship transverse is $Y_d(t)$. According to reference [22], the second-order wave drift can be divided into ECEF (earth-centered earth-fixed) east and ECEF north. Thus, from t_0 to t, the wave drift force induced drift distance in ECEF reference frame are $\Delta x_d(t)$ and $\Delta y_d(t)$, respectively, and are calculated as following:

$$\begin{cases} \Delta x_d(t) = \int_{t_0}^{t}\int_{t_0}^{t}\frac{X_d(t)}{m+m_x}dt\cos\psi(t)dt - \int_{t_0}^{t}\int_{t_0}^{t}\frac{Y_d(t)}{m+m_y}dt\sin\psi(t)dt \\ \Delta y_d(t) = \int_{t_0}^{t}\int_{t_0}^{t}\frac{X_d(t)}{m+m_x}dt\sin\psi(t)dt + \int_{t_0}^{t}\int_{t_0}^{t}\frac{Y_d(t)}{m+m_y}dt\cos\psi(t)dt \end{cases} \quad (3)$$

Due to the dynamic changing of the encounter frequency, the equivalent incident wave lengths λ_{BX} and λ_{BY} are introduced and satisfy the following equation (Equation (4)). The equivalent incident wave length has been used in reference [22].

$$\begin{cases} E_X = X_d\lambda_{BX} = X_d\frac{\lambda}{-\cos\alpha_d} \\ E_Y = Y_d\lambda_{BY} = Y_d\frac{\lambda}{-\cos\alpha_d} \end{cases} \quad (4)$$

where λ is incident mean wave length of the sea area and α_d is the wave direction.

For real-time requirements, the Daidola formula [18] is used to calculate second-order wave drift force. The Daidola method has been applied in reference [19]. The surge and sway second-order wave drift force $X_{\mathrm{d}}(t)$ and $Y_{\mathrm{d}}(t)$ are as the following:

$$\begin{cases} X_{\mathrm{d}}(t) = \frac{\rho g L^2 \zeta^2}{2}\left[0.05 - 0.2\left(\frac{\lambda_{\mathrm{BX}}}{L}\right) + 0.75\left(\frac{\lambda_{\mathrm{BX}}}{L}\right)^2 - 0.51\left(\frac{\lambda_{\mathrm{BX}}}{L}\right)^3\right]\cos(\alpha_{\mathrm{d}}) \\ Y_{\mathrm{d}}(t) = \frac{\rho g L^2 \zeta^2}{2}\left[0.46 + 6.83\left(\frac{\lambda_{\mathrm{BY}}}{L}\right) - 15.65\left(\frac{\lambda_{\mathrm{BY}}}{L}\right)^2 + 8.44\left(\frac{\lambda_{\mathrm{BY}}}{L}\right)^3\right]\sin(\alpha_{\mathrm{d}}) \end{cases} \tag{5}$$

where ζ is the mean wave height of the sea area and ζ, λ and α_{d} are calculated by wind velocity, based on the hypothesis of the fully developed wave and the hypothesis of long-crested wave. Therefore, the mean wave height of the sea area and mean wave length were estimated by wind force and direction information.

3.3. Resultant Distance Induced by Wind, Wave and Current

Based on wind load calculation and wave drift force calculation, the resultant distance induced by the wind, waves and currents is calculated as following:

$$\begin{cases} \Delta x(t) = k_1 \Delta x_{\mathrm{w}}(t) + k_3 \Delta x_{\mathrm{d}}(t) + k_5 \Delta x_{\mathrm{c}}(t) \\ \Delta y(t) = k_2 \Delta y_{\mathrm{w}}(t) + k_4 \Delta y_d(t) + k_6 \Delta y_{\mathrm{c}}(t) \end{cases} \tag{6}$$

where k_1~k_6 are adjusting parameters and Δx_{c} and Δy_{c} are east current and south current set as 1.0 m/s.

Suppose during time $[t_1, t_2]$, the ship position is $\left(x(t_j), y(t_j)\right), t_j \in [t_1, t_2]$. Then, the correction ship position $\left(\hat{x}(t_j), \hat{y}(t_j)\right)$ can be calculated as following:

$$\begin{cases} \hat{x}(t_j) = x(t_j) - \Delta x(t_j) \\ \hat{y}(t_k) = y(t_k) - \Delta y(t_j) \end{cases} \tag{7}$$

Based on the hypothesis of steady turning, the following equation has a solution for adjusting parameters k_i:

$$\begin{gathered} \underset{k_i | i=1,2,3,4,5,6}{\operatorname{argmin}} \sum_{t_j = t_1}^{t_2} \left\| \left(\hat{x}(t_j), \hat{y}(t_j)\right) - (x_0, y_0) \right\|_2 \\ \text{subject} : \begin{cases} (x_0, y_0) = f_C\left(\hat{x}(t_j), \hat{y}(t_j)\right) \\ k_i \in [-10, +10] \end{cases} \end{gathered} \tag{8}$$

where (x_0, y_0) is center of a circle and can be solved by f_C; f_C is a function from Kasa [25]. Meanwhile, $k_i \in [-10, +10]$ is a restrict condition for abnormal current. The value of these coefficients, k_i, are estimated by the optimization algorithm called pattern search. This function is established based on steady turning. The steady turning is a hypothesis condition from the IMO, ITTC and SNAME methods. Based on this hypothesis, the correction will form the final stage of turning in a circle.

To sum up this section, the illustration is shown in Figure 2. Figure 2 introduces the drift distances induced by the wind, waves and sea surface currents. The distances are divided into their east and north components. This distances also consist of the total drift distance in order to correct the ship track.

At the end of this section, based on the corrected ship position $\left(\hat{x}(t_j), \hat{y}(t_j)\right)$ and heading angle $\psi(t)$, the velocities of surge, sway and yaw are derived. These velocities are called identified ship velocities, and are written as u_{T}, v_{T} and r_{T}, where "T" stands for identified ship.

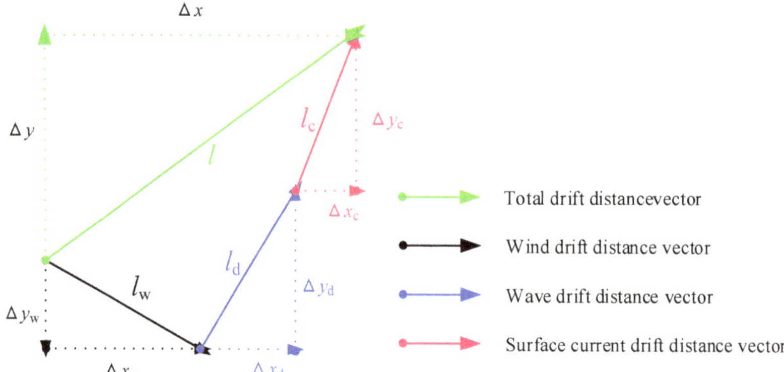

Figure 2. The drift distances and components induced by the wind, waves and surface currents.

4. Maneuver Modeling Method

In this section, the RM-SVM whole ship model is established. Firstly, the RM-SVM model is identified with trial data. Based on the prediction of RM-SVM, the acceleration data and velocity data are reproduced. Then, the data is used to identify whole ship model by the least square algorithm.

The data from RM-SVM is smooth and does not have noise. These advantages will make the least square result much more precise than the trials data. In addition, the hydrodynamic coefficients are stable and not over-fitting.

4.1. RM-SVM Model

In this section, the corrected sea trial data is applied for ship maneuver modeling. This identification modeling method determines maneuverability aspects at rough sea and poor weather conditions, which is an important function used in order avoid collisions at actual sea conditions [26]. As introduced by Mei et al. [23], the reference model support vector machine (RM-SVM) method is utilized. Although modeling cannot describe wake information, as shown by Niu et al. [27], the model prediction precision is outstanding. Taking surge acceleration as example, suppose there existed an n identified ship measurement sample. The kth sample surge, sway and yaw velocity are $u_{\mathrm{T}}(k)$, $v_{\mathrm{T}}(k)$ and $r_{\mathrm{T}}(k)$, and the kth surge acceleration function is $H_{\mathrm{T}}(u_{\mathrm{T}}(k), v_{\mathrm{T}}(k), r_{\mathrm{T}}(k), \delta_{\mathrm{T}}(k))$. For the RM, the kth sample surge, sway and yaw velocity are $u_{\mathrm{R}}(k)$, $v_{\mathrm{R}}(k)$ and $r_{\mathrm{R}}(k)$, and the kth surge acceleration function is $H_{\mathrm{R}}(u_{\mathrm{R}}(k), v_{\mathrm{R}}(k), r_{\mathrm{R}}(k), \delta_{\mathrm{R}}(k))$, where "R" stands for reference model. The concept and selection method of the reference model is introduced in reference [23]. Based on the identified ship trials sample and reference model, the surge SVM can be written as the following:

$$L_{\mathrm{D}} = \sum_{k=1}^{n} (\widetilde{\alpha}_k - \alpha_k)\Delta H(k) - \frac{1}{2}\sum_{k=1}^{n}\sum_{\ell=1}^{n} (\widetilde{\alpha}_k - \alpha_k)(\widetilde{\alpha}_\ell - \alpha_\ell)\mathbf{W}_k{}^{\mathrm{T}}\mathbf{W}_\ell - \varepsilon\sum_{k=1}^{n}(\widetilde{\alpha}_k + \alpha_k), \quad (9)$$

subject to:

$$\begin{cases} \sum\limits_{k=1}^{n} (\alpha_k - \widetilde{\alpha}_k) = 0 \\ 0 \le \alpha_k, \widetilde{\alpha}_k \le \tau \\ \alpha_k\left[\xi_k + \varepsilon - \boldsymbol{w}^{\mathrm{T}}\mathbf{W}_k - l_1 + \Delta H(k)\right] = 0 \\ \widetilde{\alpha}_k\left[\widetilde{\xi}_k + \varepsilon + \boldsymbol{w}^{\mathrm{T}}\mathbf{W}_k + l_1 - \Delta H(k)\right] = 0 \\ \alpha_k\widetilde{\alpha}_k = 0, \xi_k\widetilde{\xi}_k = 0 \\ (\tau - \alpha_k)\xi_k = 0, (\tau - \widetilde{\alpha}_k)\widetilde{\xi}_k = 0 \end{cases} \quad (10)$$

where $\ell = 1, 2, \cdots, n$ is the order of sample data, $\boldsymbol{\alpha}$, $\tilde{\boldsymbol{\alpha}}$, $\boldsymbol{\theta}$ and $\tilde{\boldsymbol{\theta}}$ are the Lagrangian multiplier vector of SVM hyper-plane, $\boldsymbol{\xi}$ and $\tilde{\boldsymbol{\xi}}$ are the slack variable vector of SVM hyper-plane, $\boldsymbol{w}$ is the normal vector of SVM hyper-plane, l_1 is constant bias of SVM hyper-plane, τ is the regularization constant and ε is the Insensitive-band parameter. $\boldsymbol{W}_k$ is the SVM input vector, as following:

$$\boldsymbol{W}_k = (u_{\mathrm{T}}(k), v_{\mathrm{T}}(k), r_{\mathrm{T}}(k), \delta_{\mathrm{T}}(k))^{\mathrm{T}} \tag{11}$$

Substituting sample data into Equations (9) and (11), the surge SVM is solved.

In the same way as the surge SVM, the sway and yaw SVM can be calculated. In addition, the identified ship accelerations can be predicted as the following:

$$\begin{cases} \dot{u}_{\mathrm{T}}(t) = \dot{u}_{\mathrm{R}}(t) + \boldsymbol{w}^{\mathrm{T}}(u_{\mathrm{T}}(t), v_{\mathrm{T}}(t), r_{\mathrm{T}}(t), \delta_{\mathrm{T}}(t))^{\mathrm{T}} + l_1 \\ \dot{v}_{\mathrm{T}}(t) = \dot{v}_{\mathrm{R}}(t) + \boldsymbol{p}^{\mathrm{T}}\left(v_{\mathrm{T}}(t), r_{\mathrm{T}}(t), \dot{v}_{\mathrm{T}}(t), \dot{r}_{\mathrm{T}}(t), \delta_{\mathrm{T}}(t)\right)^{\mathrm{T}} + l_2 \\ \dot{r}_{\mathrm{T}}(t) = \dot{r}_{\mathrm{R}}(t) + \boldsymbol{q}^{\mathrm{T}}\left(v_{\mathrm{T}}(t), r_{\mathrm{T}}(t), \dot{v}_{\mathrm{T}}(t), \dot{r}_{\mathrm{T}}(t), \delta_{\mathrm{T}}(t)\right)^{\mathrm{T}} + l_3 \end{cases} \tag{12}$$

where $\dot{u}_{\mathrm{T}}$, $\dot{v}_{\mathrm{T}}$ and $\dot{r}_{\mathrm{T}}$ are the identified ship sway and yaw accelerations, $\dot{u}_{\mathrm{R}}$, $\dot{v}_{\mathrm{R}}$ and $\dot{r}_{\mathrm{R}}$ are the RM sway and yaw accelerations, $\boldsymbol{p}$ and $\boldsymbol{q}$ are the normal vector of sway and yaw SVM hyper-plane and l_2 and l_3 are constant bias of sway and yaw SVM hyper-plane, respectively. The Equation (12) can be solved by Runge–Kutta integration.

4.2. Whole Ship Model

Based on the prediction of Equation (12), the identified ship accelerations $\dot{u}_{\mathrm{T}}$, $\dot{v}_{\mathrm{T}}$ and $\dot{r}_{\mathrm{T}}$ are reproduced by the RM-SVM model. Following that the input vector is $[u_{\mathrm{T}}, v_{\mathrm{T}}, r_{\mathrm{T}}]$, the output is $[\dot{u}_{\mathrm{T}}, \dot{v}_{\mathrm{T}}, \dot{r}_{\mathrm{T}}]$. Once the input vector and output vector are submitted into Equation (13), the whole ship model from reference [28] is identified with the least square method. The whole model structure and parameters are list as Equation (13).

$$\begin{cases} \left(m' - X'_{\dot{u}}\right)\dot{u}'_{\mathrm{T}} = X'_{\eta}(1-\eta_{\mathrm{T}}) + X'_{\eta\eta}(1-\eta_{\mathrm{T}})^2 + X'_{\eta\eta\eta}(1-\eta_{\mathrm{T}})^3 + X'_{vv}{v'_{\mathrm{T}}}^2 \\ +(X'_{rr} + m'x'_{\mathrm{G}}){r'_{\mathrm{T}}}^2 + X'_{\delta\delta}{\delta'_{\mathrm{T}}}^2 + (X'_{vr} + m')v'_{\mathrm{T}}r'_{\mathrm{T}} + X'_{vv\eta}{v'_{\mathrm{T}}}^2(1-\eta_{\mathrm{T}}) + X'_{\delta\delta\eta}{\delta'_{\mathrm{T}}}^2(1-\eta_{\mathrm{T}}) \\ \left(m' - Y'_{\dot{v}}\right)\dot{v}_{\mathrm{T}} + \left(m'x'_{\mathrm{G}} - Y'_{\dot{r}}\right)\dot{r}_{\mathrm{T}} = Y'_0 + Y'_v v'_{\mathrm{T}} + Y'_{vvv}{v'_{\mathrm{T}}}^3 + Y'_{vrr}v'_{\mathrm{T}}{r'_{\mathrm{T}}}^2 + (Y'_r - m')r'_{\mathrm{T}} + Y'_{rrr}{r'_{\mathrm{T}}}^3 \\ +Y'_{vvr}{v'_{\mathrm{T}}}^2 r'_{\mathrm{T}} + Y'_{\delta}\delta'_{\mathrm{T}} + Y'_{\delta\delta\delta}{\delta'_{\mathrm{T}}}^3 + Y'_{\eta}(1-\eta_{\mathrm{T}}) + Y'_{\eta\eta}(1-\eta_{\mathrm{T}})^2 + Y'_{\delta\eta}\delta'_{\mathrm{T}}(1-\eta_{\mathrm{T}}) + Y'_{\delta\eta\eta}\delta'_{\mathrm{T}}(1-\eta_{\mathrm{T}})^2 \\ \left(m'x'_{\mathrm{G}} - N'_{\dot{v}}\right)\dot{v}'_{\mathrm{T}} + \left(I'_z - N'_{\dot{r}}\right)\dot{r}'_{\mathrm{T}} = N'_0 + N'_v v'_{\mathrm{T}} + N'_{vvv}{v'_{\mathrm{T}}}^3 + N'_{vrr}v'_{\mathrm{T}}{r'_{\mathrm{T}}}^2 + (N'_r - m'x'_{\mathrm{G}})r'_{\mathrm{T}} + N'_{rrr}{r'_{\mathrm{T}}}^3 \\ +N'_{vvr}{v'_{\mathrm{T}}}^2 r'_{\mathrm{T}} + N'_{\delta}\delta'_{\mathrm{T}} + N'_{\delta\delta\delta}{\delta'_{\mathrm{T}}}^3 + N'_{\eta}(1-\eta_{\mathrm{T}}) + N'_{\eta\eta}(1-\eta_{\mathrm{T}})^2 + N'_{\delta\eta}\delta'_{\mathrm{T}}(1-\eta_{\mathrm{T}}) + N'_{\delta\eta\eta}\delta'_{\mathrm{T}}(1-\eta_{\mathrm{T}})^2 \end{cases} \tag{13}$$

where the $\eta_{\mathrm{T}} = u_{\mathrm{T}}/u_{0\mathrm{T}}$, $u_{0\mathrm{T}}$ is the ship service speed.

5. The Case of Trial Correction

In this section, the improved trial correction is applied to calculate the influences of wind, waves and currents. In addition, the drift distance and velocity of the turning circle test are solved. Then, the trial track and velocity for the full-scale ship are corrected.

5.1. General Details of Sea Trial

The study object of this paper is a motor vessel called Yukun; Table 1 and Figure 3 note Yukun particulars. The sea trial time was from 08:00 to 14:00 on 24 August 2012. The sea trial site is located in the northwest of the Yellow Sea, about 14 nautical miles from Dalian Port. The sea trials were carried out in open and deep water in clear and well weather conditions, as shown in Figure 4.

Table 1. Ship particulars of Yukun motor vessel.

Particulars	Values	Particulars	Values
Displacement	5710.2	Rudder area	11.8 m^2
Length overall	116 m	Rudder height	4.8 m
Length between perpendiculars	105 m	Propeller diameter	3.8 m
Designed waterline length	106.5 m	Blade number	4
Ship breadth	18 m	Blade area ratio	0.67
Full-load draft	5.4 m	Maximum rudder rate	2.8°/s
Block coefficient	0.56	Prismatic coefficient	0.58

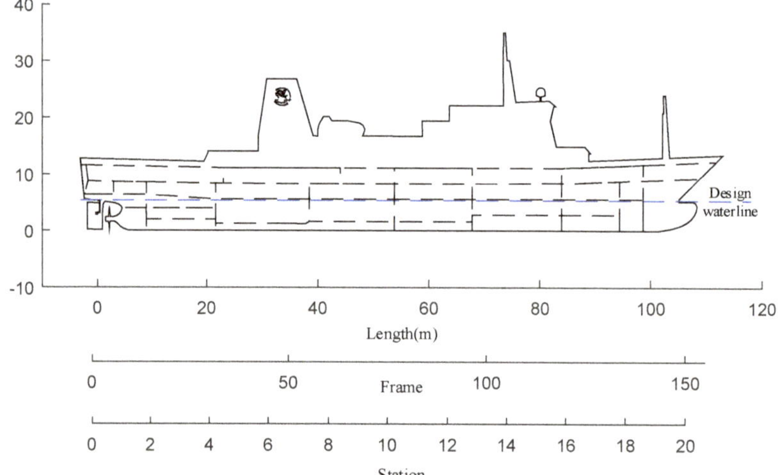

Figure 3. A general arrangement of Yukun motor vessel [29,30] (the figure permission has been achieved from Dalian Maritime University).

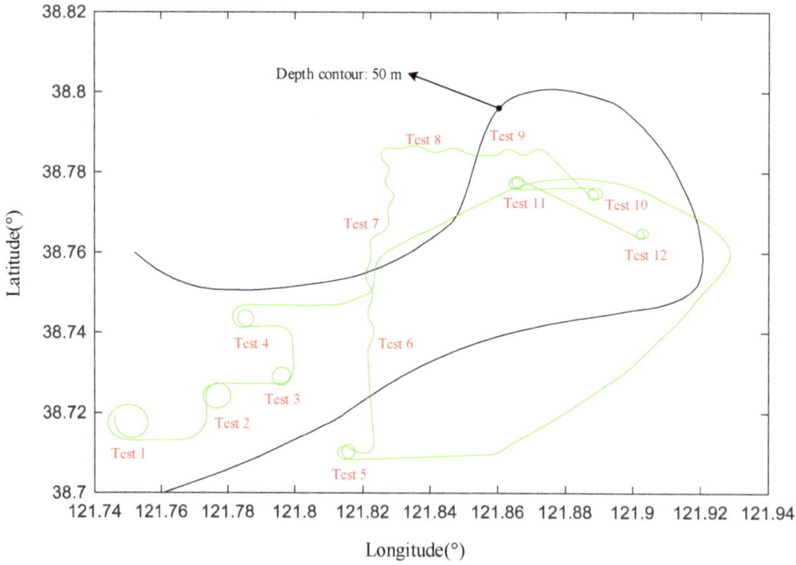

Figure 4. Sea trial area and water depth of **Yukun** test.

Figure 3 is part material from Dalian Maritime University and has been published in references [29,30]. From Figure 4, the maneuvers tests are listed in Table 2.

Table 2. Maneuver type and rudder angle details for all of the trials.

NO.	Time Points (s)	Maneuvers Type	Rudder Angle (°)	Sample Points
1	38213–38769	Turning circle	5	556
2	39232–39635	Turning circle	10	403
3	39898–40231	Turning circle	15	333
4	40619–40915	Turning circle	20	296
5	44577–44985	Turning circle	25	408
6	45509–45946	zigzag	10/−10	437
7	45975–46542	zigzag	20/−20	567
8	46606–46910	zigzag	10/−10	304
9	46999–47263	zigzag	20/−20	264
10	47517–47912	Turning circle	25	395
11	48178–48566	Turning circle	−30	388
12	49070–49450	Turning circle	34	380

As shown in Figure 5, the wind measuring system, differential global positioning system (DGPS), fiber-optic gyro and speed log are installed on the mast, bridge, gyro deck and ship bow, respectively. The DGPS position has a higher data update frequency than an automatic identification system [31].

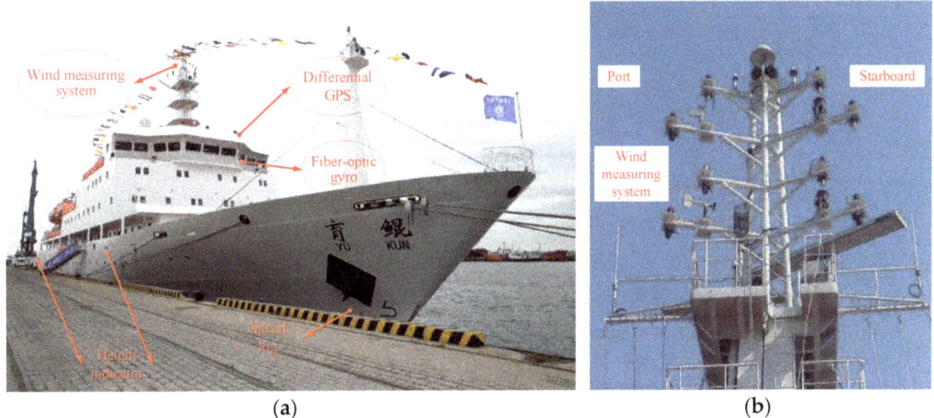

Figure 5. Instrument installation on board of Yukun. (**a**) The position, velocity and heading angle measuring instrument; (**b**) the wind measuring system installation on the mast.

5.2. Wind Load and Wave Drift Force Results

The time history subjected to the wind load and wave drift force for the +20° turning circle test were solved by the improved method of Equation (2). As shown in Figure 6, the surge and sway forces induced by wind and waves are calculated. The wind load shows dynamic fluctuations changing with time. As shown in Figure 5, the wind measuring system is shielded by the mast. Therefore, wind fluctuations included mast shielding, gusty components and random wind components. The details of the fluctuations also enhance the judgment of wind force and direction.

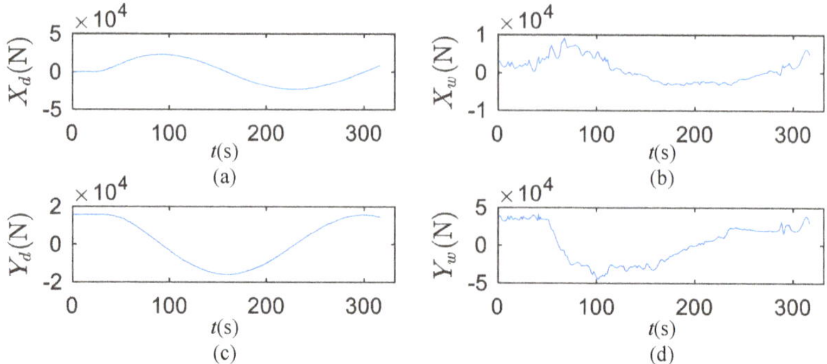

Figure 6. Wind load and wave drift force for +20°turning circle test. (**a**) Surge force induced by wave; (**b**) surge load induced by wind; (**c**) sway force induced by wave; (**d**) sway load induced by wind.

In the calculation of Figure 6, the ship front projected area, A_{fw}, is 297 m^2, and lateral projected area, A_{lw}, is 1304.6 m^2. The longitude centroid position of A_{lw} is 2.46 m and the vertical centroid position is 6.8 m.

5.3. Wind- and Wave-Induced Acceleration Results

As shown in Figure 7, the surge and sway accelerations induced by wind and waves are calculated by Equation (5).

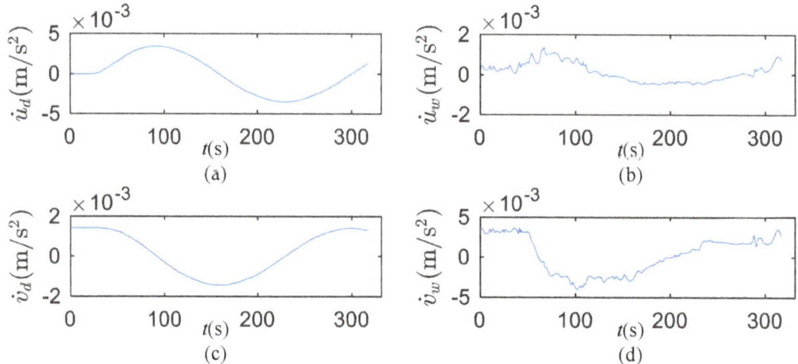

Figure 7. Wind- and wave-induced accelerations of +20° turning circle test. (**a**) Surge acceleration induced by wave; (**b**) surge acceleration induced by wind; (**c**) sway acceleration induced by wave; (**d**) sway acceleration induced by wind.

5.4. Wind- and Wave-Induced Distance Results

As shown in Figure 8, the time history of Yukun being subjected to the wind- and wave-induced distance in the +20° turning circle test are solved by the improved method of Equations (6) and (8), respectively. The results of adjusting parameters k_1~k_6 are −2.46, 1.02, −1.24, 1, −0.21 and −0.37, respectively.

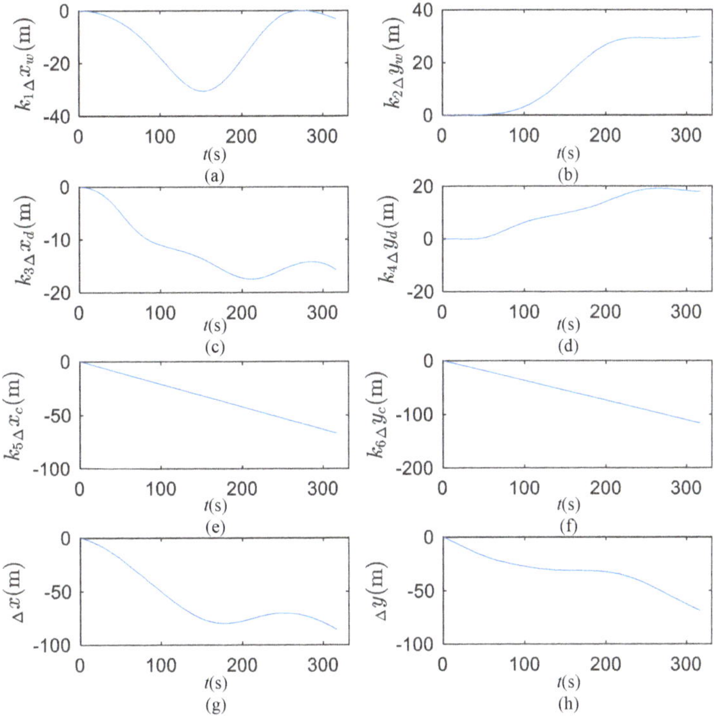

Figure 8. Wind, waves and sea surface currents induced earth-centered earth-fixed (ECEF) distances of +20° turning circle test. (**a**) Transverse drift distance induced by wind; (**b**) longitude drift distance induced by wind; (**c**) transverse drift distance induced by wave; (**d**) longitude drift distance induced by wave; (**e**) transverse drift distance induced by current; (**f**) longitude drift distance induced by current; (**g**) transverse resultant drift distance induced by wind, wave and current; (**h**) longitude resultant drift distance induced by wind, wave and current.

Figure 8 depicts the drift distance components induced by wind, waves and currents. The distances present the same order of magnitude of the wind, wave and sea surface current influence, none of distance can be ignored. The summery drift distances will be used to correct the ship track and calculate the surge sway and yaw velocities.

5.5. Track and Velocity Correction Results

The wind- and wave-induced distances have been used to correct the ship track; the comparison of the original turning circle and the corrected turning circle are presented in Figure 9. Figure 9 shows that the original turning circle moves significantly in the ECEF negative direction when under the influence of the wind and wave currents.

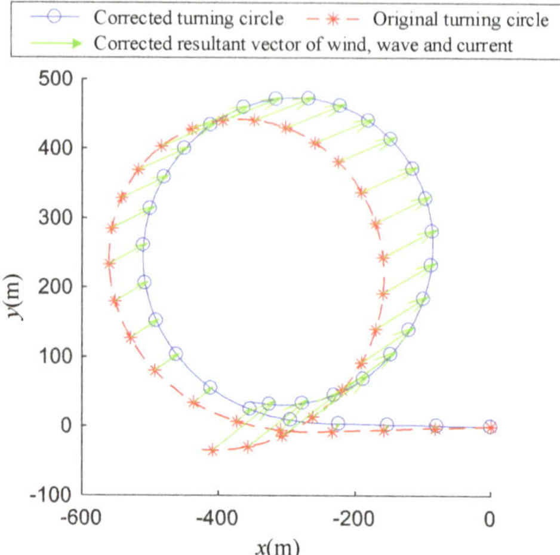

Figure 9. Correction of drift distance induced by the wind, waves and currents for +20° turning circle test.

As shown in Figure 10, the surge sway and yaw velocities are corrected. As the turning reaches a steady stage, the corrected longitudinal velocity decreases and gradually converges the stable value. However, the log velocity increases at the stage of 200 s–250 s. The uncorrected sway velocity is stable at 0 m/s, while the corrected sway velocity increased rapidly, within 0 s–60 s, and converges to −1.52 m/s at 57 s. Since the yaw velocity is not corrected, the lines overlap.

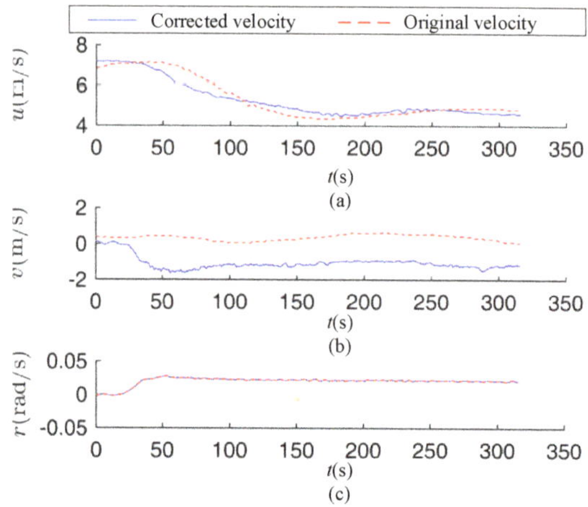

Figure 10. Correction of ECEF reference frame velocities for +20° turning circle test. (**a**) Ship surge velocity; (**b**) ship sway velocity; (**c**) ship yaw velocity.

All of the trials in Figure 4 and Table 2 have been corrected. In the previous manuscript, we only selected the 20° zigzag test for correction and modelling. Currently, the others are present in the Appendix B. These corrected cases of trials indicate that the improved correction method is valid for sea trials. These sea trials will be used for modelling in the next section.

6. The Case of Maneuver Modeling

In this section, the ship maneuvering model is established by the zigzag tests 6, 7, 8 and 9 from Figure 4. Identification model is one data driven-based method, and it is a common method in the maritime field [32]. In addition, the prediction model of +20° turning circle test is trained by Equations (9)–(11). Based on the method proposed in reference [22], the S175 ship is selected as the reference model of Yukun. In addition, the SVM is trained by zigzag test. Therefore, the RM-SVM of Yukun maneuvering model is established.

6.1. +20° Turning Circle Test

The +20° turning circle test is predicted by Equation (12), as shown in Figure 11.

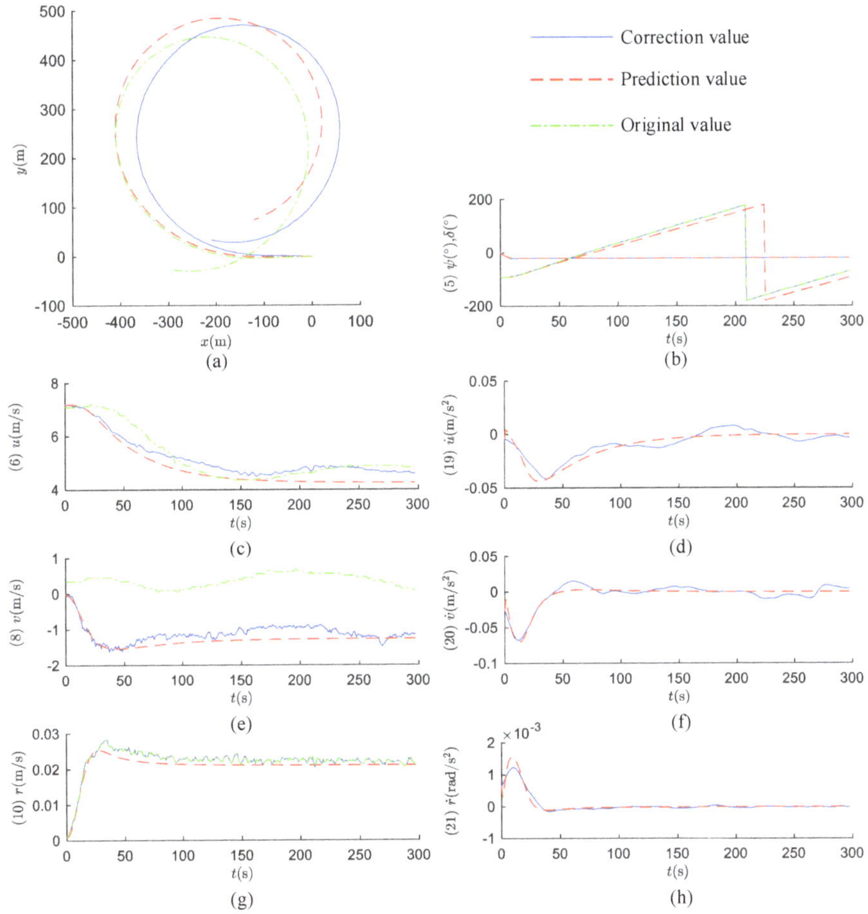

Figure 11. Prediction for +20° turning circle test. (**a**) Ship track; (**b**) ship heading angle and rudder angle; (**c**) surge velocity; (**d**) surge acceleration; (**e**) sway velocity; (**f**) sway acceleration; (**g**) yaw velocity; (**h**) yaw acceleration.

The following conclusions can be drawn from Figure 9. According to the overall prediction results, the values of RM-SVM are relatively stable, without significant numerical anomalies and fluctuations, which indicates the stability of the identification model established by RM-SVM. According to the velocity prediction results, the values of RM-SVM are close to the corrected values. Since the yaw moment is not considered in the solution of influence, the yaw velocity is not corrected, so the corrected yaw velocity is equal to the original value.

6.2. The Ship Hydrodynamic Coefficient Result and −35° Turning Circle Validation

Based on the Equation (12), the RM-SVM model is acquired. Then, the RM-SVM is used to produce the accelerations data [$\dot{u}_T$, $\dot{v}_T$ and $\dot{r}_T$], velocities data [u_T, v_T, r_T] and rudder angle data δ_T. These data consist of input and output samples. Using the least-square linear regression algorithm and submitting the sample into Equation (13), the ship hydrodynamic coefficients are estimated, as shown in Table 3. The details of ship hydrodynamic coefficients are noted in [28].

Table 3. Identification results of non-dimensional surge, sway and yaw hydrodynamic coefficients for Yukun ship model.

Surge Coefficients (×10^5)		Sway Coefficients (×10^5)		Yaw Coefficients (×10^5)	
$X'^{\delta\delta}_{r}$	−130.4	Y'_0	−69.1	N'_0	−3.8
		Y'_δ	1079.1	N'_δ	−124.0
		$Y'_{\delta\delta\delta}$	−808.7	$N'_{\delta\delta\delta}$	104.3
X'_η	630.3	Y'_η	622.3	N'_η	38.4
$X'_{\eta\eta}$	−1637.8	$Y'_{\eta\eta}$	−1397.1	$N'_{\eta\eta}$	−80.0
$X'_{\eta\eta\eta}$	3135.2	$Y'_{\delta\eta}$	−1711.9	$N'_{\delta\eta}$	194.2
$X'_{\delta\delta\eta}$	−981.7	$Y'_{\delta\eta\eta}$	5978.4	$N'_{\delta\eta\eta}$	−670.6
X'_{vv}	−1159.0	Y'_v	283.8	N'_v	−59.9
$X'_{vv\eta}$	0.0	Y'_{vvv}	−12,927.7	N'_{vvv}	3647.3
X'_{rr}	59.5	$Y'_r - m'$	249.2	$N'_r - m'x'_G$	−53.8
		Y'_{rrr}	−4829.9	N'_{rrr}	418.3
$X'_{vr} + m'$	419.2	Y'_{vrr}	−23,264.0	N'_{vrr}	2774.4
		Y'_{vvr}	−23,891.8	N'_{vvr}	4561.3

Based on the hydrodynamic coefficients in Table 3, the −35° turning circle is predicted. The ship mass and inertia moment are known. The added mass and added moment are estimated by reference [3,16,17] as $m' - X'_{\dot{u}} = 0.010249$, $m' - Y'_{\dot{v}} = 0.017853$, $m'x'_G - Y'_{\dot{r}} = 0.00071412$, $m'x'_G - N'_{\dot{v}} = 0.001086$ and $I'_z - N'_{\dot{r}} = 0.000015514$. The −35° turning circle test prediction results are as shown in Table 4.

Table 4. Validation for −35° turning circle test of Yukun full-scale ship and comparison with IMO standard.

Method		Advance	Tactical Diameter
IMO standard for ship maneuverability		4.5L_{PP}	5L_{PP}
Sea trial result	Value	3.21L_{PP}	2.98L_{PP}
	Percentage	71%	60%
Prediction by RM-SVM in this paper	Value	2.99L_{PP}	3.08L_{PP}
	Percentage	66%	62%
	Prediction accuracy	93%	97%

Where L_{PP} is ship length between perpendiculars.

As presented in the Table 4, the advance from the sea trial and prediction by RM-SVM are both smaller than the limit of the IMO standard for ship maneuverability 4.5L, as is the tactical diameter 5L. It is found that the full-scale ship complies with the IMO standard. On the other hand, the advance prediction accuracy of RM-SVM is 93% of the sea trial result, and the tactical diameter is 97%. This accuracy shows the high precision of the maneuver modeling.

7. Discussion

For the correction method, the improved method calculates the impact of the wind, waves and currents, but the traditional method takes wind and waves as a uniform current. Thus, the improved method proposed in this paper is a general form, whereas the traditional method is a special form. However, this does not mean that the improved one is perfect. Generally speaking, sea trial requires a buoy or radar wave system to measure the wave height. The Yukun test does not have this device. As described in the methodology, the improved method supposes that the wind and waves have been fully developed, and the sea state and induced ship motion are taken as stationary processes. The wave height and wave length are predicted via wind force. On the other hand, in the calculation of ship drift, the yaw induced by wind and wave has been ignored. Therefore, future works may consider correcting the yaw on the actual sea.

For the maneuvering modeling part, the ship motion with the constant engine setting is predicted, and the precision is good. The engine setting condition of the sea trials satisfies the IMO standard for ship maneuverability. However, as in the Maritime Autonomous Surface Ship (MASS), the requirement of ship maneuvering will be much more technically demanding. The other conditions, such as engine RPM changing and ballasted loading, will be common in future research. It is foreseeable that the ship maneuver modeling will be associated with MASS for sophisticated ship path planning, tracking and collision avoidance.

8. Conclusions

In this paper, the measurement data of the installation equipment of the full-scale motor vessel were checked, wind and wave influences were solved and eliminated and sea trial track and velocity were corrected. Based on the corrected free running sea trials, the maneuvering model of the full-scale ship was established. Zigzag tests were used as training data to predict the turning circle test. Based on the identification model, the accelerations were reproduced. Finally, the whole ship model was identified and the modeling performance of +35 turning circle test was verified. To sum up the above work, the following conclusions can be drawn:

(1) Due to the sea trial track and velocity being difficult to use for modeling directly, based on the assumption of the full developed wind and wave, an improved sea trial correction method was proposed. In this method, the wind, wave and current drift influences were calculated separately, and the adjusting parameters for the optical drift distances were solved by pattern search algorithm. The corrected track and velocity vectors were applied to modify the original data. The correction results of all trials illustrated the effectiveness of the proposed method.

(2) According to the prediction example of the Yukun +20° turning circle test, it can be concluded that the maneuver model was precise. On the basis of the estimation results, the ship hydrodynamic coefficients in whole ship model were identifiable. From the track prediction of a −35° turning circle, the Yukun satisfies the IMO standard for ship maneuverability. In addition, the accuracy of the advance and tactical diameter reached 93% and 95%.

(3) It will be much more convincing to validate this manuscript in several ships. However, it is not easy to obtain sea trials, as only the Yukun motor vessel test was organized and collected. In the future, there will be a new motor vessel built for maritime autonomous surface ship (MASS) research at Dalian Maritime University. The public building project has been approved. The correction and modeling of the new MASS will appear soon, once the trials are carried out.

(4) Nowadays, ship maneuvering in waves is a tough and hot issue for the researcher as presented by ITTC 2017. Full-scale maneuvering in waves, including the rolling, heaving and pitch of ships, will be included in future works as soon as possible.

Author Contributions: Conceptualization, B.M. and L.S.; methodology, B.M.; software, B.M.; validation, G.S.; formal analysis, B.M.; investigation, B.M.; resources, B.M.; data curation, B.M.; writing—original draft preparation, B.M.; writing—review and editing, L.S.; visualization, G.S.; supervision, L.S. All authors have read and agreed to the published version of the manuscript.

Funding: This research was funded by the National Natural Science Foundation of China, grant numbers: 51579025; the Natural Science Foundation of Liaoning Province, grant number: 20170540090.

Acknowledgments: The authors would like to acknowledge Dalian Maritime University and X.K. Zhang for their support.

Conflicts of Interest: The authors declare no conflict of interest.

Appendix A

As the prismatic coefficient Cp is one of the key main particulars, the estimation is proposed in this appendix. In the Figure A1, Lpp is the length between perpendiculars, s_AM is the midship section area and Δ is ship displacement.

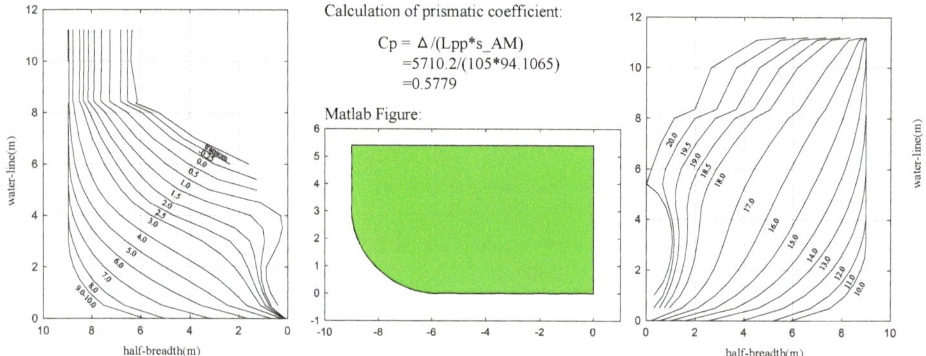

Figure A1. Molded lines of amidships area [29,30] (the figure permission has been achieved from Dalian Maritime University).

Appendix B

As shown in Figures A2–A4, all of the sea trials are corrected.

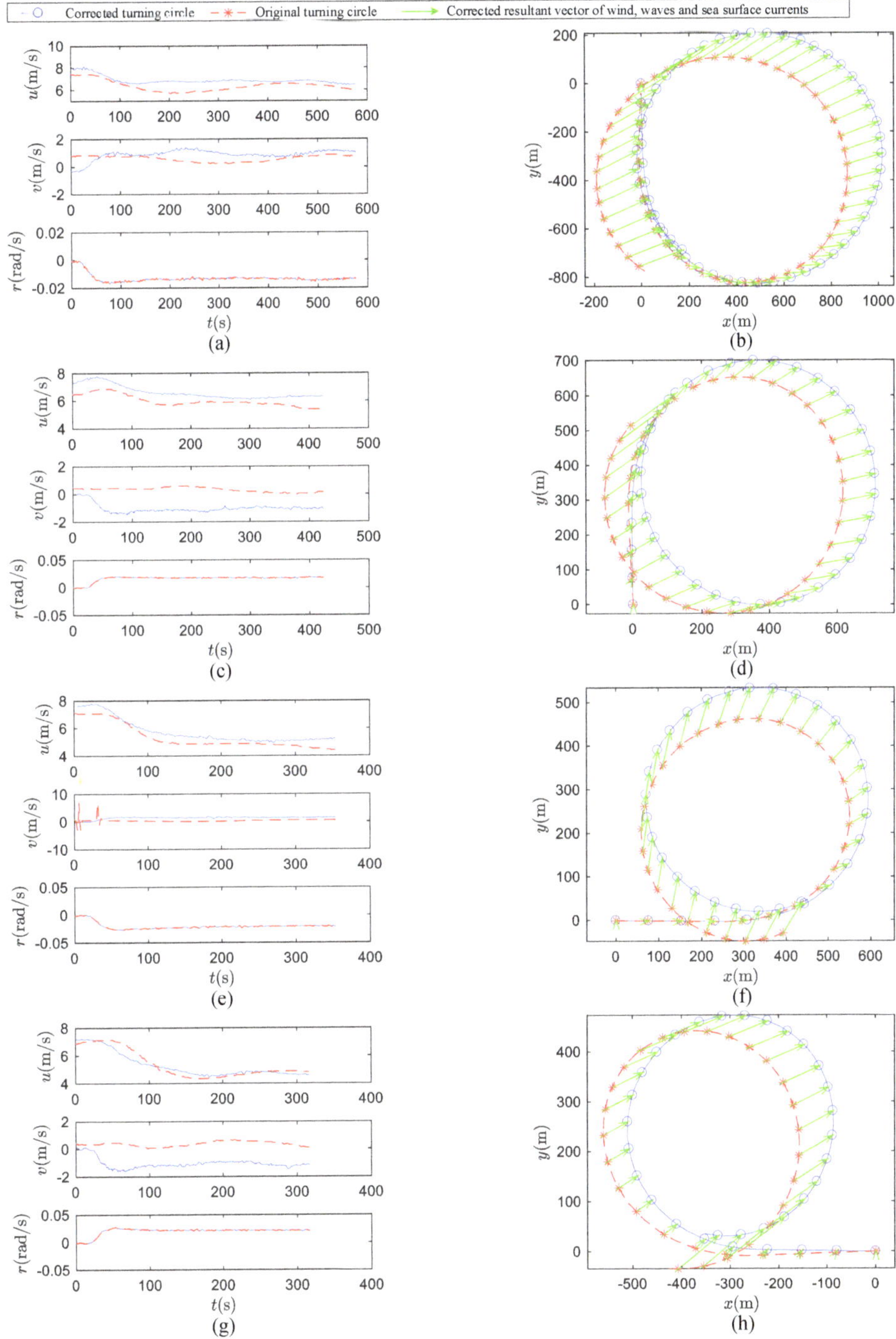

Figure A2. Correction result of track and velocities for test NO. 1–4. (**a**) Track of NO.1; (**b**) velocities of NO.1; (**c**) track of NO.2; (**d**) velocities of NO.2; (**e**) track of NO.3; (**f**) velocities of NO.3; (**g**) track of NO.4; (**h**) velocities of NO.4.

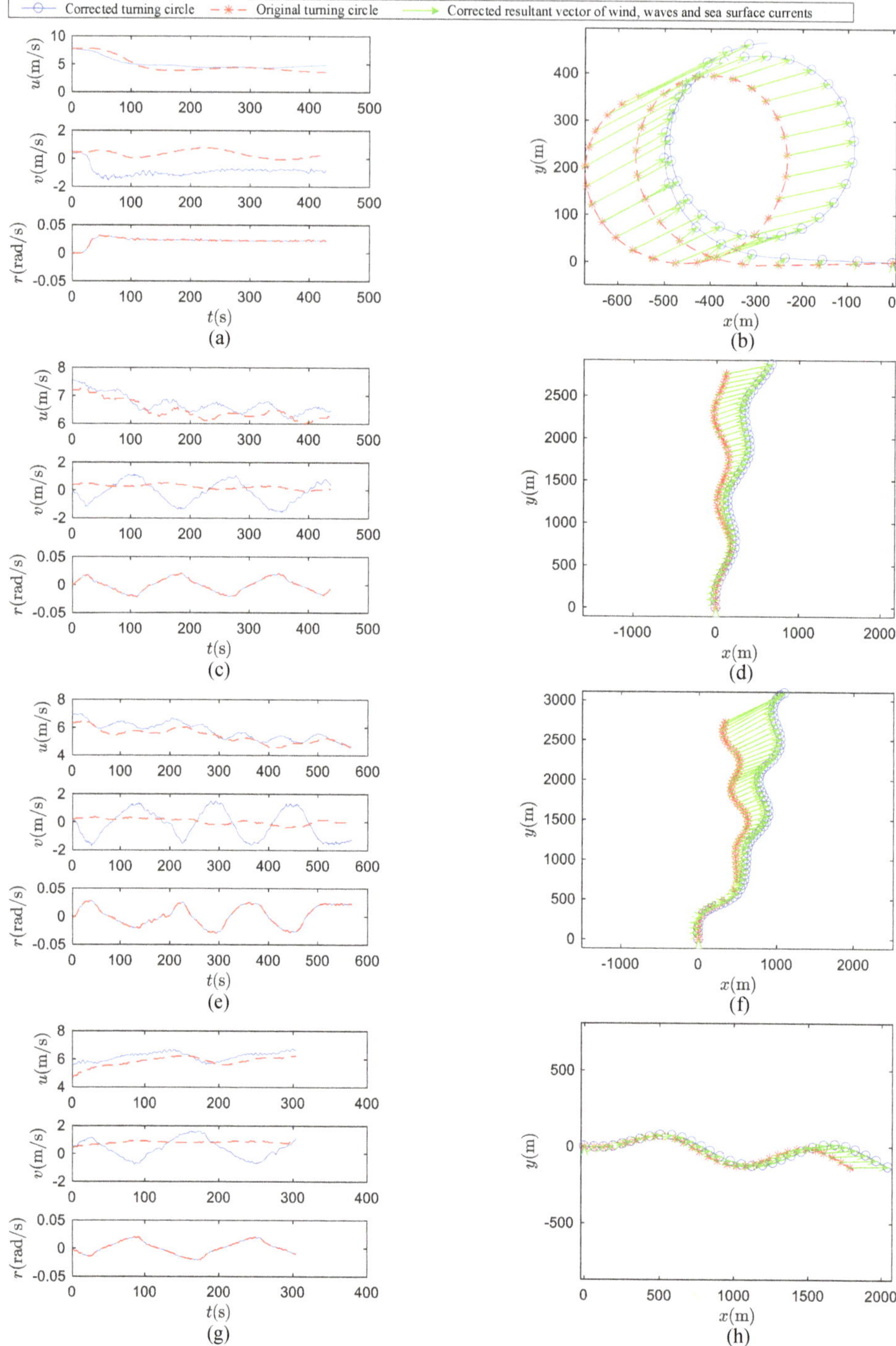

Figure A3. Correction result of track and velocities for test NO. 5–9. (**a**) Track of NO.5; (**b**) velocities of NO.5; (**c**) track of NO.6; (**d**) velocities of NO.6; (**e**) track of NO.7; (**f**) velocities of NO.7; (**g**) track of NO.8; (**h**) velocities of NO.8.

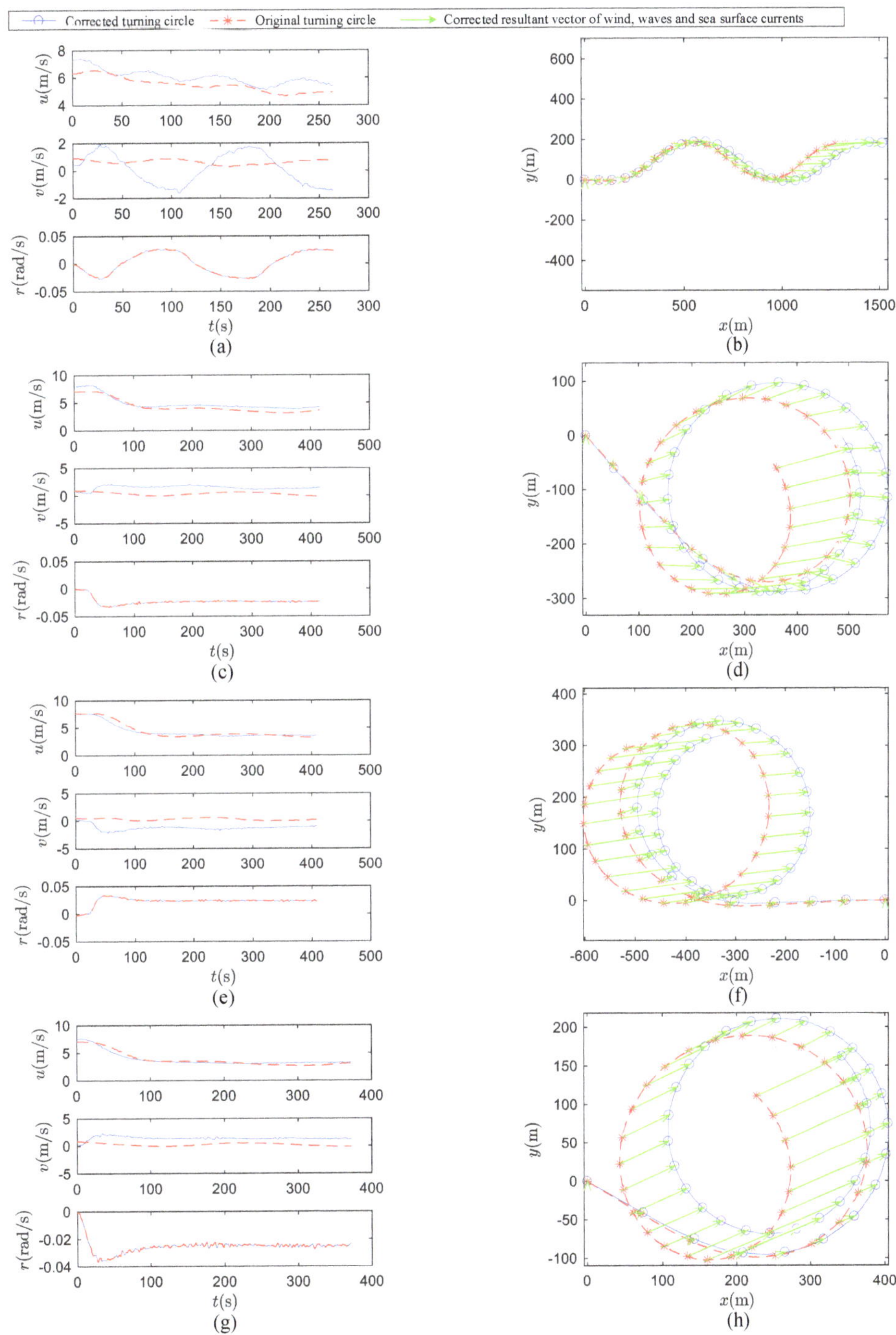

Figure A4. Correction result of track and velocities for test NO. 9–12. (**a**) Track of NO.9; (**b**) velocities of NO.9; (**c**) track of NO.10; (**d**) velocities of NO.10; (**e**) track of NO.11; (**f**) velocities of NO.11; (**g**) track of NO.12; (**h**) velocities of NO.12.

References

1. Lee, H.W.; Roh, M.I. Review of the multibody dynamics in the applications of ships and offshore structures. *Ocean Eng.* **2018**, *167*, 65–76. [CrossRef]
2. Abkowitz, M.A. Measurement of hydrodynamic characteristics from ship maneuvering trials by system identification. *Trans. SNAME* **1980**, *88*, 283–318.
3. Zhang, G.; Zhang, X.; Pang, H. Multi-innovation auto-constructed least squares identification for 4 DOF ship maneuvering modelling with full-scale trial data. *ISA Trans.* **2015**, *56*, 75–85. [CrossRef] [PubMed]
4. Bai, W.W.; Ren, J.S.; Li, T.S. Multi-innovation gradient iterative locally weighted learning identification for a nonlinear ship maneuvering system. *China Ocean Eng.* **2018**, *32*, 288–300. [CrossRef]
5. Kim, D.W.; Knud, B.; Mathias, P. Estimation of hydrodynamic coefficients from sea trials using a system identification method. *J. Korean Soc. Mar. Environ. Saf.* **2017**, *23*, 258–265. [CrossRef]
6. Maritime Safety Committee. The standards for ship maneuverability MSC 137(76). Available online: http://www.imo.org/en/KnowledgeCentre/IndexofIMOResolutions/Maritime-Safety-Committee-%28MSC%29/Documents/MSC.137%2876%29.pdf (accessed on 13 July 2020).
7. Society of Naval Architects and Marine Engineers (SNAME). *Guide for Sea Trials (Progressive Speed, Maneuvering and Endurance) Technical and Research Bulletin*; SNAME: Alexandria, VA, USA, 2015; pp. 3–47.
8. International Towing Tank Conference (ITTC). *Recommended Procedures and Guidelines-Full Scale Maneuvering Trials*; ITTC: Universitätsstrasse, Zürich, Switzerland, 2017; pp. 14–15.
9. Bishop, R.C.; Belknap, W.; Turner, C. *Parametric Investigation on the Influence of GM Roll Damping and Above-Water Form on the Roll Response of Model 5613*; Naval Surface Warfare Center: West Bethesda, MD, USA, 2005; pp. 11–17.
10. National Maritime Research Institute (NMRI). *Development of an Innovative Tank Model Test Methodology for Measuring Actual Sea Performance of Ships*; NMRI: Tokyo, Japan, 2016; pp. 13–21.
11. National Maritime Research Institute (NMRI). *Research on the Development of Energy Saving Device in Actual Sea*; NMRI: Tokyo, Japan, 2017; pp. 79–84.
12. Isherwood, R.M. Wind resistance on merchant ships. *Trans. RINA* **1973**, *115*, 327–338.
13. Blendermann, W. Parameter identification of wind loads on ships. *J. Wind Eng. Ind. Aerod.* **1994**, *51*, 339–351. [CrossRef]
14. Fujiwara, T.; Ueno, M.; Nimura, T. Estimation of wind forces and moments acting on ships. *J. Soc. Nav. Archit. Jpn.* **1988**, *183*, 77–90. [CrossRef]
15. Andersen, V.I.M. Wind loads on post-panamax container ship. *Ocean Eng.* **2013**, *58*, 115–134. [CrossRef]
16. Motora, S. On the measurement of added mass and added moment of inertia for ship motions. *J. Zosen Kiokai* **1959**, *105*, 83–92. [CrossRef]
17. Zhou, Z.M.; Sheng, Z.Y.; Fen, W.S. On maneuverability prediction for multipurpose cargo ship. *Ship Eng.* **1983**, *6*, 21–36.
18. Daidola, J.C.; Graham, D.A.; Chandrash, L. A simulation program for vessel's maneuvering at slow speeds. In Proceedings of the 11th Ship Technology and Research Symposium, Portland, OR, USA, 21–23 May 1986; pp. 156–161.
19. Li, Z.; Sun, J.; Beck, R.F. Evaluation and modification of a robust path following controller for marine surface vessels in wave fields. *J. Ship Res.* **2010**, *54*, 141–147.
20. Yasukawa, H. Simulations of ship maneuvering in waves 1st report turning motion. *J. Soc. Nav. Archit. Jpn.* **2006**, *4*, 127–136. [CrossRef]
21. Zhang, W.; Zou, Z.J. Time domain simulations of the wave induced motions of ships in maneuvering condition. *J. Mar. Sci. Technol. Jpn.* **2016**, *21*, 154–166. [CrossRef]
22. Hong, L.; Zhu, R.C.; Miao, G.P.; Fan, J.; Li, S. An investigation into added resistance of vessels advancing in waves. *Ocean Eng.* **2016**, *123*, 238–248. [CrossRef]
23. Mei, B.; Sun, L.; Shi, G. Ship maneuvering prediction based on grey-box model identification via adaptive RM-SVM with small rudder angle. *Pol. Marit. Res.* **2019**, *26*, 115–127. [CrossRef]
24. Yabuki, H.; Yoshimura, Y.; Ishiguro, T.; Ueno, M. *Turning Motion of a Ship with Single CPP and Single Rudder during Stopping Maneuver under Windy Condition*; International Conference on Ship Manoeuvrability and Maritime Simulation: Terschelling, The Netherlands, 2006; pp. 4–5.
25. Kasa, I. A curve fitting procedure and its error analysis. *IEEE Trans. Instrum. Meas.* **1976**, *25*, 8–14. [CrossRef]

26. Bakdi, A.; Glad, I.K.; Vanem, E.; Engelhardtsen, Ø. AIS-based multiple vessel collision and grounding risk identification based on adaptive safety domain. *J. Mar. Sci. Eng.* **2020**, *8*, 5. [CrossRef]
27. Niu, J.; Liang, X.; Zhang, X. Time-Varying Kelvin Wake Model and Microwave Velocity Observation. *Sensors* **2020**, *20*, 1575. [CrossRef]
28. Sung, Y.J.; Park, S.H.; Ahn, K.S. Evaluation on deep water maneuvering performances of KVLCC2 based on PMM test and RANS simulation. In *SIMMAN 2014*; FORCE Technology: Lyngby, Denmark, 2014; pp. 1–6.
29. Shan, X.F. Probing the Cyclicity of "YUKUN" under the Influence of Wind and Current. Masters' Thesis, Dalian Maritime University, Dalian, China, 2013.
30. Su, Z.J. The Numerical Simulation and Analysis of Parametric Rolling for Vessel "YUKUN". Masters' Thesis, Dalian Maritime University, Dalian, China, 2011.
31. Jiang, Y.; Zheng, K. The Single-Shore-Station-Based position estimation method of an Automatic Identification System. *Sensors* **2020**, *20*, 1590. [CrossRef]
32. Kim, D.; Lee, S.; Lee, J. Data-driven prediction of vessel propulsion power using support vector regression with onboard measurement and ocean data. *Sensors* **2020**, *20*, 1588. [CrossRef] [PubMed]

Article

Remanufacturing System with Chatter Suppression for CNC Turning

Karol Miądlicki *, Marcin Jasiewicz, Marcin Gołaszewski, Marcin Królikowski and Bartosz Powałka

Department of Mechanical Engineering and Mechatronics, West Pomeranian University of Technology, Szczecin, al. Piastów 19, 70-310 Szczecin, Poland; marcin.jasiewicz@zut.edu.pl (M.J.); gm37274@zut.edu.pl (M.G.); marcin.krolikowski@zut.edu.pl (M.K.); bartosz.powalka@zut.edu.pl (B.P.)

* Correspondence: karol.miadlicki@zut.edu.pl; Tel.: +48-91-494-4338

Received: 5 August 2020; Accepted: 4 September 2020; Published: 7 September 2020

Abstract: The paper presents the concept of a support system for the manufacture of machine spare parts. The operation of the system is based on a reverse engineering module enabling feature recognition based on a 3D parts scan. Then, a CAD geometrical model is generated, on the basis of which a machining strategy using the CAM system is developed. In parallel, based on the geometric model, a finite element model is built, which facilitates defining technological parameters, allowing one to minimize the risk of vibrations during machining. These parameters constitute input information to the CAM module. The operation of the described system is presented on the example of machining parts of the shaft class. The result is a replacement part, the accuracy of which was compared by means of the iterative closest point algorithm obtaining the RMSE at the level of scanner accuracy.

Keywords: feature recognition; geometry recognition; 3D scanning; chatter; machining assistance; machining stability; receptance coupling; finite element model; refactoring

---

1. Introduction

In the modern industry, in line with the latest trends and the idea of Industry 4.0, the objective is to increase automation and autonomy of production. The aim of these activities is to reduce the dependence of production plants on qualified machine operators and technologists, and to increase production efficiency. Therefore, the importance of intelligent production support systems and operators is growing. New machines are increasingly being equipped with support systems that significantly simplify operation and allow one to avoid costly errors, which are often due to a lack of operator experience. This approach reduces training costs, among other things, and makes the plant independent of qualified personnel. Owing to this approach, it is possible to assign employees to operate the machine, even if they do not have specialist knowledge of how to produce workpieces. It may turn out to be particularly important in the maintenance departments of industrial companies or in onboard ships applications, in the engine department. An integral part of the tasks performed by these units is the damaged parts replacement. If the spare part is available, the repair can be performed instantly. Otherwise, it is necessary to manufacture the parts by carrying out machining. It should be done as quickly as possible; therefore, any mistakes made or unforeseen difficulties such as vibrations during machining are unacceptable. In such situations, the assistance systems can be invaluably helpful.

Available tools' supporting production can be divided into two groups. The first group are tools used mainly by managers, executives, and technologists. The available production support tools include production management systems, production planning tools, parts design support systems, production quality control systems and reverse engineering tools, supporting, for example, CAD model building. These tools usually include advanced software installed on dedicated PCs. The second group is software and HMI interfaces dedicated to machine and equipment operators. Most often, they are integrated directly with machines. To this group we may include overlays on user interfaces or support systems based on virtual and augmented reality.

The machinery/machining industry in particular requires qualified staff, both at the stages of geometry design and development of the parts manufacturing process (technologists) and at the time of manufacture (machine operators). Therefore, tools for designers and technologists have developed over the years, along with the development of computer aided design and manufacturing techniques for machine parts. These tools include CAD systems to support the design of parts, CAM systems to facilitate the generation of machining programs and reverse engineering tools used, for example, to identify structural features. Most of these systems are integrated in commercial programs such as Dassault Systèmes SOLIDWORKS, Dassault Systèmes Catia, Siemens NX or PTC Creo. Reverse engineering systems are particularly advanced. Some commercial programs such as Ansys SpaceClaim have been equipped with a module supporting basic identification of structural features based on cross-sections in parallel planes. In parallel, newer and newer techniques are being developed to open up the geometry of parts [1], both in two and three dimensions [2]. V. B Sunil et al., in their work, present an intelligent system for recognizing prismatic part machining features from CAD models using an artificial neural network [3]. X. Lin et al. present a similar approach in [4], where a propose intelligent hybrid strategy is proposed for edge inconsistent feature detection by machine vision. The two deep neural networks are employed together in series, to first detect and then recognize polishing workpieces in an industrial environment; these were used by F. Liu et al. in [5].

Techniques are also being developed to identify structural features and dynamic properties for steel machine parts [6,7], as well as for modern composite parts [8,9].

In the machining industry, the key stage is to make a part from the blank. For this process, qualified CNC operators with extensive knowledge of G-code machining and programming technology are required, especially in view of the increasing demands concerning: machining time, surface quality [10], topographic control and continuous miniaturization, for which micrometers accuracy is required [11]. The machining of new polymer materials injected [12] or printed [13] is also a major challenge for operators. For many companies, the cost of hiring a qualified operator is too high and the time needed for training is too long. Therefore, the control systems of machine tools are extended with operator support systems, such as: compensation of temperature errors of ball screws [14] or 3D (three-dimensional) scanning vision system for positioning the workpiece [15]. More extensive CNC systems have additional built-in options, which include extensive graphical interfaces to facilitate the generation and analysis of G-code i.e., Siemens: Shopturn/ShopMill, Fanuc: Manual Guide. New graphic solutions with 3D elements, touch screens, remote controls and even gesture support are introduced to increase operator comfort. The latest systems also support users in the selection of technological parameters [16,17]. All systems mentioned above support an inexperienced operator in the machining process, owing to which he makes fewer mistakes and does not have to undergo expensive training.

However, despite the existence of systems to support designers/technologists and operators, so far, no solution has been developed to copy/manufacture a part without specialist knowledge including part design, development of machining technology, and manufacturing the part using a CNC machine. Currently, due to the development of the idea of Industry 4.0 and the Internet of Things, systems combining the tasks of a designer, a technologist and an operator are the object of increased research [18,19]. The popularity of these systems will continue to grow. This will be supported by the growing computing power of CNC systems and the increasing number of sensors integrated in machines.

This paper proposes an innovative system based on reverse engineering that allows for simple and intuitive copying of shaft type elements. The following procedures have been implemented in the system: scanning of parts, geometry reproduction, CAD model generation, simulation, selection of machining parameters (reduction of self-excited vibrations) and the generation of machining technology with the G-code. The main novelty presented in the article concerns the integration of known computational methods with the innovative feature recognition algorithm. The presented methodology may contribute to the development of manufacturing support systems. This can be particularly useful on ships, where access to qualified specialists and spare parts is significantly limited. Section 2 discusses and explains the various stages of system operation. Section 3 presents the results of the system operation and their discussion. Section 4 provides a summary and further plan for the development of the system.

2. CNC Machining Assistance System

The following subsections present the concept of a system in which a fully parametric CAD model is built with the use of reverse engineering, and then, on its basis, analyses supporting the technological process are conducted.

2.1. System Concept

The idea of operation of the developed system is based on the use of reverse engineering, in which a parametric CAD model is built on the basis of a 3D scan of the part. Then, the processing technology is developed on its basis, using the CAM module. One of the elements determining its effectiveness is the selection of appropriate technological parameters. In the presented system, this selection is supported by a module allowing one to minimize the risk of self-excited vibrations. This is possible on the basis of the analysis of dynamic properties of the workpiece. This analysis is carried out using the FEM model based on the CAD model. As a result of the proposed system, a part machining program is obtained with technological parameters that allow one to avoid vibrations during machining. The block diagram showing the system concept with the data flow is presented in Figure 1.

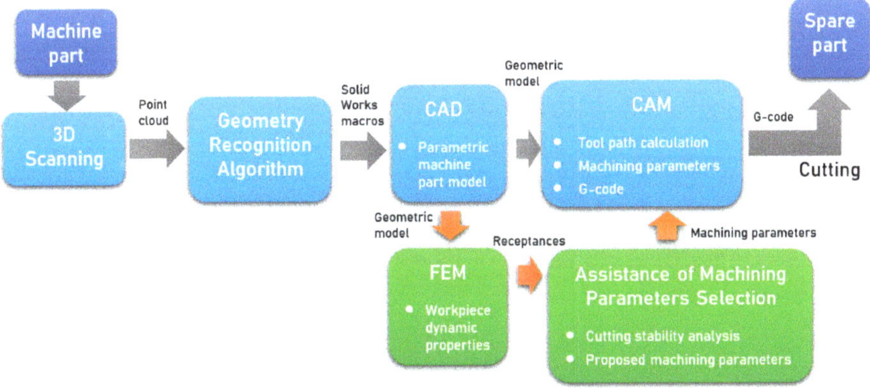

Figure 1. Block diagram showing the system concept.

2.2. 3D Scanning

The part on the basis of which the operation of the system will be presented is a 183 mm long steel shaft with a maximum diameter of 37 mm made of steel, as shown in Figure 2. This is a part with a low degree of complexity, but due to the high average length/diameter ratio (L/D), there is a high risk of self-excited vibration during machining.

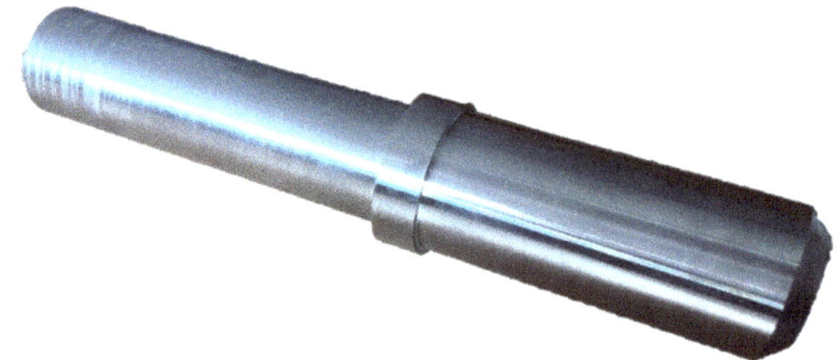

Figure 2. The part under consideration.

Scanning was performed in rotary mode on the PICZA-LPX1200 scanner manufactured by Roland (1-6-4 Shinmiyakoda, Kita-ku, Hamamatsu-shi, Shizuoka-ken, 431-2103 Japan). After being covered with an anti-reflective white film, the object is placed in the axis of the rotary table, as shown in Figure 3. The following parameters were used during the scanning process: angular pitch—0.90 deg, lace cut in the axis of the object every 0.1 mm. XYZ point clouds with native size of 11.244 KB and 11.126 KB (0.1 mm axial lace cut) were obtained.

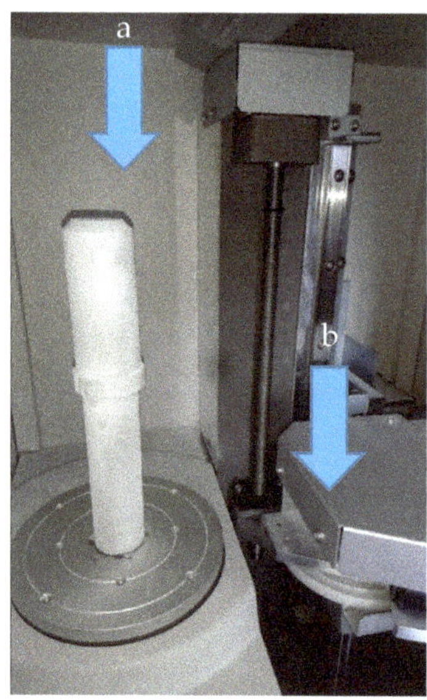

Figure 3. Experimental setup—shaft prepared for scanning; (**a**)—scanned object, (**b**)—scanning head.

2.3. Geometry Recognition Algorithm

The next step was to convert the point cloud to parametric geometry. The implemented algorithms (Figure 4) identify the geometry of the shaft class parts based on cross-sections. The geometry of the identified part is then imported into SOLIDWORKS for further processing.

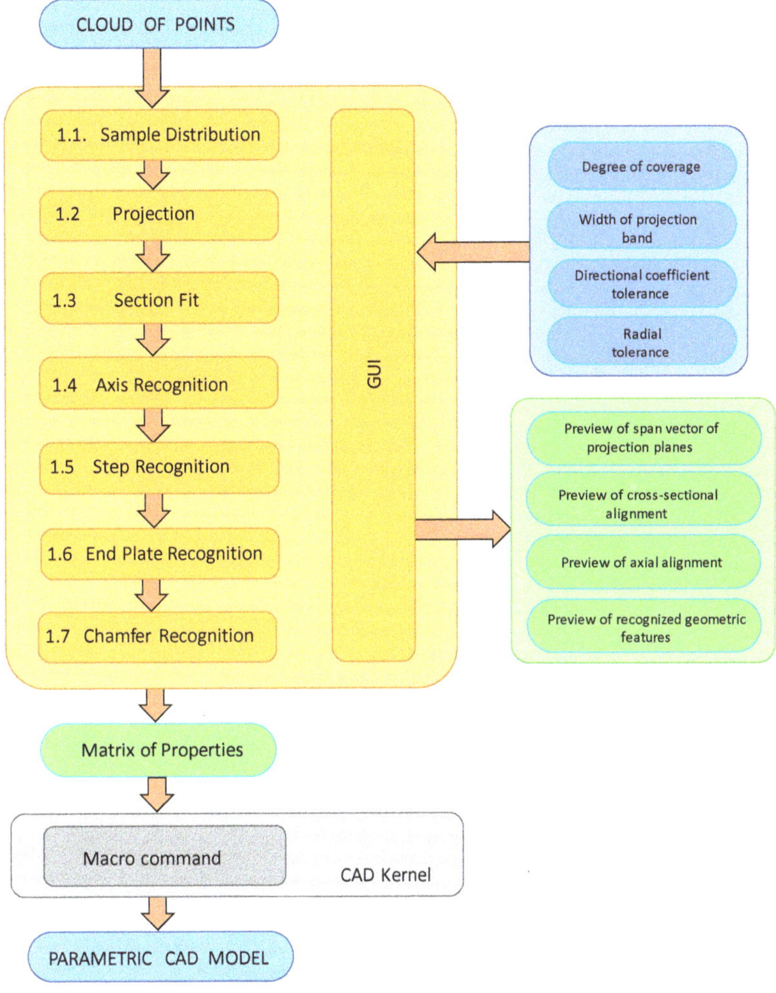

Figure 4. Diagram of cloud of points conversion to parametric geometry.

The operation of the geometry identification algorithm began with the determination of the vector of the span of planes (Figure 5), which was performed using the 'Sample Distribution' function. This function determines the vector based on the projection bandwidth and degree of coverage. The width of the projection band determines the symmetrical area around the section plane from which points from the cloud are projected. The degree of coverage is, on the other hand, a percentage parameter, which determines the total width of the projection bands. against the background of the Z-axis point cloud span (1).

$$p = \frac{k \cdot s}{r} \cdot 100\% \tag{1}$$

where:

p—coverage parameter
k—number of scanning planes
s—width of a single projection band
r—span of the point cloud in the Z axis

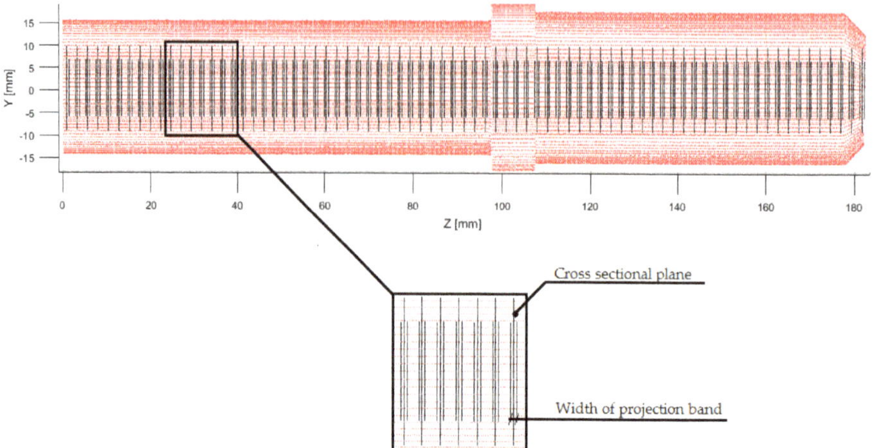

Figure 5. Span vector of cross-section planes with marked projection bands.

Hence:

$$k = \frac{p \cdot r}{s \cdot 100\%} \tag{2}$$

Then, on the basis of the designated span vector, using the 'Projection' function, points were projected on the cross-section planes. The circular cross-sections were adjusted using the 'Section Fit' function.

The 'Section Fit' function approximates a circular cross-section using the smallest squares method. To determine the coordinates of the center and the value of the section radius, a canonical equation of a circle (3) has been formulated, where x_i, y_i are the coordinates of the scanned i-th point, x_c, y_c are the coordinates of the center of the circle and R is the radius of the circle (Figure 6).

$$(x_i - x_c)^2 + (y_i - y_c)^2 = R^2 \tag{3}$$

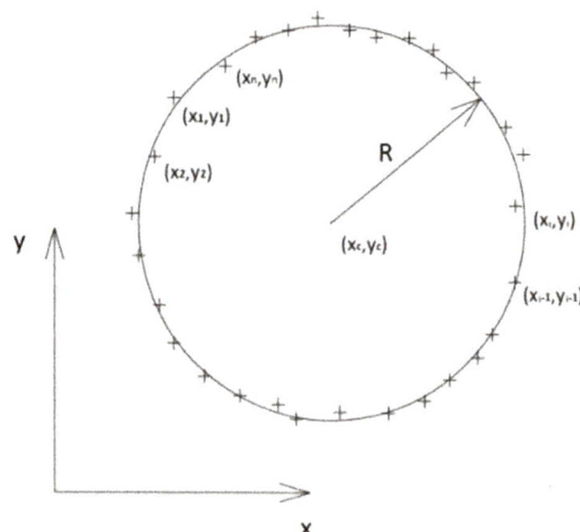

Figure 6. Approximation of the circular cross-section.

After the transformation of Equation (3), the general form of the circle equation is obtained:

$$x_i^2 + y_i^2 = Ax_i + By_i + C \tag{4}$$

where: the constants A, B, C Equations (5)–(7) have been introduced to simplify the notation.

$$A = 2x_c \tag{5}$$

$$B = 2y_c \tag{6}$$

$$C = -\left(A^2 + B^2 + R^2\right) \tag{7}$$

For each point (Figure 6) projected on the cross-sectional plane, the equation was formulated in the determined general form Equations (8)–(12).

$$x_1^2 + {y_1}^2 = Ax_1 + By_1 + C \tag{8}$$

$$x_2^2 + {y_2}^2 = Ax_2 + By_2 + C \tag{9}$$

$$x_i^2 + {y_i}^2 = Ax_i + By_i + C \tag{10}$$

$$x_{n-1}^2 + y_{n-1}^2 = Ax_{n-1} + By_{n-1} + C \tag{11}$$

$$x_n^2 + {y_n}^2 = Ax_n + By_n + C \tag{12}$$

Next, the system of Equations (8)–(12) was transformed into a matrix notation Equation (13).

$$\begin{bmatrix} x_1 & y_1 & 1 \\ x_2 & y_2 & y \\ \cdots & \cdots & \cdots \\ x_i & y_i & 1 \\ \cdots & \cdots & \cdots \\ x_n & y_n & 1 \end{bmatrix} \begin{bmatrix} A \\ B \\ C \end{bmatrix} = \begin{bmatrix} x_1^2 + y_1^2 \\ x_2^2 + y_2^2 \\ \cdots \\ x_i^2 + x_i^2 \\ \cdots \\ x_n^2 + y_n^2 \end{bmatrix} \tag{13}$$

After solving the linear system of Equation (13), the values of constants A, B and C Equations (5)–(7) were obtained. Next, from their values, the coordinates of the circle center x_c, y_c Equations (14) and (15) and the radius R of the circle Equation (16) were determined.

$$x_c = -\frac{A}{2} \tag{14}$$

$$y_c = -\frac{B}{2} \tag{15}$$

$$R = \sqrt{\frac{A^2 + B^2 + 4C}{4}} \tag{16}$$

The circles were matched in the local coordinate system of the section plane. On the basis of the determined coordinates of centers using the 'Axis Recognition' function, the shaft axis in the global system was matched. Having regard to the defined directional coefficient tolerance, the matched axis has been corrected (Figure 7).

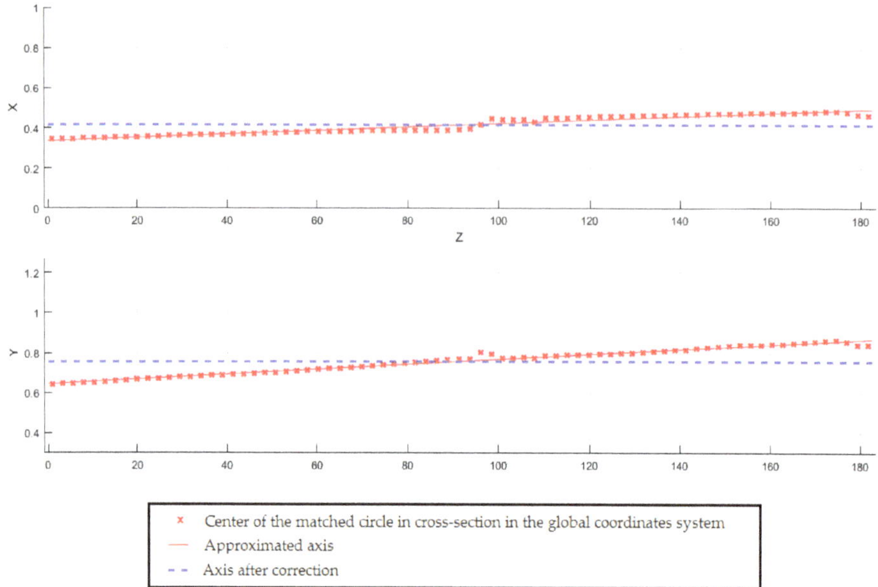

Figure 7. Overview of the shaft axis in XZ plane and YZ plane.

The identification of the shaft steps was made on the basis of the values of the cross-sectional radii. The radial tolerance parameter was the condition contained in the 'Step Recognition' module to recognize the belonging of successive cross sections to one shaft step. When *i*-th cross-section radius did not deviate within the radial tolerance from the cross-section radius *i*1, both cross-sections were considered to belong to one shaft step. The detected sequence of less than three cross-sections is considered as an apparent degree, which is not taken into account in further proceedings (Figure 8). The appearance of the apparent degree results from the distortion of the radius by projecting points from the surfaces closing the shaft steps.

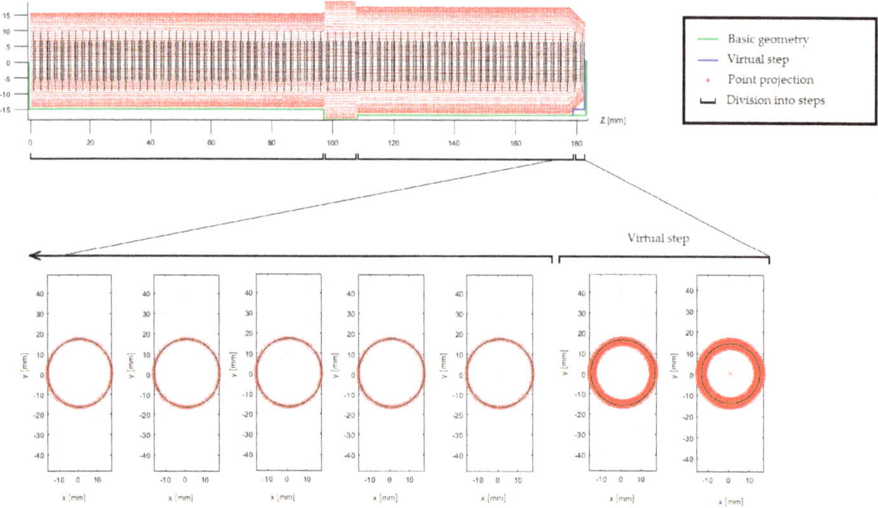

Figure 8. Overview of matched cross-sections with the apparent degree shown.

Using the 'End Plate Recognition' module, fragmented axial cross-sections were created in the transition areas of the shaft steps. The closing surfaces were adjusted using the 'polyfit' function from the Matlab library.

In order to detect chamfering, fragmented axial cross-sections have been created in the end areas of the shaft steps. Using the 'Chamfer Recognition' module, a chamfer was determined by detecting the distance of projection points from the recognized basic geometry (Figure 9). The identified geometry with division into the base body and technological features has been recorded in the geometric properties matrix (Table 1).

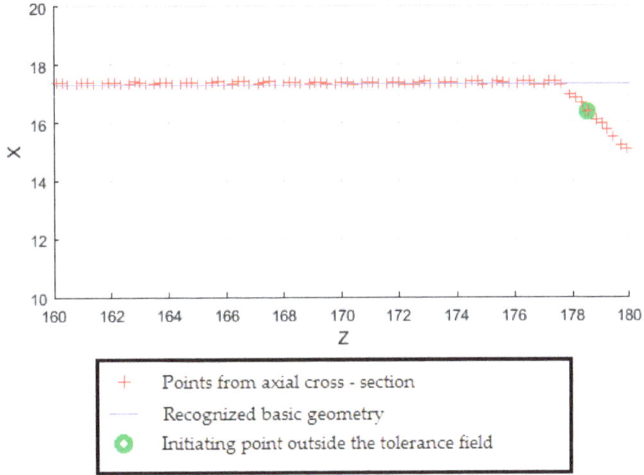

Figure 9. Chamfer detection based on axial cross-section.

Table 1. Matrix of recognized geometric properties.

No.	X	Y	R	Z Start	Z End	Chamfer Start	Chamfer End
1	0	0	15	0	98	0.0	0.0
2	0	0	18.5	98	108	0.0	0.0
3	0	0	16.9	108	183	0.0	5.0

The geometric properties matrix has been taken over by the macro command in Dassault Systèmes SOLIDWORKS 2018. The pseudo-code of the algorithm for importing geometry into SOLIDWORKS is shown in Table 2. First, the basic geometry was created by adding/extracting by rotation. Then, technological operations were added to the model. This approach allowed to reproduce the operations tree. The reconstructed geometry is shown in Figure 10.

Table 2. Pseudocode of geometry import algorithm to SOLIDWORKS software.

Input: Matrix of recognized geometric properties **Output: Parametric CAD model**
1. Determine the number of steps (based on Matrix of properties) 2. Create 'New Part' 3. Open Sketch on XZ plane 4. Draw contour of the basic geometry 5. Create the basic geometry using 'Revolved Boss/Base' function 6. Find edges to chamfer 7. For i = number of edges to chamfer 8. Specify the type of chamfer 9. Create sketch of chamfer on XZ plane 10. Create chamfer using 'Recolved Boss/Base' or 'Revolved Cut' (type of function dependent on the specified type of chamfer) 11. Save CAD model in specified format

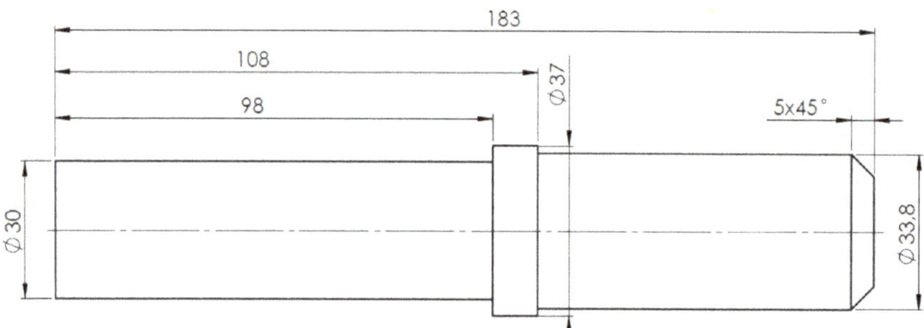

Figure 10. Identified part geometry.

2.4. Finite Element Model

Then, on the basis of the geometric model, a finite element model of the workpiece was built, in order to determine its characteristic frequency transition functions. The described frequency transition functions were the information necessary to determine the area of stability of machining using the CNC assistance module.

The finite element model is built using Midas NFX 2018 R1 software (Midas Information Technology Co. Ltd., Seongnam, Korea). In the first step, the geometric model was discretized using eight node, cubic, isoparametric finite elements (CHEXA) and six node, five-walled, isoparametric finite elements (CPENTA). The applied finite elements were characterized by linear shape functions and three translation degrees of freedom in each node. As a result of the discretization, a model consisting of 2.103 finite elements and 6.237 degrees of freedom was obtained. The discrete model is shown in Figure 11.

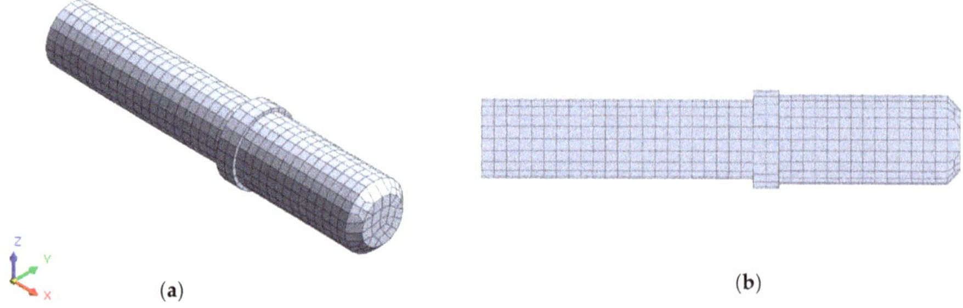

Figure 11. Discretized model of analyzed shaft, (**a**) isometric view and (**b**) cross-sectional view.

Due to the receptance method used in the CNC assistance module, on the basis of which the stable machining area is determined, an underdetermined model was adopted for further calculations.

In the next step, using the Nastran Solver processor (SOL108), a set of receptance functions in the X and Z direction in the frequency ranging from 50 to 5000 Hz with a 1 Hz step was determined. Examples of calculation results as frequency response functions are presented in Figure 12.

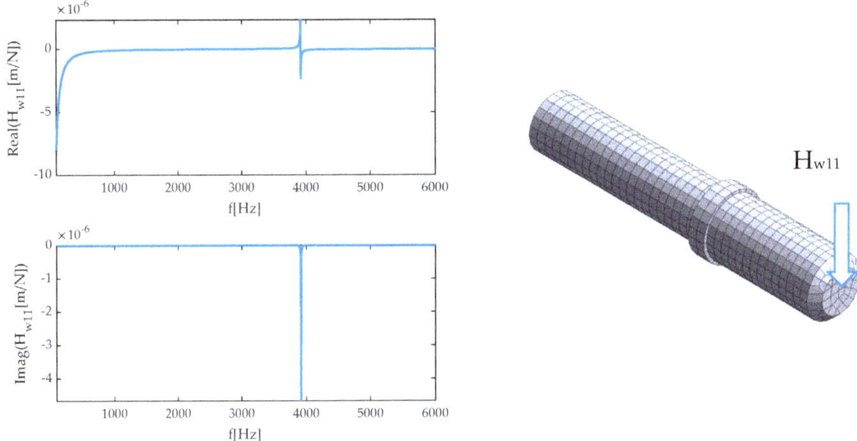

Figure 12. An example of FRF for the workpiece.

2.5. Assistance of Machining Parameters Selection

In order to determine the process stability conditions for assessing the risk of self-excited vibrations, it is necessary to know the dynamic properties (as frequency response functions—FRFs) of the machine tool—workpiece system [20]. In the developed system, the system consists of a lathe spindle with a three-jaw chuck and a mounted workpiece. The dynamic properties of a given system can be determined with the application of modal synthesis, using the receptance coupling approach (RCA) [21,22]. This method allows the FRF function of a combined system to be determined (Figure 13a), having the dynamic properties of the components of which the system is composed (Figure 13b).

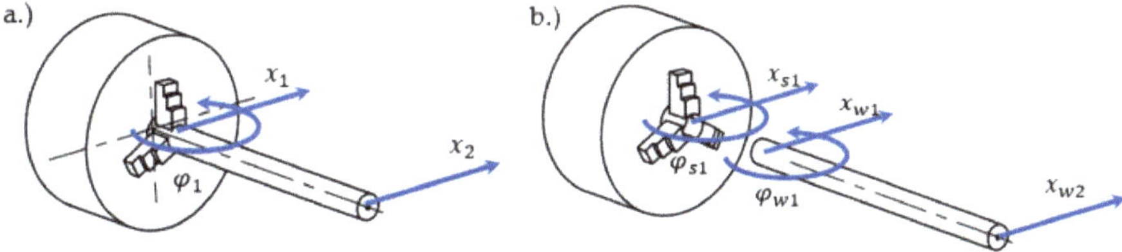

Figure 13. Receptance coupling system: (**a**) components, (**b**) coupled system.

The dynamic properties of the machine tool are reduced to the properties of the spindle with the three-jaw chuck, and the experimental extended inverse receptance coupling (EIRC) method, described in detail in [23]. This method allows one to take into account the rotational degrees of freedom (RDOF) of the system, necessary to properly model the way the workpiece is clamped in the three-jaw chuck. Moreover, while modelling the properties of the spindle, it is possible to add an extra length of the machined part resulting from the use of a longer workpiece. The dynamic properties of the spindle remain unchanged over time; thus, they can be determined once for a given machine tool, by performing a series of impulse tests. The variable element in the system is the workpiece. Most often, for RCA applications, the workpiece is modelled analytically as a Timoshenko beam, however, due to a more complex geometry, in the presented example, the workpiece is modelled using the finite element method, as described in Section 2.4. For RCA modal synthesis, the dynamic properties of the components (w index—workpiece, s index—spindle) are noted as matrix equations:

$$\begin{bmatrix} x_{s1} \\ \varphi_{s1} \end{bmatrix} = \begin{bmatrix} H_{s11} & L_{s11} \\ N_{s11} & P_{s11} \end{bmatrix} \cdot \begin{bmatrix} F_{s1} \\ M_{s1} \end{bmatrix} \tag{17}$$

$$\begin{bmatrix} x_{w1} \\ \varphi_{w1} \\ x_{w2} \end{bmatrix} = \begin{bmatrix} H_{w11} & L_{w11} & H_{w12} \\ N_{w11} & P_{w11} & N_{w12} \\ H_{w21} & L_{w21} & H_{w22} \end{bmatrix} \cdot \begin{bmatrix} F_{w1} \\ M_{w1} \\ F_{w2} \end{bmatrix} \tag{18}$$

where: x—translational displacement in direction x, φ—angle of rotation, F—force, M—torque, transfer functions: translational H[m/N] and rotational L[m/Nm], N[rad/N], P[rad/Nm].

The matrix equation describing the dynamic properties of the combined system shall be noted as:

$$\begin{bmatrix} x_{1} \\ \varphi_{1} \\ x_{2} \end{bmatrix} = \begin{bmatrix} H_{11} & L_{11} & H_{12} \\ N_{11} & P_{11} & N_{12} \\ H_{21} & L_{21} & H_{22} \end{bmatrix} \cdot \begin{bmatrix} F_{1} \\ M_{1} \\ F_{2} \end{bmatrix} \tag{19}$$

Having both a model of the dynamic properties of the machine tool and the workpiece, boundary conditions and force balance conditions between the components are noted as follows:

$$\begin{cases} x_{s1} = x_{w1} = x_1 \\ \varphi_{s1} = \varphi_{w1} = \varphi_1 \end{cases}, \quad \begin{cases} F_{s1} + F_{w1} = F_1 \\ M_{s1} + M_{w1} = M_1 \end{cases} \tag{20}$$

By making appropriate transformations using Equations (17), (18), (20), it is possible to determine the dynamic properties matrix of the combined system:

$$\begin{bmatrix} H_{11} & L_{11} & H_{12} \\ N_{11} & P_{11} & N_{12} \\ H_{21} & L_{21} & H_{22} \end{bmatrix} = \begin{bmatrix} 1+\frac{H_{w11}P_{s11}-H_{w12}L_{s11}}{H_{s11}P_{s11}-L_{s11}N_{s11}} & \frac{H_{w12}H_{s11}-H_{w11}L_{s11}}{H_{s11}P_{s11}-L_{s11}N_{s11}} & 0 \\ \frac{H_{w11}P_{s11}-P_{w11}L_{s11}}{H_{s11}P_{s11}-L_{s11}N_{s11}} & 1+\frac{P_{w11}H_{s11}-N_{w11}L_{s11}}{H_{s11}P_{s11}-L_{s11}N_{s11}} & 0 \\ \frac{H_{w21}P_{s11}-L_{w21}L_{s11}}{H_{s11}P_{s11}-L_{s11}N_{s11}} & \frac{L_{w21}H_{s11}-H_{w21}L_{s11}}{H_{s11}P_{s11}-L_{s11}N_{s11}} & 1 \end{bmatrix}^{-1} \cdot \begin{bmatrix} H_{w11} & L_{w11} & H_{w12} \\ N_{w11} & P_{w11} & N_{w12} \\ H_{w21} & L_{w21} & H_{w22} \end{bmatrix} \tag{21}$$

For the stability analysis, the translational function of the transition H_{22} to the x (Figure 14a) direction is used, as it is the highest susceptibility at the end of the workpiece, which is equivalent to the highest risk of vibration occurrence during machining. The tool for predicting system stability is the stability lobe diagram (SLD). The formation of self-excited vibrations during machining is associated with exceeding the cutting depth limit a_{lim}, which is presented as follows:

$$a_{lim} = -\frac{1}{2K_r Re(H(j\omega))} \tag{22}$$

where $Re(H(j\omega)$—the real part of the translational transfer function for a susceptible component of the system, K_r—coefficient of the cutting forces. As it results from formula (14), positive values of the machining depth limit are obtained for negative values of the function $Re(H(j\omega)$, which are used to build the SLD diagram. The reference of the machining depth limit to the spindle speed is realized by replicating the lobe with the following relationship:

$$N_c = \frac{60 \cdot f_c}{k}, \quad \text{for } k = 1, 2 \ldots, n \tag{23}$$

where f_c—is the chatter frequency, k—consecutive integers denoting the lobe number.

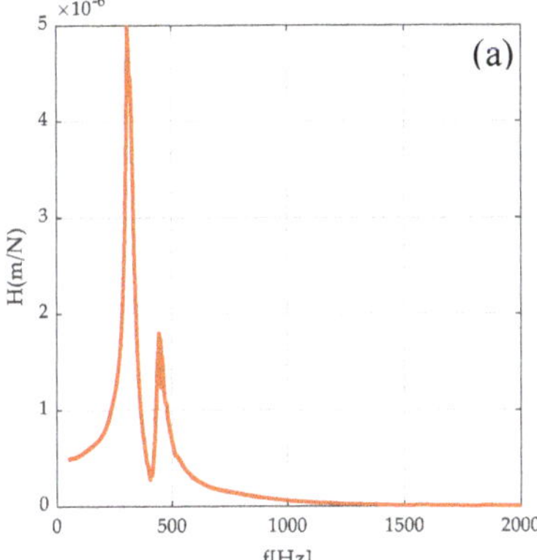

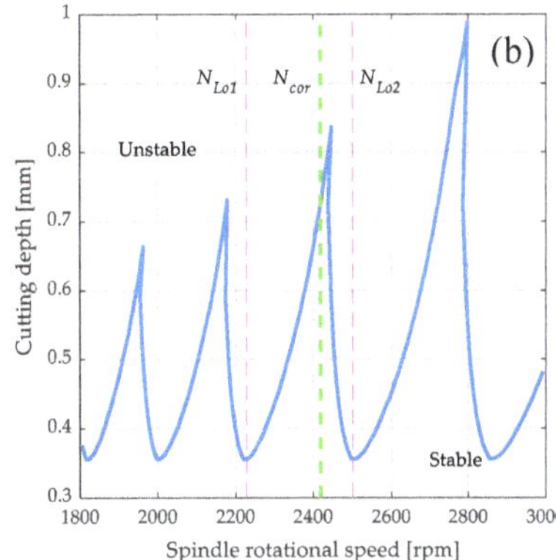

Figure 14. Stability analysis: (**a**) FRF at the end of the workpiece based on RCA (**b**) Stability lobes.

The stability lobes generated in the speed range from 1800 to 3000 rpm for the transfer function under consideration H_{22} are shown in Figure 14a. The stability lobes delineate the areas of technological parameters for which vibrations do not occur ('Stable' area) and the area in which self-excited vibrations will develop ('Unstable' area).

When analyzing the stability lobes, it should be noted that the selected rotational speeds are more resistant to the occurrence of vibrations during machining. The idea of the presented support for the selection of technological parameters assumes proposing the correction of arbitrarily adopted spindle speeds to the nearest values allowing one to obtain stable machining. The authors experience shows that indicating peaks of the stability lobes as the recommended speeds does not give the expected results due to the proximity of the stability limit. In the proposed approach, first, rotational speeds N_{Lo1} and N_{Lo2} are selected, which define the single lobe range, as shown in Figure 14b. Then, the recommended spindle speed correction N_{cor} for the lobe is determined as:

$$N_{cor} = (N_{Lo2} - N_{Lo1}) \cdot P_{sh} + N_{Lo1} \tag{24}$$

where P_{sh} is a peak shift coefficient, in the presented system, with the value $P_{sh} = 0.7$. The precise determination of the cutting depth is possible with the value of the specific cutting force coefficient determined experimentally for a given material and tool configuration. In the presented system, it was decided to adopt an average value of this coefficient for steel, which significantly simplifies the procedure of supporting the selection of rotational speed, and at the same time does not affect the position of 'stable' rotational speeds.

2.6. Computer-Aided Manufacturing (CAM)

The technological parameters for which stable conditions and favorable machining performance have been achieved while maintaining the recommended blade life have been imported into the Solidworks CAM system lathe module. Solidworks CAM enables the development of 2.5 and 3 axis turning and milling processes. The software is based on the CAMWorks system. The CAM module implemented in Solidworks provides full support for configuration and parts, which enables the use of geometric features recognized on the basis of the model and the corresponding CAD operations.

Solidworks CAM uses rule-based machining principles, allowing to program the most important machining strategies in the system to be used as standard. These rules can be automatically applied depending on the material type and geometry of the operation.

The system is equipped with an operation recognition function, so it identifies standard geometric primitives such as ruled figures—holes, cylinders. This makes it possible to automate the process of tool path generation by identifying properties from the project tree (feature). For rotating parts, these features are even expected to be optimal. Most importantly, in the application to the research work described in this paper, Solidworks CAM is fully parametrically integrated into the Solidworks system's graphical kernel, so all modifications to parts are taken into account when rebuilding tool paths and post-processing.

3. Results

The first step in verifying the effectiveness of the proposed approach was to carry out the machining of the part without any additional support system for the selection of technological parameters. The machining program was generated in the CAM system, and the technological parameters were selected arbitrarily, taking into account the catalogue data provided by the tool manufacturer and the experience of the technologist.

The assumption was that the object was to be made on a CNC lathe in one clamping. The prefabricated product used for processing was a bar made of steel 1.0715 (11SMn30), with a circular cross-section of a diameter of $D = 40$ mm and overhang $L = 200$ mm. The Figure 15 shows the orientation of the part in relation to the prefabricated product.

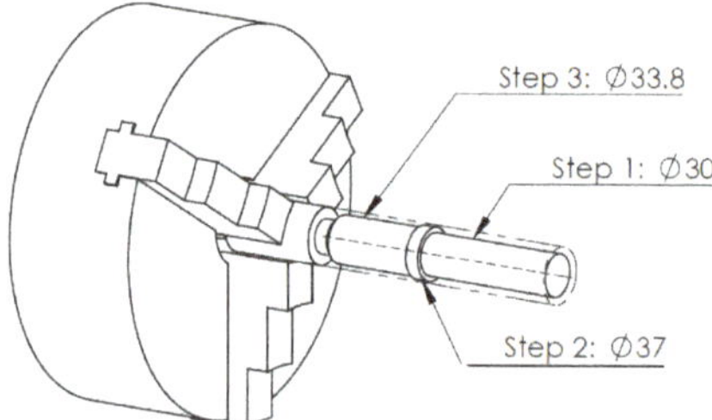

Figure 15. Orientation of the parts to be machined in relation to the prefabricated item.

For shaping, we used SCLCL/R 2020K09P tools equipped with Sandvik CCMT 09T304 PF 4325 plates (Sandvik Coromant, Sandviken, Sweden). Machining was performed on an AFM TAE-35N CNC lathe Andrychowska Fabryka Maszyn DEFUM, Andrychów, Poland), equipped with a Fanuc control system (Fanuc Robotics Ltd., Oshino, Japan).

Due to the geometry of the machined part in Steps 1 and 2 (Figure 15), the left-hand tool is used for machining. The highest susceptibility and therefore the highest risk of vibration occurs in Step 1, for which the diameter of the workpiece is to be 30 mm. The machining is carried out at a constant machining speed $v_c = 190$ m/min, depth of cut $a_p = 1.0$ mm and feed rate $f_n = 0.1$ mm/rev. The surface after machining for Steps 1 and 2 is shown in Figure 16.

Figure 16. Surface after machining without the system supporting the selection of technological parameters.

The surface machined in Step 1 shows the characteristic trace of vibration occurrence during the machining. When machining the last layer, on the diameter of $D = 30$ mm for the selected machining speed, the spindle speed was $N = 2016$ rpm. When analyzing the course of the stability lobes shown in Figure 14b, it should be noted that there is an increased risk of vibrations during machining at a given rotational speed.

The next step was to make a part, taking into account the conducted process stability analysis. The application of the module supporting the selection of machining parameters allowed one to introduce a correction of the spindle speed to the nearest 'stable' speeds, in order to minimize the risk of vibration during machining. In the selected case, the corrected spindle speed was $N_{cor} = 2154$ rpm. Figure 17 presents the finished part with the rotational speeds proposed by the developed support system. No vibrations were observed during the machining, which is confirmed by the obtained surface condition.

Figure 17. The part after machining with the support of technological parameters selection.

Figure 18 presents a comparison of the surface condition in Step 1 for the part manufactured without analysis of the stability of the cutting process, (a) and for the part where the proposed system of technological parameters support was used (b). The photographs showing the machined surfaces were taken using the Hawk Elite measuring microscope (Vision Engineering Ltd., Woking, UK) with a 20× magnification lens.

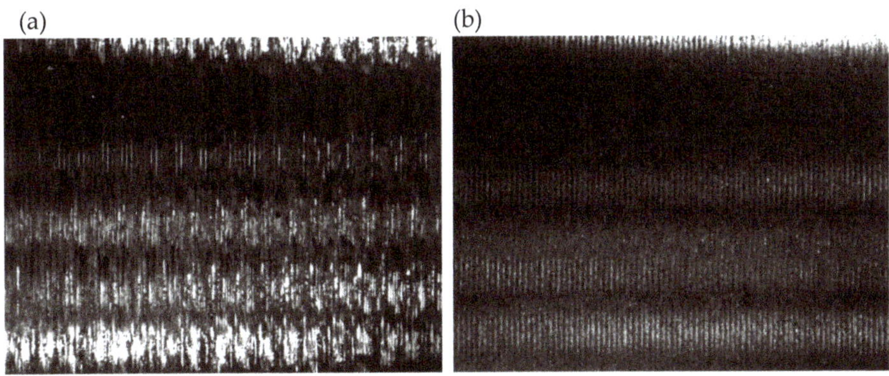

Figure 18. Surface comparison in Step 1: (**a**) without technological parameters selection support (**b**) with the support.

The comparison shows a clear improvement in surface condition due to the rotational speed correction. No signs of vibration were observed on surface (b), while these are clearly visible on surface (a). It is particularly important to note that the observed improvement was achieved using a software solution only, by changing the set rotational speed.

The last step of the verification was a 3D scan of the machined spare part and a comparison with the geometry of the original part. Overlapping geometries are shown in Figure 19.

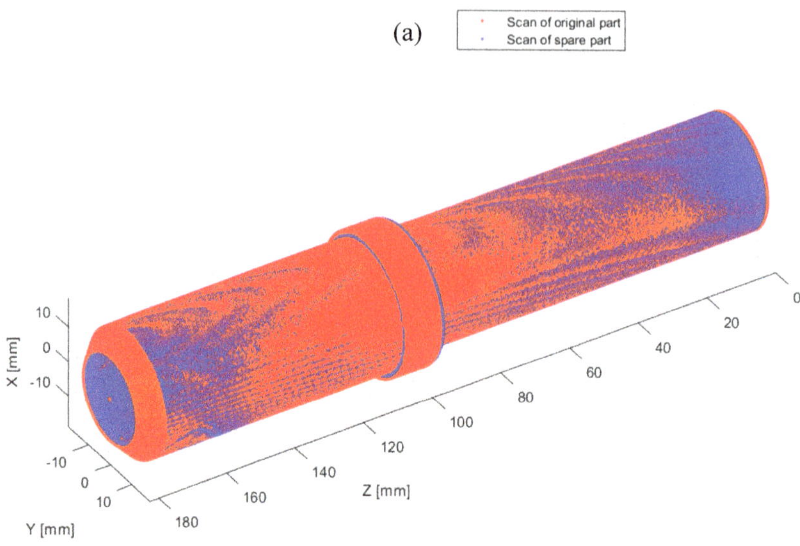

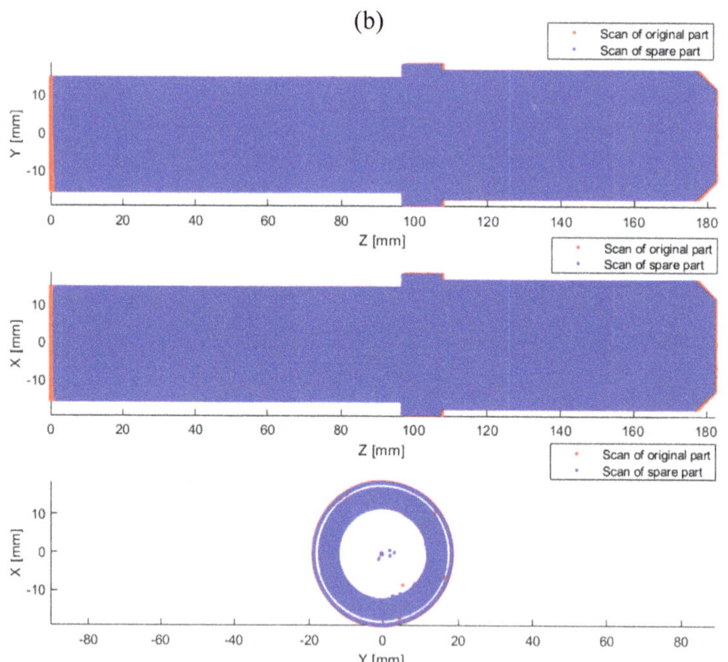

Figure 19. Comparison of 3D scans of the original part and a manufactured spare part, (**a**) 3D view (top), (**b**) side view (bottom).

In Figure 19, the blue color indicates the original part, while the red color indicates the scan of the spare part produced using the proposed system. The resulting discrepancies expressed in the form of RMSE are 0.15 mm, which corresponds to the accuracy of the scanner used.

4. Conclusions

The paper presents the concept of a system, allowing one to simplify the procedure of manufacturing spare parts. The presented concept allows one to manufacture a spare part on the basis of the original part scan. Such an approach is particularly useful when manufacturing spare parts for special machines, where access to technical documentation or finished parts is difficult or impossible.

The feature recognition module presented is characterized by the high accuracy of cylindrical elements and chamfers recognition. The result of this module is a parametric CAD model. On the basis of the obtained model, technological strategies and parameters of the processing are determined to minimize the risk of vibration occurrence during processing. Such an approach enables obtaining the required dimensional and shape accuracy and high quality of the machined surface.

The main limitation of the presented system is the heterogeneity of computing environments. Therefore, the direction of system development is its unification. Moreover, it is planned to extend the functionality of the system by recognizing further geometric features (e.g., roundness, conical surfaces, cavities) and to introduce the possibility of generating a CAD model based on damaged surfaces.

To sum up, the presented system allows for the highly automated production of copies of special machine parts, thus reducing the need to engage qualified staff and specialized measuring equipment, while providing technological processing parameters beneficial from the point of view of process stability.

Author Contributions: Conceptualization, K.M. and M.J.; Data curation, M.J. and M.G.; Formal analysis, M.J. and M.K.; Investigation, K.M.; Methodology, K.M. and M.J.; Resources, M.J., M.G., M.K. and B.P.; Software, M.J. and M.G.; Supervision, K.M. and B.P.; Validation, K.M., M.J. and B.P.; Visualization, K.M. and M.G.; Writing—original draft, K.M., M.J., M.G. and M.K.; Writing—review and editing, K.M. and M.J. All authors have read and agreed to the published version of the manuscript.

Funding: The research was financed by Smart Growth Operational Program in the project POIR.04.01.02-00-0078/16 "Vertical lathe of light construction". Research carried out on research apparatus purchased as part of the project No. RPZP.01.03.00-32-0004/17. Project co-financed by the European Union from the European Regional Development Fund under the Regional Operational Program of the West Pomeranian Voivodeship 2014-2020. Project co-financed by the Ministry of Science and Higher Education.

Conflicts of Interest: The authors declare no conflict of interest.

References

1. Chybowski, L.; Nozdrzykowski, K.; Grządziel, Z.; Dorobczyński, L. Evaluation of Model-Based Control of Reaction Forces at the Supports of Large-Size Crankshafts. *Sensors* **2020**, *20*, 2654. [CrossRef] [PubMed]
2. Zhang, K.; Yan, M.; Huang, T.; Zheng, J.; Li, Z. 3D reconstruction of complex spatial weld seam for autonomous welding by laser structured light scanning. *J. Manuf. Process.* **2019**, *39*, 200–207. [CrossRef]
3. Sunil, V.B.; Pande, S.S. Automatic recognition of machining features using artificial neural networks. *Int. J. Adv. Manuf. Technol.* **2009**, *41*, 932–947. [CrossRef]
4. Lin, X.; Wang, X.; Li, L. Intelligent detection of edge inconsistency for mechanical workpiece by machine vision with deep learning and variable geometry model. *Appl. Intell.* **2020**, *50*, 2105–2119.
5. Liu, F.; Wang, Z. PolishNet-2d and PolishNet-3d: Deep Learning-Based Workpiece Recognition. *IEEE Access* **2019**, *7*, 127042–127054. [CrossRef]
6. Niesterowicz, B.; Dunaj, P.; Berczyński, S. Timoshenko beam model for vibration analysis of composite steel-polymer concrete box beams. *J. Theor. Appl. Mech.* **2020**, *58*, 799–810. [CrossRef]
7. Pajor, M.; Marchelek, K.; Powałka, B. Method of Reducing the Number of DOF in the Machine Tool-Cutting Process System from the Point of View of Vibrostability Analysis. *J. Vib. Control* **2002**, *8*, 481–492. [CrossRef]
8. Dunaj, P.; Berczyński, S.; Chodźko, M.; Niesterowicz, B. Finite Element Modeling of the Dynamic Properties of Composite Steel–Polymer Concrete Beams. *Materials* **2020**, *13*, 1630. [CrossRef]
9. Dunaj, P.; Berczyński, S.; Chodźko, M. Method of modeling steel-polymer concrete frames for machine tools. *Compos. Struct.* **2020**, *242*, 112197. [CrossRef]

10. Chybowski, L.; Nozdrzykowski, K.; Grządziel, Z.; Jakubowski, A.; Przetakiewicz, W. Method to Increase the Accuracy of Large Crankshaft Geometry Measurements Using Counterweights to Minimize Elastic Deformations. *Appl. Sci.* **2020**, *10*, 4722. [CrossRef]
11. Wojciechowski, S.; Mrozek, K. Mechanical and technological aspects of micro ball end milling with various tool inclinations. *Int. J. Mech. Sci.* **2017**, *134*, 424–435. [CrossRef]
12. Irska, I.; Paszkiewicz, S.; Gorący, K.; Linares, A.; Ezquerra, T.A.; Jędrzejewski, R.; Rosłaniec, Z.; Piesowicz, E. Poly(Butylene terephthalate)/polylactic acid based copolyesters and blends: Miscibility-structure-property relationship. *Express Polym. Lett.* **2020**, *14*, 26–47. [CrossRef]
13. Dunaj, P.; Berczyński, S.; Miądlicki, K.; Irska, I.; Niesterowicz, B. Increasing Damping of Thin-Walled Structures Using Additively Manufactured Vibration Eliminators. *Materials* **2020**, *13*, 2125. [CrossRef] [PubMed]
14. Zapłata, J.; Pajor, M. Piecewise compensation of thermal errors of a ball screw driven CNC axis. *Precis. Eng.* **2019**, *60*, 160–166. [CrossRef]
15. Pajor, M.; Grudziński, M. Intelligent machine tool—Vision based 3D scanning system for positioning of the workpiece. *Solid State Phenom.* **2015**, *220–221*, 497–503. [CrossRef]
16. Aslan, D.; Altintas, Y. On-line chatter detection in milling using drive motor current commands extracted from CNC. *Int. J. Mach. Tools Manuf.* **2018**, *132*, 64–80. [CrossRef]
17. Jasiewicz, M.; Powałka, B. Prediction of turning stability using receptance coupling. In Proceedings of the AIP Conference Proceedings, Maharashtra, India, 5–6 July 2018; American Institute of Physics Inc.: College Park, MD, USA, 2018; Volume 1922, p. 100005.
18. Urbikain, G.; Olvera, D.; López de Lacalle, L.N.; Beranoagirre, A.; Elías-Zuñiga, A. Prediction Methods and Experimental Techniques for Chatter Avoidance in Turning Systems: A Review. *Appl. Sci.* **2019**, *9*, 4718. [CrossRef]
19. Yu, B.F.; Chen, J.S. Development of an analyzing and tuning methodology for the CNC parameters based on machining performance. *Appl. Sci.* **2020**, *10*, 2702. [CrossRef]
20. Dunaj, P.; Marchelek, K.; Chodźko, M. Application of the finite element method in the milling process stability diagnosis. *J. Theor. Appl. Mech.* **2019**, *57*, 353–367. [CrossRef]
21. Park, S.S.; Altintas, Y.; Movahhedy, M. Receptance coupling for end mills. *Int. J. Mach. Tools Manuf.* **2003**, *43*, 889–896. [CrossRef]
22. Schmitz, T.L.; Duncan, G.S. Three-component receptance coupling substructure analysis for tool point dynamics prediction. *J. Manuf. Sci. Eng. Trans. ASME* **2005**, *127*, 781–790. [CrossRef]
23. Jasiewicz, M.; Powałka, B. Receptance coupling for turning with a follower rest. *Adv. Mech. Theor. Comput. Interdiscip. Issues* **2016**, 245–248. [CrossRef]

Article

An Attitude Prediction Method for Autonomous Recovery Operation of Unmanned Surface Vehicle

Yang Yang, Ping Pan, Xingang Jiang, Shuanghua Zheng, Yongjian Zhao, Yi Yang *, Songyi Zhong and Yan Peng

School of Mechatronic Engineering and Automation, Shanghai University, Shanghai 200444, China; yangyang_shu@shu.edu.cn (Y.Y.); parkin@shu.edu.cn (P.P.); jackong@shu.edu.cn (X.J.); 18717927360@163.com (S.Z.); zhaoyongjian@shu.edu.cn (Y.Z.); zhongsongyi@shu.edu.cn (S.Z.); pengyan@shu.edu.cn (Y.P.)

* Correspondence: yiyangshu@t.shu.edu.cn; Tel.: +86-021-6613-6396

Received: 28 July 2020; Accepted: 30 September 2020; Published: 3 October 2020

Abstract: The development of launch and recovery technology is key for the application to the unmanned surface vehicle (USV). Also, a launch and recovery system (L&RS) based on a pneumatic ejection mechanism has been developed in our previous study. To improve the launch accuracy and reduce the influence of the sea waves, we propose a stacking model of one-dimensional convolutional neural network and long short-term memory neural network predicting the attitude of the USV. The data from experiments by "Jinghai VII" USV developed by Shanghai University, China, under levels 1–4 sea conditions are used to train and test the network. The results show that the stabilized platform with the proposed prediction method can keep the launching angle of the launching mechanism constant by regulating the pitching joint and rotation joint under the random influence from the wave. Finally, the efficiency and effectiveness of the L&RS are demonstrated by the successful application in actual environments.

Keywords: unmanned surface vehicle (USV); launch and recovery system (L&RS); attitude prediction; convolutional neural network (CNN); long short-term memory (LSTM) neural network

---

1. Introduction

Many countries are developing intelligent unmanned maritime equipment for marine exploitation and protection of maritime rights and interests owing to the rich biological and mineral resources. Unmanned surface vehicle (USV) is a kind of unmanned surface platform with autonomous navigation and obstacle avoidance ability, and it can independently complete tasks such as marine environment information perception, inshore island mapping, and disaster rescue, which is suitable for dangerous and human-unsuited missions instead of manned surface boats vessels [1–5].

Limited by endurance, the USV is usually carried by the mother ship to the mission area and then placed on the surface for autonomous operation. After the mission is completed, it would be recovered to the mother ship's deck. Therefore, the launch and recovery technology is a key technology for the application of USV [6]. The launch and recovery system (L&RS) for the manned surface boat can be divided into two types. The first one is the stern ramp type, which is mainly composed of a slide and a winch. Kern et al. have designed a device with an inclined chute and traction mechanism that can recover both autonomous underwater vehicle and remote-operated vehicle [7]. Hayashi et al. developed a set of devices composed of an obstacle avoidance system, slope L&RS, and a matching remote control for reducing the number of operators in the recovery process [8]. However, the requirement of slideway is limited to the application range, and thus, the davit system is more widely used compared with the slide system. The davit is generally installed on both sides of the mother ship's deck. During the operation, the USV is out of the ship's side and is lowered to or hoisted from the sea surface [9]. The RHP L&RS developed by Global davit gmbH can be used for the launch and recovery of boats with mass from 1000 to 3500 kg [10]. Marine Equipment Pellegrini, a company from Italy, has developed a marine L&RS capable of operating at level 6 sea conditions and has strong adaptability and large load capacity. The existing L&RS, however, requires human intervention to operate the boat, which is a difficulty when this is used for USV.

One of the challenges of the launch and recovery technology is the connection of USV to the mother ship's recovery system, dealing with the uncertainty and randomness under the impact of the waves, especially during high seas conditions. Thus, predicting the movement trend of USVs is important during launch and recovery operations. Consequently, several methods have been used, such as statistical forecast, Kalman filter, and time series [11]. Wiener et al. proposed an optimal linear prediction method based on the minimum mean square error [12]. The method can obtain better prediction results just within 5–6 s; however, the prediction error significantly increased with the prolonged prediction time [13]. Furthermore, Dodin and Sidar obtained the ship's motion state equation based on the force analysis and then deduced the multi-step prediction [14]. On the basis of the autoregressive model, Peng et al. proposed a real-time modeling and prediction method for predicting attitude motion of large ships under random wave action. The method is found suitable for application under non-stationary motion conditions, and the prediction time takes only 7–10 s [15]. Khan et al. combined the autoregressive model and the moving average model with an artificial neural network for predicting ship motion to achieve better prediction accuracy [16]. On the other hand, methods of attitude prediction were focused on large tonnage ships. The amplitude and frequency of the attitude are smaller than that of common USVs because of the inertia. In improving the maneuverability, the mass of the USV should be generally small. However, because of the influence of the wind and waves, large and high-frequency changes in attitudes have occurred.

For previously developed L&RS based on the pneumatic ejection mechanism, this study presents the attitude prediction for USV to improve the operation success rate. The rest of the paper is organized as follows. Section 2 introduces the concept and mechanism of the L&RS. In Section 3, the USV attitude prediction algorithm for stacking one-dimensional convolutional neural network (1D CNN) and long short-term memory (LSTM) neural network is proposed for improving the aiming accuracy of the L&RS under the influence of the waves. The experiments in Section 4 verify the validity of the algorithm, and efficiency of the automatic L&RS. The conclusions and plans for future studies are presented in Section 5.

2. Launch and Recovery System

2.1. Mechanism

A L&RS for USV based on pneumatic projectile has been developed in our previous study [17]. As shown in Figure 1, it is composed of a launching mechanism, a 2-degree-of-freedom-stabilized platform mechanism and a docking mechanism. During the launching process, the USV was lowered to the sea surface by the davit, and then the locking mechanism separated the USV from the conical butt joint.

The recovery process after the completion of a mission is as follows (Figure 2).

1. First, by automatic regulation of the pitching and rotation joints of the stabilized platform mechanism, the launching mechanism is aimed to the mother ship's deck (Figure 2a).
2. After the launching switch was acted, the air projectile was separated from the catapult mechanism and was driven by the high-pressure gas. It drives the guide rope to drop on the mother ship's deck (Figure 2b).
3. Then, the mother ship's crew passes the guide rope through the hole of the conical butt joint, and the conical butt joint slides along with the guide rope into the docking mechanism (Figure 2c).

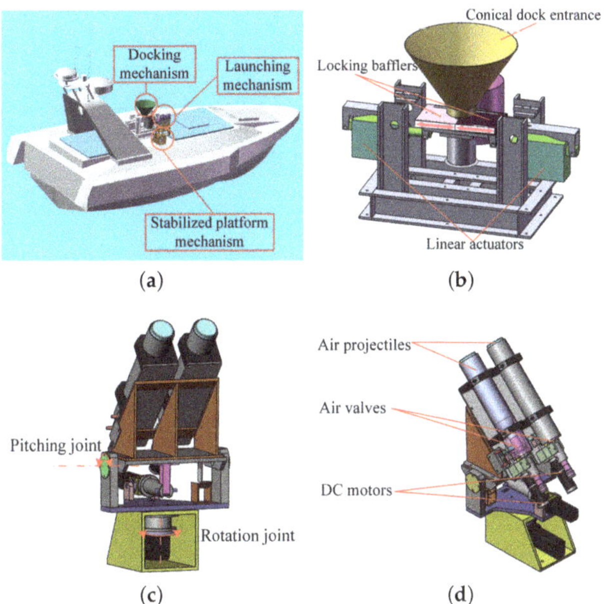

Figure 1. Schematic of an automated L&RS. (**a**) Concept. (**b**) Docking mechanism. (**c**) Stabilized platform mechanism. (**d**) Launching mechanism.

4. Finally, the docking mechanism locks the conical butt joint, and the USV is lifted from the sea surface by the davit (Figure 2d).

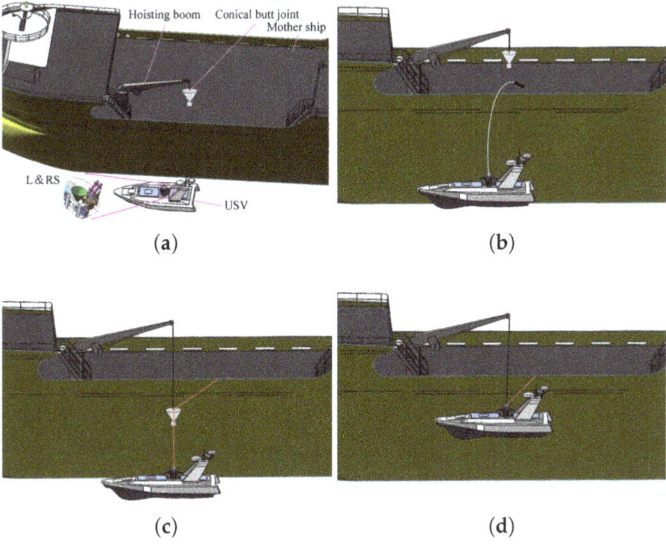

Figure 2. Operation process in recovering a USV. (**a**) Homing and aiming phase. (**b**) Launching phase. (**c**) Docking phase. (**d**) Lifting phase.

2.2. Launching Angle

The accurate launching of the air projectile to the mother ship is important for the recovery operation of USV. When the launching angle is too large, it could result in high elevation and short-range, causing the air projectile to fall into the sea. Hence, when the launching angle is too small, it could result in low elevation and longer range, causing the air projectile to hit the mother ship's sidewall. Therefore, during the recovery process, it is required to regulate the launching angle of the stabilized platform according to the measured distance and direction information between the USV and the mother ship.

A world coordinate system $O_0X_0Y_0Z_0$ is established, as shown in Figure 3a. The O_0Y_0 axis points to the bow; the O_0Z_0 axis is perpendicular to the sea level, opposite to the direction of gravity, and direction of the O_0X_0 is determined by the right-hand rule. The coordinates of the port of the canister launcher O_p and the target landing point of the air projectile O_h are denoted by (x_p,y_p,z_p) and (x_h,y_h,z_h), respectively. On the basis of the aerodynamics, the desired angle of the azimuth angle φ and the elevation angle η in the world coordinate system (Figure 3b) can be calculated as follows [18]:

$$\varphi = \arctan\frac{y_h - y_p}{x_h - x_p} = \arctan\frac{\operatorname{sgn}(x_h - x_p)\cdot s_1}{s_0 + s_j} \tag{1}$$

$$\eta = \arcsin\left[\frac{cW(\frac{c^2-1}{e}e^{\frac{b^2 l_h}{m^2 g}})}{c^2 - 1 - W(\frac{c^2-1}{e}e^{\frac{b^2 l_h}{m^2 g}})}\right] \tag{2}$$

$$c = \frac{bv_0}{mg} \tag{3}$$

where s_0 is the distance between the USV and mother ship's side, s_1 is the distance between the USV and the landing point along the direction of the O_0Y_0 axis, h_0 is the altitude difference between the mother ship and the USV, s_j is the distance from the mother ship's side to the landing point, l_h and v_0 are the elevation and initial velocity of the air projectile, respectively, m is the mass of air projectile, and b is the damping coefficient."sgn" is a symbolic function, which returns an integer variable indicating the sign of the argument:

$$sgn(x) = \begin{cases} 1, & x > 0 \\ 0, & x = 0 \\ -1, & x < 0 \end{cases} \tag{4}$$

The Lambert W function is a multivalued complex function:

$$W(x)e^{W(x)} = x \tag{5}$$

where x is a complex number. The detailed relationship between x and the Lambert $W(x)$ function can be found in [19].

By coordinate transformation, the corresponding angles of the pitching and rotation joints of the stabilized platform can be derived. Thus, the launching angle can be kept constant by regulating the stabilized platform in real time to compensate for the attitude change of the USV due to the influence of sea waves.

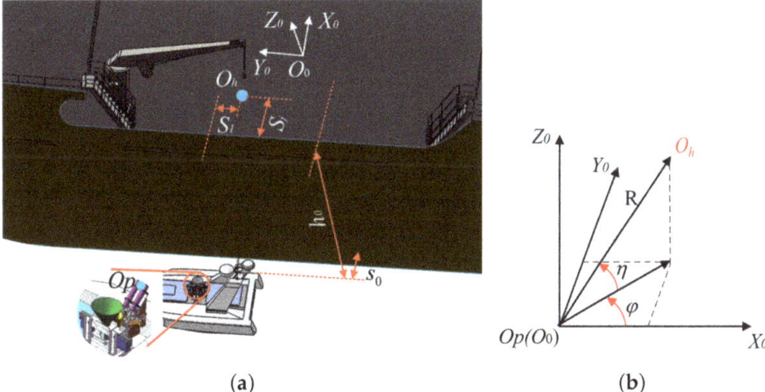

Figure 3. Coordinate systems. (**a**) Overview. (**b**) The azimuth angle φ and the elevation angle η in the world coordinate system.

3. USV Attitude Prediction

Having the features of being small in size and lightweight, the USV has superior maneuverability over the common manned surface vehicle. However, the changes in attitude due to the influence of waves will affect the landing point accuracy of air projectile in a recovery operation. By reducing the tracking error of the catapult mechanism due to time delays, we predict the attitude of the USV based on the previous state information. In this study, the LSTM neural network model is used to predict the attitude of the USV at sea in real time. To improve the performance of the prediction model, 1D CNN is superimposed on the LSTM. It can reduce the fluctuation range of prediction results and prediction errors, resulting in a successful recovery operation.

3.1. LSTM Neural Network

LSTM is an advanced recurrent neural network (RNN) structure that can learn and predict time series data [20,21]. For an ordinary RNN network, the output at the time t is as follows:

$$Y_t = \delta(W_o X_t + U_o S_t + b_o) \tag{6}$$

where X_t is the input at the current moment; S_t is the state of the network at time t, which is derived from the output of the network at the previous moment (i.e., $S_t = Y_{t-1}$); W_o is the weight matrix of the input; U_o is the weight matrix for the states; b_o is the bias; and δ is the activation function of the network. After Y_t is inputted to the softmax layer, the final prediction result can be obtained as follows:

$$prediction = softmax(V_s Y_t + b_s) \tag{7}$$

where V_s and b_s are the weights of the softmax layer. Because state S_t is a recursive variable, the derivative term increases with the time step when calculating the gradient, causing the gradient to disappear. To solve this problem, we improved the LSTM neural network on the basis of the original RNN network.

Figure 4 shows the internal structure of the LSTM, which adds a memory cell state C_t on the original basis. During the training process, the signal is not only controlled by the input and output but also passes the forgetting control unit. The forget gate F_t enables the network to delete some memory cell state information according to the previous training feedback without changing the weight, and it also selects certain neurons to update the weight. Also, an input gate I_t and an output gate K_t are added to make the model nonlinear. The input gate determines the amount of current input information used to calculate the carrying value. The output gate determines the amount of output from the carrying value to the final state. The state information of the three gates is calculated as follows:

$$\begin{cases} I_t = \delta_i(W_i X_t + U_i S_t + b_i) \\ F_t = \delta_f(W_f X_t + U_f S_t + b_f) \\ K_t = \delta_k(W_k X_t + U_k S_t + b_k) \end{cases} \quad (8)$$

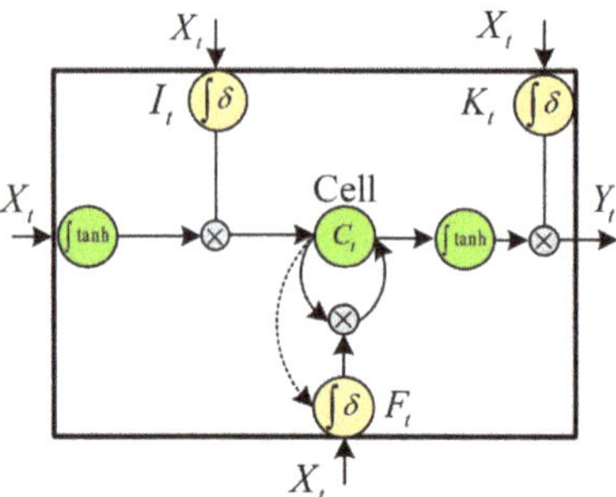

Figure 4. The internal structure of the LSTM neurons.

The memory cell state information carried by the network at the next moment is as follows:

$$\tilde{C}_t = \tanh(W_c X_t + U_c S_t + b_c) \quad (9)$$

$$C_t = F_t * C_{t-1} + I_t * \tilde{C}_t \quad (10)$$

where $\tilde{C}_t$ is the candidate value used to calculate the memory cell state information. The output of the network is as follows:

$$Y_t = K_t * \tanh(C_t) \quad (11)$$

where Y_t is the input to the softmax layer, obtaining the final prediction result. The carrying information can still be retained even after several time steps. The final output can derive long-term dependencies from the carrying information, thus solving the problem of gradient disappearance. At the same time, the input gate and the output gate can also adjust the influence of the output of different timing on the model, so it can effectively solve the problem of the gradient explosion of RNN.

3.2. One-Dimensional Convolutional Neural Network

Although LSTM has a suitable performance for processing time series, it is difficult to apply to a huge number of input data, which significantly reduced the calculation efficiency. 1D CNN is an effective method in dealing with sequence objects. It can extract high-level features from local input data through convolution operations, which can efficiently use data and reduce the input dimension. As a result, computational cost can be significantly reduced [22].

As shown in Figure 5, a local one-dimensional sequence segment is extracted from the original sequence in a 1D CNN. It is dotted with the weights in the convolution kernel to generate a shorter one-dimensional sequence. The sequence is trained as the input to the LSTM layer. As the input sequence length is shortened, the input dimension of the LSTM layer and the required training parameters are reduced. Therefore, the computing load can be effectively reduced, and the training time of the network can be shortened. Also, the 1D CNN extracts more advanced and abstract features from the original sequence in advance, so that the data use is high and the performance of the network can be effectively improved.

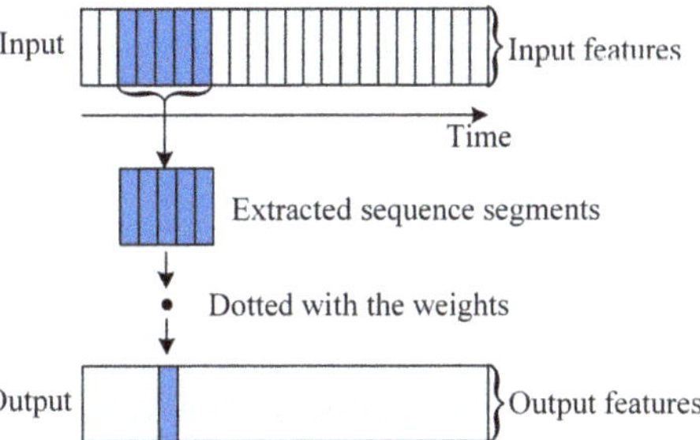

Figure 5. The working principle of 1D CNN. Each output time step is obtained using a small segment of the input sequence in the time dimension.

3.3. USV Attitude Data

Owing to the strong correlation between the attitude motion sequences of USV, the attitude in the next moment can be predicted by the LSTM neural network based on the attitude sequence value within the past time. In this paper, the training data are measured by a six-axis gyroscope mounted on the "Jinghai VII" USV developed by Shanghai University. The attitude data include posture angle and angular velocity in the heading, pitch, and roll directions of the USV. The measurement frequency of the sensing system is 10 Hz, which is far faster than that of the large tonnage ships. The experimental data with a time step of 0.1 s can be converted into dimensionless data through standardized formulas:

$$x^*(t) = \frac{x(t) - x_{min}(t)}{x_{max}(t) - x_{min}(t)} \tag{12}$$

where t is time; $x(t)$ is input data at time t; $x_{max}(t)$ and $x_{min}(t)$ are maximum and minimum values of the input data at time t, respectively; and $x^*(t)$ is the normalized value of the input data in which the range is from 0 to 1.

3.4. Determination of Hyper-Parameters

In this study, we applied a neural network model with two hidden layers and ten hidden units. The input layer inputs the attitude angle and angular velocity information of the USV in one direction. The output layer uses the tanh function as the activation function to output the predicted result. The uniform distribution randomly generates the network weights. The range is (−limit, limit):

$$limit = \sqrt{\frac{6}{n_j + n_{j+1}}} \tag{13}$$

where n_j is the number of units in the layer j network. The bias is initialized at 0. The square sum of the difference between the predicted value and the actual value of the output is selected as the loss function. By minimizing the loss function, all weights and bias parameters in the network can be obtained:

$$loss = \frac{\sum_{j=1}^{n} (y_i - y_i{}^p)^2}{n} \tag{14}$$

where y_i is the actual value, $y_i{}^p$ is the predicted value, and n is the total number of data.

The loss function can also be used to evaluate the accuracy of the neural network training model. The trained heading, pitch, and roll neural network models of the USV can be respectively derived from the posture angle and angular velocity in three directions.

3.5. Training Process

As shown in Figure 6, first, the weights and biases of the stacked LSTM network are initialized. After entering the normalized attitude data, the forecasted value of heading, roll, and pitch angles can be obtained. Equation (14) is used to calculate the loss between the desired value and the real value. Finally, the gradient descent method is used to constantly adjust the weight and bias until the loss value converges or the number of iterations reaches the peak.

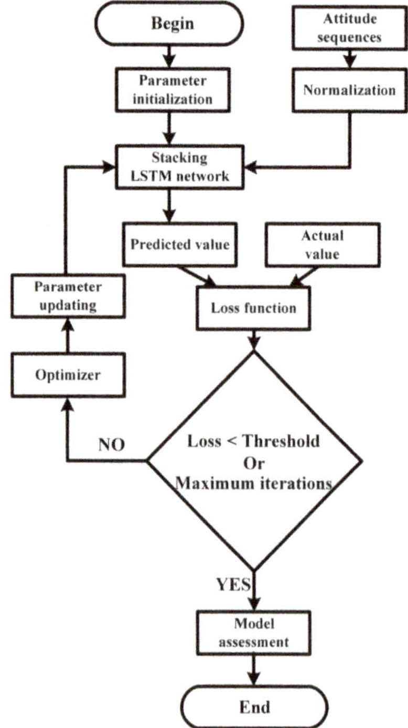

Figure 6. The training process of the 1D CNN-LSTM neural network.

It is noted that a large size of database takes a long training time. Thus, a reasonable batch size needs to be set to reduce the number of iterations required for the training model. To obtain the suitable batch size, it was set to 1 at the beginning, and then the value increased until the improvement in training accuracy was no longer apparent in this study.

4. Experiments and Discussion

4.1. Experimental Overview

To verify the validity of the proposed prediction algorithm, we used the experimental data measured from the "Jinghai VII" USV under levels 1–4 sea conditions for training the testing. Table 1 lists the specifications of the USV. According to [23,24], the sea level conditions were generally defined by the ranges of significant wave heights as listed in Table 2. The data set contains the posture angles and angular velocities of the USV in the heading, pitch, and roll directions. Each set has approximately 3000 sets of data, in which 60%, 20%, and 20% were taken as training, verification, and test sets, respectively. The batch size is 128 in both training process and validation process.

Table 1. Specifications of "Jinghai VII" USV.

Parameters	Values	Unit
Length	8.2	m
Width	2.45	m
Height	1.84	m
Mass	3000	kg
Depth of immersion	0.5	m

Table 2. Definition of sea level conditions [23,24].

Sea Condition Level	Sea States	Significant Wave Height (m)
0	Calm (glassy)	0
1	Calm (ripples)	0–0.1
2	Smooth (wavelets)	0.1–0.5
3	Slight	0.5–1.25
4	Moderate	1.25–2.5
5	Rough	2.5–4.0
6	Very rough	4.0–6.0
7	High	6.0–9.0
8	Very high	9.0–14.0
9	Phenomenal (Extreme)	Over 14.0

To prevent model overfitting, we took samples 100 times per turn for a total of 50 rounds. The neural network model is trained by inputting the angle and angular velocity information simultaneously. To demonstrate the performance of the proposed method, we compared the results with that predicted by the original LSTM, Nonlinear Autoregressive Exogenous Model (NARX) network, and Time Delay Neural Network (TDNN) under 1–4 sea levels.

4.2. Results and Discussion

To evaluate the performance of the 1D CNN-LSTM neural network, we compared the effect of that LSTM neural network model, in which the parameter settings and training test data are the same for the two models. The results of the heading angle of USV are taken as an example. As shown in Figure 7a, although the sea condition is at level 1, the heading angle fluctuates greatly because of the slight weight of the USV. Both neural network models can predict the trend of the USV attitude, but the predicted results by LSTM neural network have a rougher degree of agreement with the actual curve. It is deduced that as the data set is small, the original LSTM neural network is difficult to learn the changing characteristics of the heading angle, reducing the accuracy of the prediction model. It can be seen from the enlarged area of Figure 7a that when the heading angle suddenly changed, the prediction error of the LSTM neural network increased. However, the predicted results by 1D CNN-LSTM neural network can achieve higher accuracy. It is attributed that after the addition of CNN, the model can effectively extract the features of the USV attitude data and reduce the redundant information input to the LSTM neural network. CNN enables LSTM neural network to mine the deeper features of heading angle variation so that the predicted results are closer to the actual test data. Table 3 shows the comparison of the training speed and various losses of the model. It can be confirmed that the training speed of the LSTM neural network is 22 ms/step, whereas the training speed of the 1D CNN-LSTM network model is increased by 55% to 10 ms/step. The LSTM test loss was 0.0511, whereas the test loss of the 1D CNN-LSTM network model was reduced by 59% to 0.0212. The 1D CNN-LSTM neural network has higher training efficiency and higher prediction accuracy.

The proposed prediction algorithm is required to guarantee suitable prediction accuracy for various sea conditions and environments. To demonstrate the performance of the proposed network model under unknown complex sea conditions and higher noise disturbances, we showed the results of the heading angle under level 2–4 sea conditions in Figure 7b–d, respectively. As shown in Table 3, the test error of the LSTM neural network increased with the increase of sea level. It means that the complexity of high sea conditions has a great influence on the accuracy of the LSTM neural network prediction result. However, the predicted results by the 1D CNN-LSTM network model after training agreed with the test data well. The proposed network model test error is approximately 0.015–0.025.

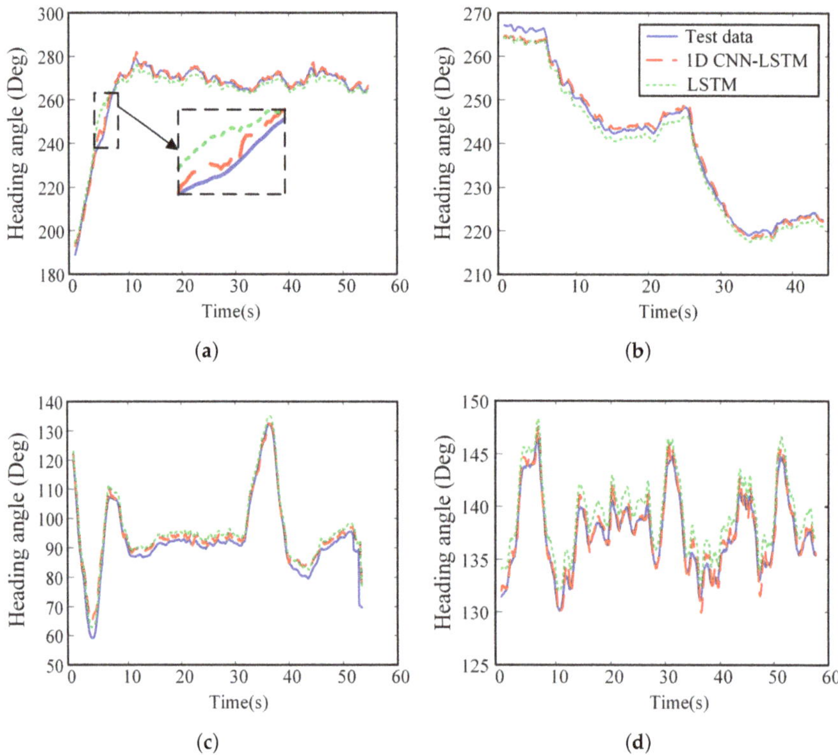

Figure 7. Comparison of the predict results and actual results by LSTM with CNN and LSTM without CNN at different sea level. (**a**) Level 1 sea condition. (**b**) Level 2 sea condition. (**c**) Level 3 sea condition. (**d**) Level 4 sea condition.

As shown in Figure 8, all the neural network models can predict the trend of the USV attitude, in which the 1D CNN-LSTM can achieve the higher prediction accuracy than the other 3 networks. Table 3 demonstrated the comparison of the training speed and various losses of the four networks at different sea level. Compared with NARX, the training speed of 1D CNN-LSTM increased by 25% (level 1), 18% (level 2), 17% (level 3) and 21% (level 4), and test loss decreased by 4% (level 1), 43% (level 2), 16% (level 3) and 6% (level 4). Although the training speed of 1D CNN-LSTM has slight increase with TDNN, the test loss was 25% (level 1), 39% (level 2), 29% (level 3) and 26% (level 4) lower than that of TDNN. In addition, it can be found that the training speed of 1D CNN-LSTM in levels 1–4 sea conditions is increased by 45%, 30%, 25% and 29% respectively, compared with the test loss of LSTM. The test loss of 1D CNN-LSTM in levels 1–4 sea conditions is reduced by 59%, 45%, 30%, and 65%, respectively, compared with the test loss of LSTM. It is noticed that the degree of the improvement in the prediction accuracy has no relationship with the sea level conditions.

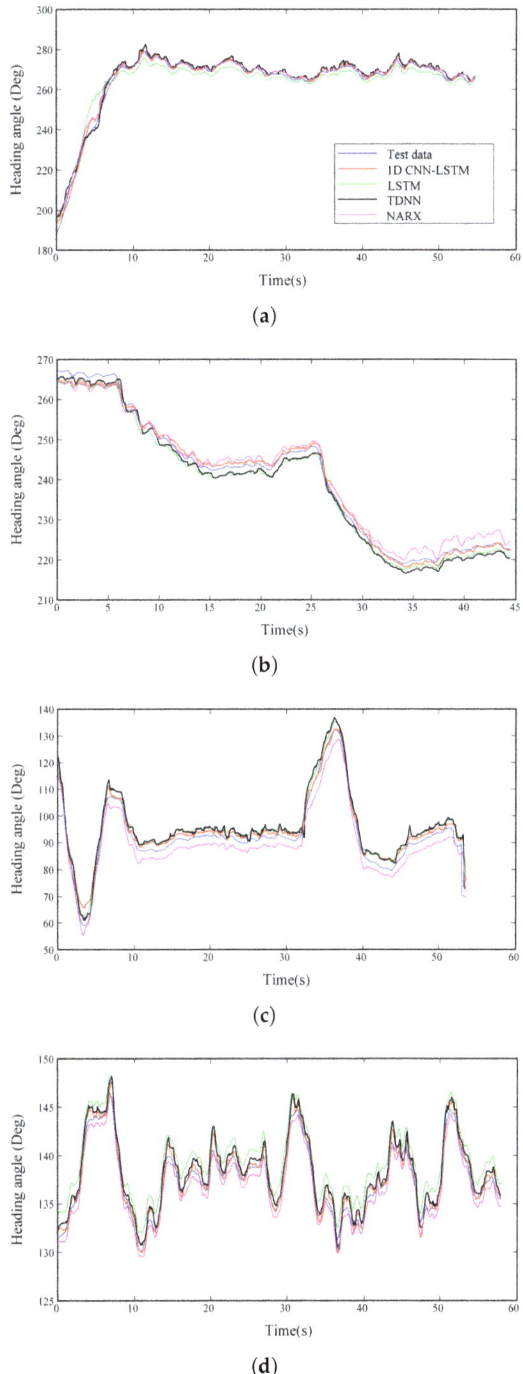

Figure 8. Comparison of the heading angle by 1D CNN-LSTM network, LSTM network, NARX network and TDNN network at different sea level. (**a**) Level 1 sea condition. (**b**) Level 2 sea condition. (**c**) Level 3 sea condition. (**d**) Level 4 sea condition.

Table 3. Comparison of the results by LSTM, 1D CNN-LSTM, NARX and TDNN.

Sea Condition	Network Model	Training Speed (ms/step)	Training Loss	Validation Loss	Test Loss
Level 1	LSTM	22	0.1013	0.1232	0.0511
	1D CNN-LSTM	12	0.0289	0.0128	0.0212
	NARX	16	0.0301	0.0120	0.0222
	TDNN	13	0.0247	0.0416	0.0283
Level 2	LSTM	20	0.0402	0.0264	0.0240
	1D CNN-LSTM	14	0.0284	0.0201	0.0132
	NARX	17	0.0308	0.0244	0.0232
	TDNN	14	0.0222	0.0188	0.0216
Level 3	LSTM	20	0.0309	0.0107	0.0310
	1D CNN-LSTM	15	0.0279	0.0211	0.0218
	NARX	18	0.0305	0.0186	0.0261
	TDNN	14	0.0212	0.0221	0.0308
Level 4	LSTM	21	0.0765	0.0461	0.0381
	1D CNN-LSTM	15	0.0415	0.0125	0.0133
	NARX	19	0.0377	0.0239	0.0141
	TDNN	16	0.0286	0.0232	0.0179

It is noticed that compared with the common online prediction algorithms, 1D CNN-LSTM requires no calculation of the model parameters in real time. It ensures the high speed, real-time and reliability of the usv's attitude data prediction. The fluctuation range of model prediction error under sea conditions is relatively stable regardless of the time, and the prediction error keep stable in the prolonged prediction time. Therefore, the 1D CNN-LSTM network is superior to the other three networks in terms of prediction accuracy, adaptability, and efficiency. It can achieve suitable prediction accuracy under various sea conditions, which is important for the recovery operation of the USV.

4.3. Field Application

The autonomous launch and recovery operation of "Jinghai VII" USV was performed in the East China Sea to demonstrate the effect of the proposed attitude prediction method in the practical application. After the mission was completed, the USV returned to the vicinity of the mother ship in preparation for recovery. The mother ship was stopped, and the position was continuously changing due to the influence of sea waves. By the guidance of the navigation system, the USV finally stopped approximately 3 m away from the port side of the mother ship with the same heading.

After the propeller was stalled, the USV switched to the aiming phase. Owing to the influence of the wave, the attitude of the USV continuously changes. From Equations (1) and (2), the desired launching angle of the air projectile can be calculated from a distance between the stabilized platform and the target landing point in the world coordinate system measured by the laser rangefinder and GPS positioning system mounted on the USV.

Figure 9 demonstrates the effect of the proposed prediction algorithm on the stability control of the stabilized platform. It can be seen that the tracking error of the azimuth and elevation angles can be improved by using the proposed prediction algorithm. Without the attitude prediction, the stabilized platform regulated the azimuth and elevation angles based on the tracking error by the PID algorithm. As shown in Figure 9a,c, although the desired values and actual values have similar trends, it has the obvious time delay. It is attributed that the attitude of the USV changes instantly, and the launching angle of the air projectile changes synchronously with the change of the USV attitude. As the stabilized platform moves to the desired angle according to the calculated value at the previous moment, the USV attitude has changed. However, using the proposed prediction algorithm, the joint angles can follow the desired values with a slight tracking error. On the basis of the predicted results, the stabilized platform can respond in advance. It compensated the influence of the wave and increased the accuracy of the air projectile launch.

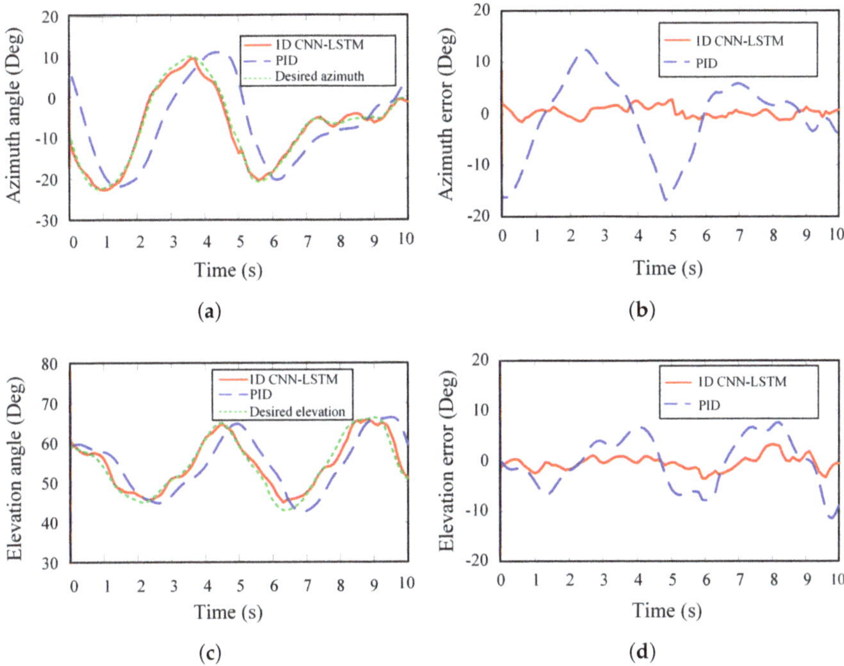

Figure 9. Comparison of the predict results and actual results in the field application. (**a**) Azimuth angle. (**b**) Error of azimuth angle. (**c**) Elevation angle. (**d**) Error of elevation angle.

By applying the proposed prediction method, we can obtain an accurate projectile on the mother ship's deck. As shown in Figure 10, following the operation process in Section 2.1, the mother ship's crew operated the hoisting boom to drop the conical butt joint to slide into the conical dock entrance along with the guide rope. Then, the locking mechanism locked the conical butt joint, and the USV was lifted and placed on the mother ship's deck.

During the operation, reducing the sway of the USV was time-consuming. Moreover, in reducing the sway of the USV during the lifting process, the operation required eight operators, a crane operator, a commander, and other people that could pull the USV in the head and stern directions through ropes. All people were working on the mother ship's deck, and no operator is required onboard the USV. It significantly improved the safety of the USV recovery operation.

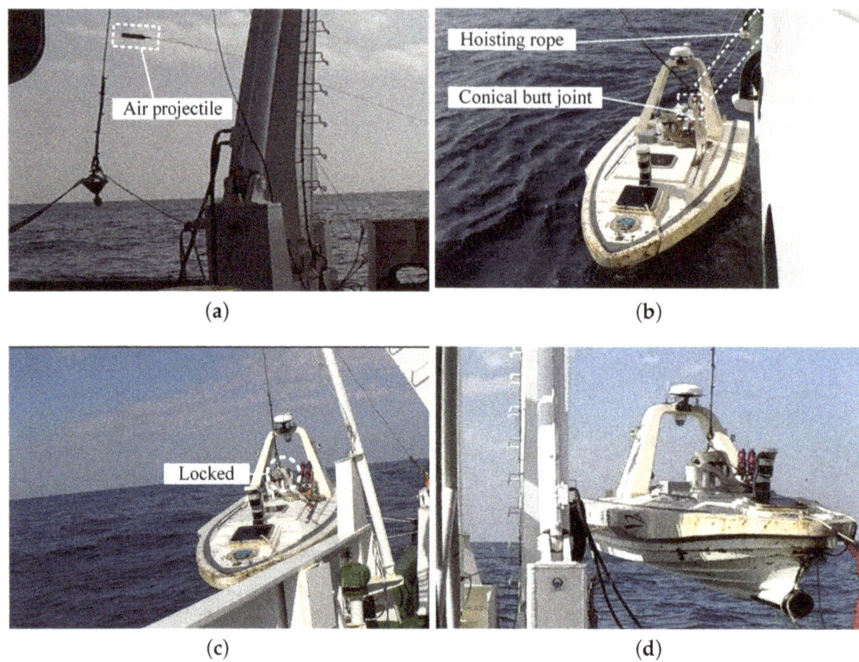

Figure 10. From (**a–d**), recovery process in field application. The recovery process took approximately 23 min. The efficiency of the aiming and launching of L&RS is suitable.

5. Conclusions and Future Work

To improve the launching accuracy of the developed pneumatic ejection mechanism-based L&RS of USV, we proposed a stacking model of 1D CNN and LSTM network in this study to predict the attitude of the USV. On the basis of the predicted results, the pitching joint and rotation joint of the stabilized platform can be regulated in real time to compensate for the disturbance of the waves, keeping the launching angle stable. The data from the experiments by "Jinghai VII" USV were used to train and test the proposed network. The results demonstrated that the algorithm has suitable prediction accuracy and calculation efficiency. From the filed application, it can be confirmed that the L&RS with the trained network can successfully recover USV. It requires no operator working on the USV, and it significantly improved safety and adaptability. In the future, the proposed method will be applied to model predictive control techniques, automatic speech recognition, etc.

Author Contributions: Conceptualization, Y.Y. (Yang Yang); Writing–original draft, P.P.; Data curation, Y.Z.; Funding acquisition, Y.Y. (Yi Yang); Methodology, X.J.; Project administration, Y.P.; Software, S.Z. (Shuanghua Zheng); Manuscript revision, S.Z. (Songyi Zhong). All authors have contributed to and approved the final manuscript.

Funding: This study was supported by the National Natural Science Foundation of China under Grant 61903243/91648119/51675318/61827812, and the Project of Shanghai Municipal Science and Technology Commission under Grant 17DZ1205000.

Conflicts of Interest: The authors declare no conflict of interest.

References

1. Furfaro, T.C.; Dusek, J.; Von Ellenrieder, K.D. Design, construction, and initial testing of an autonomous surface vehicle for riverine and coastal reconnaissance. In Proceedings of the OCEANS 2009, Biloxi, MS, USA, 26–29 October 2009; pp. 1–6.
2. Breivik, M.; Hovstein, V.E.; Fossen, T.I. Straight-Line target tracking for unmanned surface vehicles. *Model. Identif. Control* **2008**, *29*, 131–149. [CrossRef]
3. Liu, Z.; Zhang, Y.; Yu, X.; Yuan, C. Unmanned surface vehicles: An overview of developments and challenges. *Annu. Rev. Control* **2016**, *41*, 71–93. [CrossRef]
4. Yang, W.; Chen, C.; Hsu, C.; Tseng, C.; Yang, W. Multifunctional inshore survey platform with unmanned surface vehicles. *Int. J. Autom. Smart Technol.* **2011**, *1*, 19–25. [CrossRef]
5. Sonnenburg, C.; Woolsey, C.A. Modeling, identification, and control of an unmanned surface vehicle. *J. Field Robot.* **2013**, *30*, 371–398. [CrossRef]
6. Fisher, N.; Gilbert, G.R. Unmanned systems in support of future medical operations in dense urban environments. *J. Article* **2016**, *4*, 48pm.
7. Kern, F.R. Launch and Recovery Devices for Water Vehicles and Methods of Use. U.S. Patent 7,581,507, 19 February 2009.
8. Hayashi, E.; Kimura, H.; Tam, C.; Ferguson, J.; Laframboise, J.; Miller, G.; Kaminski, C.; Johnson, A. Customizing an autonomous underwater vehicle and developing a launch and recovery system. In Proceedings of the 2013 IEEE International Underwater Technology Symposium (UT), Tokyo, Japan, 5–8 March 2013; pp. 1–7.
9. Wu, G.X.; Xie, Y.; Sun, H.b.; Zou, J. Modeling and simulation of capsizing automatic recovery system for unmanned surface vehicle. *J. Syst. Simul.* **2009**, *21*.
10. Bergmann, H.D. Device for a Watercraft for Picking up and Launching Boats. U.S. Patent 7,815,394, 19 October 2010.
11. Peng, X.; Zhao, X.; Xu, L. Real-time prediction algorithm research of ship attitude motion based on order selection with corner condition. In Proceedings of the 2006 1st International Symposium on Systems and Control in Aerospace and Astronautics, Harbin, China, 19–21 January 2006; IEEE: Piscataway, NJ, USA, 2006; p. 6.
12. Norbert, W. *Interpolation and Smoothing of Stationary Time Series*; Wiely: New York, NY, USA, 1949.
13. Bates, M.R.; Bock, D.; Powell, F. Analog computer applications in predictor design. *IRE Trans. Electron. Comput.* **1957**, *EC-6*, 143–153. [CrossRef]
14. Sidar, M.; Doolin, B. On the feasibility of real-time prediction of aircraft carrier motion at sea. *IEEE Trans. Autom. Control* **1983**, *28*, 350–356. [CrossRef]
15. Peng, X.Y.; Zhao, X.R.; Gao, Q.F. Research on real-time prediction algorithm of ship attitude motion. *J. Syst. Simul.* **2007**, *19*, 267–271.
16. Khan, A.; Bil, C.; Marion, K.; Crozier, M. Real time prediction of ship motions and attitudes using advanced prediction techniques. In Proceedings of the Congress of the International Council of the Aeronautical Sciences. International Council of the Aeronautical Sciences, Yokohama, Japan, 29 August–3 September 2004.
17. Zheng, S.; Yang, Y.; Peng, Y.; Cui, J.; Chen, J.; Jiang, X.; Feng, Y. An automated launch and recovery system for USVs based on the pneumatic ejection mechanism. In Proceedings of the International Conference on Intelligent Robotics and Applications, Shenyang, China, 8–11 August 2019; Springer: Berlin/Heidelberg, Germany, 2019; pp. 289–300.
18. Qian, X.; Yin, Y.; Zhang, X.; Li, Y. Influence of irregular disturbance of sea wave on ship motion. *J. Traff. Transp. Eng. China* **2016**, *16*, 116–124.
19. Hu, H.; Zhao, Y.; Guo, Y.; Zheng, M. Analysis of linear resisted projectile motion using the Lambert W function. *Acta Mech.* **2012**, *223*, 441–447. [CrossRef]
20. Ma, C.; Wang, A.; Chen, G.; Xu, C. Hand joints-based gesture recognition for noisy dataset using nested interval unscented Kalman filter with LSTM network. *Vis. Comput.* **2018**, *34*, 1053–1063. [CrossRef]
21. Yang, B.; Sun, S.; Li, J.; Lin, X.; Tian, Y. Traffic flow prediction using LSTM with feature enhancement. *Neurocomputing* **2019**, *332*, 320–327. [CrossRef]
22. Pan, H.; He, X.; Tang, S.; Meng, F. An improved bearing fault diagnosis method using one-dimensional CNN and LSTM. *J. Mech. Eng* **2018**, *64*, 443–452.

23. Nguyen, T.D.; Sørensen, A.J.; Quek, S.T. Design of hybrid controller for dynamic positioning from calm to extreme sea conditions. *Automatica* **2007**, *43*, 768–785. [CrossRef]
24. Price, W.G. *Probabilistic Theory of Ship Dynamics*; Printed in the United Kingdom; Chapman & Hall Ltd.: London, UK, 1974; ISBN 0 412 12430 0.

Article

The Effect of Deflections and Elastic Deformations on Geometrical Deviation and Shape Profile Measurements of Large Crankshafts with Uncontrolled Supports

Krzysztof Nozdrzykowski [1], Stanisław Adamczak [2], Zenon Grządziel [1] and Paweł Dunaj [3,*]

1 Faculty of Marine Engineering, Maritime University of Szczecin, 1-2 Wały Chrobrego St., 70-500 Szczecin, Poland; k.nozdrzykowski@am.szczecin.pl (K.N.); z.grzadziel@am.szczecin.pl (Z.G.)
2 Faculty of Mechatronics and Mechanical Engineering, Kielce University of Technology, Kielce, 7 Tysiąclecia Państwa Polskiego Ave., 25-314 Kielce, Poland; adamczak@tu.kielce.pl
3 Faculty of Mechanical Engineering and Mechatronics, West Pomeranian University of Technology, 19 Piastów Ave., 70-310 Szczecin, Poland
* Correspondence: pawel.dunaj@zut.edu.pl

Received: 16 September 2020; Accepted: 6 October 2020; Published: 8 October 2020

Abstract: This article presents a multi-criteria analysis of the errors that may occur while measuring the geometric deviations of crankshafts that require multi-point support. The analysis included in the paper confirmed that the currently used conventional support method—in which the journals of large crankshafts rest on a set of fixed rigid vee-blocks—significantly limits the detectability of their geometric deviations, especially those of the main journal axes' positions. Insights for performing practical measurements, which will improve measurement procedures and increase measurement accuracy, are provided. The results are presented both graphically and as discrete amplitude spectra to make a visual, qualitative comparison, which is complemented by a quantitative assessment based on correlation analysis.

Keywords: crankshaft; geometrical error; finite element model; eccentricity; Fourier series; discrete Fourier transform

---

1. Introduction

Methodological errors occur when a model fails to include the factors of the measurement method and related phenomena. This quantity is the discrepancy between the model characteristics of the method and its real characteristics. The accuracy of measurements requires for the possible errors to be analyzed and eliminated as much as possible or treated as correction factors, especially for measurements where analysis will significantly affect the total method error. Examining and analyzing individual error components facilitates the determining of how they influence the measurement accuracy [1]. Taking these errors into account during measurements creates a wide range of possibilities for using the developed method in practical measurements.

Specific procedures for measuring large machinery components have been discussed in several studies [2–4]. They include methods to measure geometrical deviations of cylindrical surfaces of such components and include a concrete, broadly understood analysis of systematic and random errors in the proposed methods [5–8]. These studies contain valuable information for understanding the discussed issues and for perfectly matching modern metrology trends [9].

The error analysis elaborated in [2,10–12] is particularly useful for issues related to a specific group of large machine components with cylindrical assemblies [13], such as large crankshafts of ship engines. Such shafts have large masses and dimensions, and are also flaccid, have low and variable rigidity, and are susceptible to flexural deformation [14,15]. These properties require the main journals to be supported at multiple points in a controlled manner during measurements [16,17].

As shown in [10], the reaction forces at the interface of the main journals and supports vary along the shaft and also depend on the shaft rotation angle at the supports. An uncontrolled support—when a crankshaft's main journals are borne by a set of fixed, rigid vee-blocks [18,19] or when any of the main journals is not supported—causes shaft deflections that cannot be eliminated. Importantly, the supports should not limit the possible journal movements, which occur when the journal axes are not mutually aligned. Misalignment occurs when the shape and geometry of manufactured items deviate from their theoretical designs, and such geometrical quantities should be correctly characterized. If shaft journals are supported by a set of fixed, rigid vee-blocks, this type of displacement is limited by unintentional preliminary deflections, resulting in incorrect measurements of geometric quantities [11].

Therefore, to accurately assess the crankshaft geometry, measurements must be performed by controlling the reaction forces at the supports, which must be articulated and susceptible, i.e., adaptable to possible mutual displacements of the main journals due to geometric deviations of the item being measured.

The aspect related to measurement inaccuracies caused by shaft deflection under its own weight was investigated in our previous study [10], in which we described an innovative method for eliminating deformation in large crankshafts during measurement of their geometric condition. The method consists of using the measuring system with active compensation for shaft deflection, by means of actuators cooperating with force transducers monitoring the deflection of individual crank journals of a crankshaft being measured. The results have shown that the system is able to effectively eliminate the deflection and elastic deformation of the crankshaft under the influence of its own weight.

The continuation of the study presented in [10] was [12]. In this study, the support reaction forces were changed to minimize the crankshaft elastic deflection as a function of the crank angle. The changes of these reaction forces were determined according to the developed algorithm. The algorithm uses a mathematical model that interpolates the values of forces calculated previously with finite element software. The supports are continuously adjusted when the shaft rotates by precision current-controlled valves that operate in feedback with the force sensors measuring the actual force at the contact of support heads and main journals.

The latest study [11] describes the use of temporary counterweights during large crankshaft measurements and presents the specifications of the measurement system and method to stabilize the forces at the supports that fix the shaft during measurements. The study showed that the proposed solution provided constant reaction forces and ensured nearly zero deflection at the supported main journals of a shaft during its rotation (during its geometry measurement).

In this paper, we investigated the effect of elastic deformations on geometrical deviation and shape profile measurements of large crankshafts with uncontrolled supports. We considered the influence of the difference in the height of the supports and the influence of the journal eccentricity on the measurement results of the shaft geometry. The main motivation of this study was to indicate the limitation of the rigid vee-blocks measuring method. The results presented in this paper confirm that the currently used conventional support method—in which the main journals of a shaft are supported by a set of fixed, rigid vee-blocks—significantly limits the detectability of geometric deviations, especially those of the journal and pin axes' positions. The results of this study also provide insights to be considered during measurements, thereby improving the measurement procedures and increasing measurement accuracy.

The structure of the article is as follows: in Section 2, the methods of fixing crankshafts to measure geometrical deviations are presented and their limitations are indicated. Next, a study plan is formulated in order to prove that the currently used measurement methods limit the detectability of geometric deviations. According to the study plan, a finite element model of an exemplary shaft was built. Based on the finite element model, the necessary calculations were carried out, the results of which are presented in Section 3, which also includes a discussion. The main conclusions are presented in Section 4.

2. Materials and Methods

2.1. Methods of Fixing Crankshafts to Measure Geometrical Deviations

As part of the methods and techniques currently used, measurements are performed with the shaft axis fixed in the horizontal or vertical plane. Large crankshafts are placed in the horizontal plane because of their large masses and dimensions. The main journals of those shafts are rested on a set of rigid vee-block supports (Figure 1a), but these conditions do not ensure the elimination of shaft elastic deflections [10,20,21]. Small and medium shafts are usually measured using precision measuring machines [22], and their axis is located in the vertical plane (Figure 1b). Those shafts are fixed and stabilized at their ends in holders or centers without additional stiffening in their middle part. With this type of stabilization, the shaft axis buckles. Both types of stabilization cause elastic deformations of the shaft that vary in sign and value due to changes in the shaft's rigidity when it rotates.

(a)

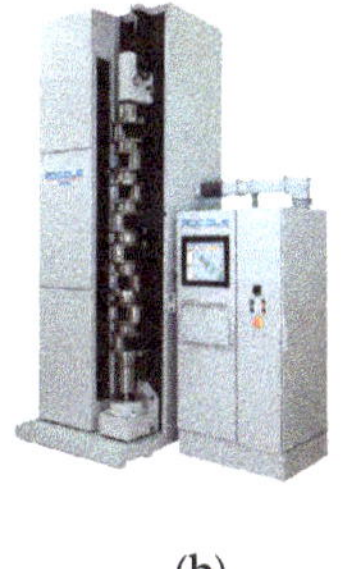

(b)

Figure 1. Geometrical deviation measurement methods: (**a**) horizontal on a set of rigid vee-block supports, (**b**) vertical using precision measuring machines (courtesy of ADCOLE).

For a horizontal shaft axis with equally elevated vee-blocks supporting all the main journals and a perfectly manufactured crankshaft (no geometrical deviations), the shaft does not undergo elastic deformations; however, the actual manufacture of machine parts always deviates from the ideal shape. Therefore, we must assume that a shaft will have geometric deviations, which qualify such shafts as usable for operation if they are within the engineering limits. Even when the geometrical deviations are within permissible limits, they cause elastic deformations of the shaft, which directly affects geometrical quantity measurements.

2.2. Tested Object

The object subjected to the analysis was the crankshaft of the main propulsion medium-speed Buckau Wolf R8 DV-136 engine (Maschinenfabrik Buckau R. Wolf AG, Magdeburg, Germany), measuring 3630 mm in length and weighing 9280 N, equipped with ten 149 mm main journals and eight 144 mm crankpins. The geometrical model with main journal numeration used in further analysis is shown in Figure 2.

Figure 2. Geometrical model of the analyzed shaft with journal numeration.

It was assumed that the material of which the shaft was made is AISI 1060-2 steel, characterized by Young modulus $E = 212$ GPa, Poisson's ratio $\nu = 0.29$, and mass density $\rho_s = 7.7{\cdot}10^{-6}\ \frac{\text{kg}}{\text{m}^3}$.

2.3. Research Plan of Crankshaft Measurements

A study plan was developed in order to prove that the currently used methods of measuring crankshaft geometrical deviations, briefly described in Section 2.1, limit the detectability of geometric deviations. It includes an analysis of deflection and reaction forces at the contact of vee-block support heads with the main journals. Four most representative cases are included in the analysis, i.e.:

- Case 1: individual main journals are perfectly coaxial, while one of the supports (of journal no. 5, counting from the timing gear end) is offset upwards by 0.03 mm relative to others;
- Case 2: the main journals of all crankshafts are perfectly coaxial, while one of the supports (of journal no. 5, counting from the timing gear end) is offset downwards by 0.03 mm relative to others;
- Case 3: the axis of one of the main journals (no. 5 counting from the timing gear end) is offset upwards by 0.03 mm from the others, while the supports are at set at the same height;
- Case 4: the axis of one of the main journals (no. 5, counting from the timing gear end), is offset downwards by 0.03 mm from the others, while the supports are set at the same height.

A graphic representation of the cases under consideration is shown in Figure 3.

2.4. Finite Element Analysis

To assess the deflections and reaction forces distribution at the contact between vee-block support heads with the main journals, a finite element model [23,24] of the crankshaft was established—Midas 2019 (Midas Information Technology Co. Ltd., Seongnam, Korea) [25,26]. The geometrical model of the analyzed shaft was discretized using four-sided solid elements (CTETRA) with three translational degrees of freedom in each node. As a result, the finite element model subjected to further analysis had 137,475 elements and 126,114 degrees of freedom. The finite element model is shown in Figure 4.

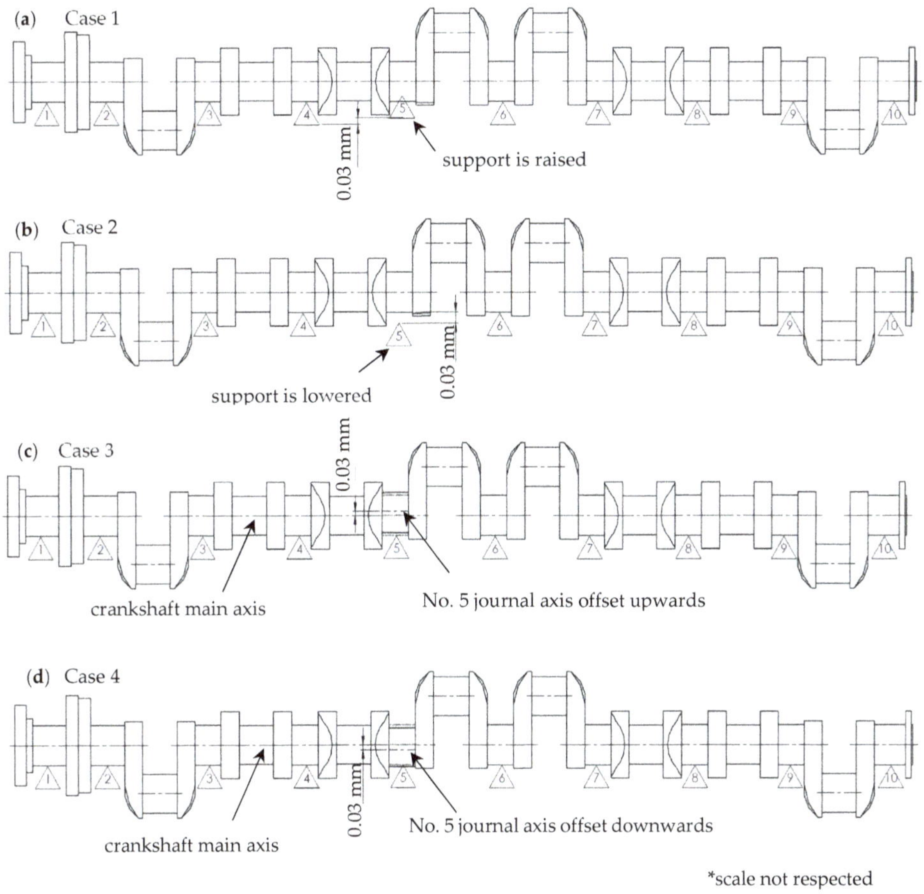

*scale not respected

Figure 3. A graphic representation of the cases under consideration.

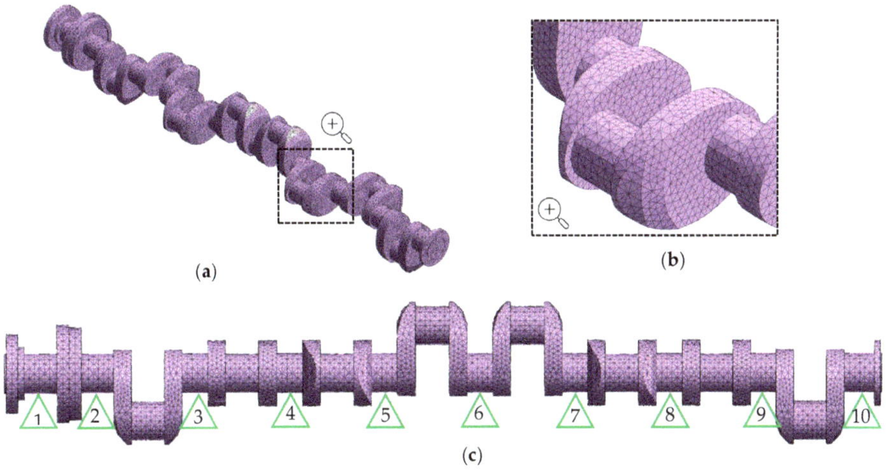

Figure 4. Finite element model of the analyzed shaft (**a**) isometric view; (**b**) mesh close-up; (**c**) main journals support.

The gravity load was applied to the model. An analysis consisting of determining shaft deflections and reaction forces acting on a supported shaft was performed using linear static Nastran solver (SOL101). Subsequent cases formulated in Section 2.3. were calculated according to the assumption that deformations and changes in the reaction forces on the supported main journals were caused by support positioning and geometrical deviations of the shaft when its journals were rested on a set of vee-block supports [16]. Subsequent angular positions of the shaft were simulated by rotating the model subjected to the force of gravity [27].

2.5. Experimental Setup

Experimental measurements were performed using a constructed system consisting of a MUK 25-600 measuring head and SAJD software, which enabled a complete qualitative evaluation of the roundness profiles (Figure 5a) [2]. The MUK 25-600 head was seated directly on the surface of the journal being tested, which evaluated the shape profile independent of the object's support conditions (Figure 5b). The roundness profile measurements were analyzed in terms of harmonics, the results of which were presented in discrete amplitude spectra [2,28].

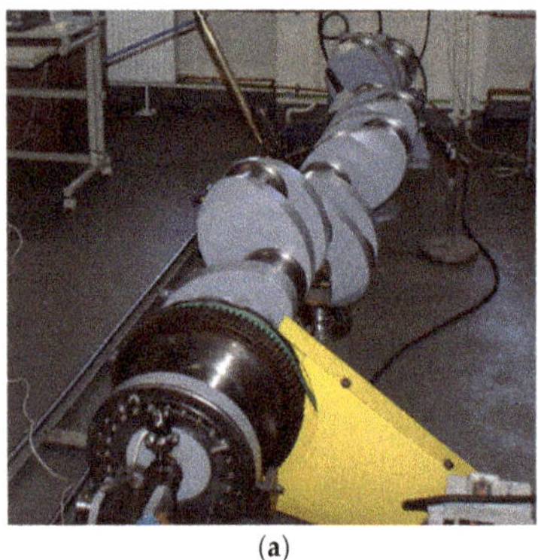

(a)

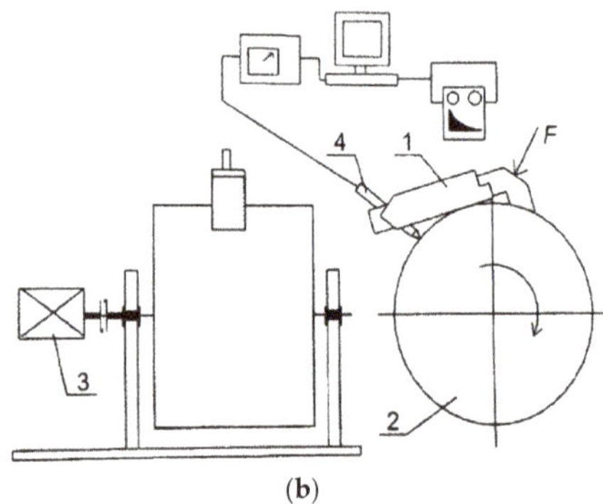

(b)

Figure 5. SAJD measurement system: (**a**) layout of the system; (**b**) measurement method [11]: 1—measuring head MUK 25-600, 2—shaft journal, 3—drive motor, 4—displacement sensor, F—measuring head pressing force.

An important advantage of this system is that measurements can be made directly in the work environment and the measured object does not need to be dismantled. Similar features have different design solutions of the measuring heads equipped with multi-contact self-adjusting vee blocks cooperating with one or more dial sensors. However, with this method it is only possible to evaluate the deviations and shape profiles, which, from the perspective of performing a comprehensive evaluation of the journal geometry, provides only a partial control of the measurement accuracy. Measuring axis deviation remains difficult.

The roundness profiles were then superimposed on the displacement profile of the center of the journal moving eccentrically and the displacement profile subject to support limitations. To completely depict the issue, the measured roundness contour was repeatedly superimposed on the displacement profile limited by the support. It was moved angularly to the defined starting position. Detailed description of the experimental system was presented in [2,11].

3. Results and Discussion

3.1. Finite Element Model Analysis

3.1.1. Case One

To implement the study plan formulated in Section 2.3, we calculated the displacements and reaction forces at the contact of support heads with main journals for the support positioning and geometric shaft deviations adopted above. The positioning of the supports at different heights generally means that the shaft will be pre-deflected, even if it is perfectly constructed. It was assumed that support no. five was offset upwards by 0.03 mm relative to the others. An exemplary finite element analysis results for case one is shown in Figure 6. Figure 7 shows the graphical interpretation of the changes in deflection values for this case when changing the shaft rotation angle by 15° at a time. The displacement of the journals not included in Figure 5 was 0 mm.

Figure 6. Exemplary finite element analysis results for case 1.

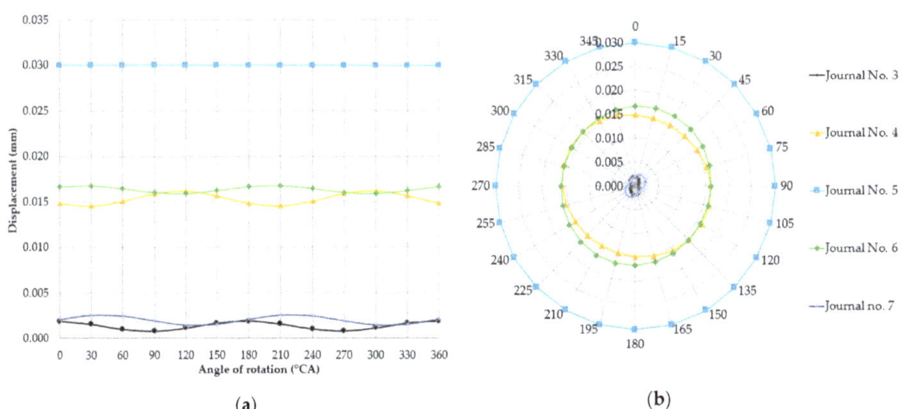

Figure 7. Changes in deflections measured in the vertical plane at individual main journals with the shaft rotated by 15° at a time and when one of the supports (of journal no. 5 counting from the timing gear end) is offset upwards by 0.03 mm relative to the others, shown in the charts in the (**a**) Cartesian and (**b**) polar coordinate systems.

For this type of support and a shaft rotation angle of 90°, Table 1 presents changes in the reaction forces at the support head/individual main journal interface.

Table 1. Changes in the reaction forces at the contact of the supports with each main journal when the shaft rotation angle was changed by 90° and one of the supports (of journal no. 5 counting from the timing gear end) was offset upwards by 0.03 mm relative to the others.

Angular Position [°CA]	Main Journal Number [-]									
	1	2	3	4	5	6	7	8	9	10
	Reaction Forces in Main Journals [N]									
0°	681.2	1545.4	0	0	3891.3	0	0	1441.7	1157.6	566.5
90°	878.8	1302.2	0	0	3928.8	0	0	1470.9	1162.4	540.7
180°	681.2	1545.4	0	0	3891.3	0	0	1441.7	1157.6	566.5
270°	878.8	1302.2	0	0	3928.8	0	0	1470.9	1162.4	540.7

In this case, the shaft is cyclically bent upwards and support no. five carries very high loads, completely relieving supports no. three, four, six, and seven. As support no. five is offset, the shaft is lifted, and the aforementioned supports lose contact with the journals.

3.1.2. Case Two

Lowering the support relative to the others removes the support from under the shaft at its location. It was assumed that support no. five was offset downwards by 0.03 mm relative to the others. The results from exemplary finite element analysis for case two are shown in Figure 8. Figure 9 shows a graphical interpretation of the changes in deflection values for this case, when the shaft rotation angle was changed by 15° at a time. The displacement of the journals not included in Figure 9 was 0 mm.

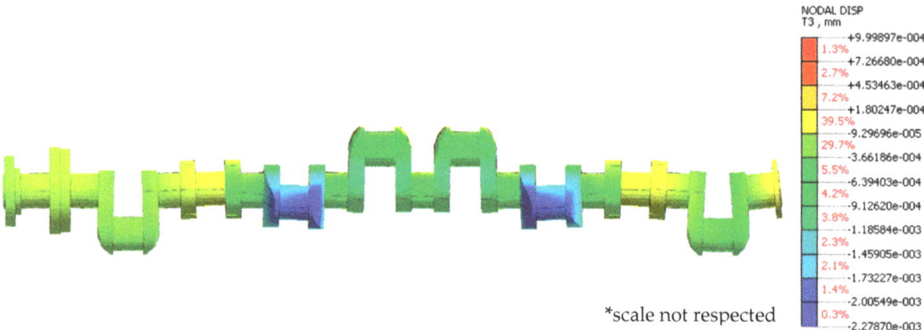

Figure 8. Exemplary finite element analysis results for case 2.

Table 2 presents the changes in reaction forces at individual main journals for this type of support at a 90° shaft rotation.

Table 2. Changes in the reaction forces where the supports contact each main journal when the shaft rotation angle was changed by 90° with one of the supports (of journal no. 5 counting from the timing gear end) offset downwards by 0.03 mm relative to the others.

Angular Position [°CA]	Main Journal Number [-]									
	1	2	3	4	5	6	7	8	9	10
	Reaction Forces in Main Journals [N]									
0°	725.4	996.9	729.4	1714.0	0	1750.0	584.3	1183.4	1025.2	572.3
90°	817.3	844.7	775.5	1773.6	0	1541.2	914.9	907.9	1168.4	540.3
180°	725.4	996.9	729.4	1714.0	0	1750.0	584.3	1183.4	1025.2	575.3
270°	817.3	844.7	775.5	1773.6	0	1541.2	914.9	907.9	1168.4	540.3

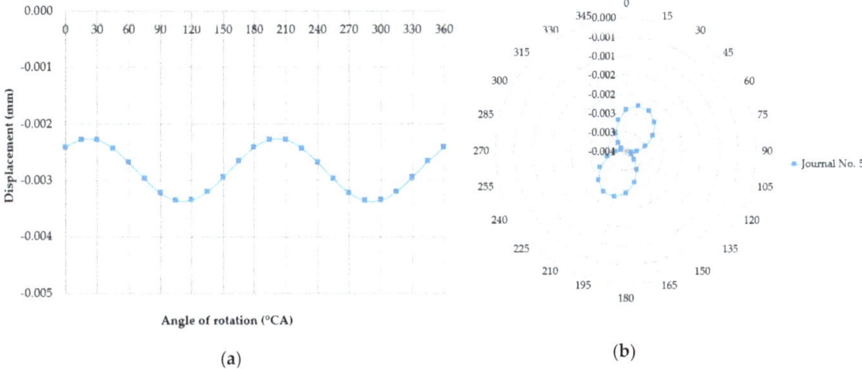

Figure 9. Changes in deflections measured in the vertical plane at individual main journals with the shaft rotated by 15° at a time and when one of the supports (of journal no. 5 counting from the timing gear end) was offset downwards by 0.03 mm relative to the others, in the (**a**) Cartesian and (**b**) polar coordinate systems.

The shaft bends down under its own weight in locations without support. Deflections at non-supported journals are insignificant (−0.00228 mm to −0.00335 mm), while the reaction forces at journals adjacent to the non-supported ones increase significantly, reaching 1713 N to 1773 N in journal no. four and 1541 N to 1749.9 N in journal no. six.

3.1.3. Case Three

The situation is slightly different when the supports are at the same height and the main journal axes positions deviate (case three and the alternative version of case four). In case three, the supports are located at the same height, while the axis of journal no. five (for the shaft's reference angular position) is offset eccentrically upwards by +0.03 mm relative to the other journals. The form of deformation of the shaft is analogous to that presented in the Figure 8. Figure 10 shows a graphical interpretation of the changes in deflection values for this case, when the shaft rotation angle was changed 15° at a time. The displacement of the journals not included in Figure 10 was 0 mm.

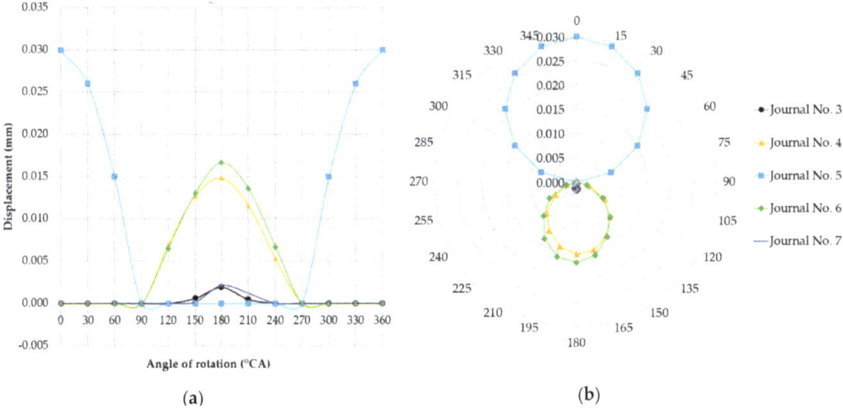

Figure 10. Changes in deflections measured in the vertical plane at individual main journals with the shaft changed by 15° at a time, and when one of the main journal axes (of journal no. 5 counting from the timing gear end) was offset eccentrically upwards by 0.03 mm relative to the others, shown in the (**a**) Cartesian and (**b**) polar coordinate systems.

Table 3 shows the reaction forces exerted by the supports for the four characteristic angular positions of the shaft in this case.

Table 3. Changes in the reaction forces where the supports contact each main journal when the shaft rotation angle changes by 90° and when one of the axis main journals (journal no. 5 counting from the timing gear end) is offset upwards by 0.03 mm relative to the others.

Angular Position [°CA]	Main Journal Number [-]									
	1	2	3	4	5	6	7	8	9	10
	Reaction Forces in Main Journals [N]									
0°	725.5	996.9	729.4	1714.0	0	1750.0	584.3	1183.4	1025.2	575.3
90°	833.3	763.1	1134.1	891.3	1126.1	809.9	1127.1	886.2	1173.4	539.5
180°	681.2	1545.4	0	0	3891.3	0	0	1441.7	1157.6	566.5
270°	833.3	763.1	1134.0	891.3	1126.1	809.9	1127.1	886.2	1173.4	539.5

3.1.4. Case Four

Changing the shaft rotation angle by 180° causes the eccentrically located journal axis to move to an extreme location opposite the reference angle used in the previous case. The considered relative positions of the journals correspond to case four and are an alternative version of case three (Figure 8.) Using the values of deflections and reaction forces for shaft rotation angles ranging from 0° to 360°, the resulting calculated quantities will take the same values as in Table 3 if an angular offset of 180° is applied. The form of deformation of the shaft is analogous to that presented in Figure 6. Figure 11 shows a graphical interpretation of the changes in deflection values for this case, when the shaft rotation angle was changed 15° at a time. The displacement of the journals not included in Figure 11 was 0 mm.

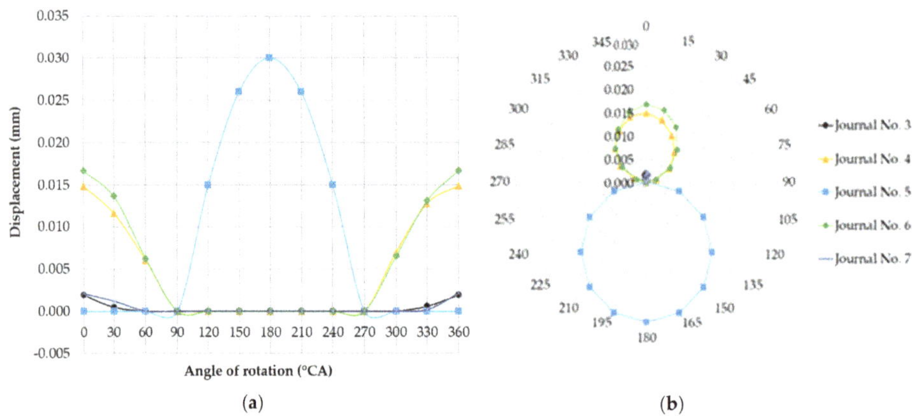

Figure 11. Changes in deflections measured in the vertical plane at individual main journals, when one of the main journal axes (no. 5 counting from the timing gear end) was offset eccentrically downwards by 0.03 mm relative to the others, shown in the (**a**) Cartesian and (**b**) polar coordinate systems.

3.2. Phenomenological Model of Detectability of Journals' Misalignment

A detailed analysis of cases three and four shows that for deviations in the position of main journal axes, the detectability of journals' misalignment was limited by supporting the shaft with a set of fixed rigid vee-blocks located at the same height. The analysis of the graphs of journal axes' deflections (Figures 7 and 8) shows that the deflections were zero in the range of angles for which journal no. five permanently contacted the support.

We analyzed the measurements of case three, in which all the main journals of the shaft were rested on equally elevated supports, and all main journals were situated coaxially, except for journal no. five, whose axis O_1 was offset upwards relative to the others by 0.03 mm in the initial angular shaft position. In case three, when the shaft was rotated in the angular range from 0 to 90°, journal no. five was unsupported, and the axis of this journal moved eccentrically with respect to the main axis O of the shaft (relative to the shaft's axis of rotation). The displacement sensor, whose probe stylus is located vertically, measures the deflection of the journal resulting from its eccentric movement during the shaft rotation (Figure 12).

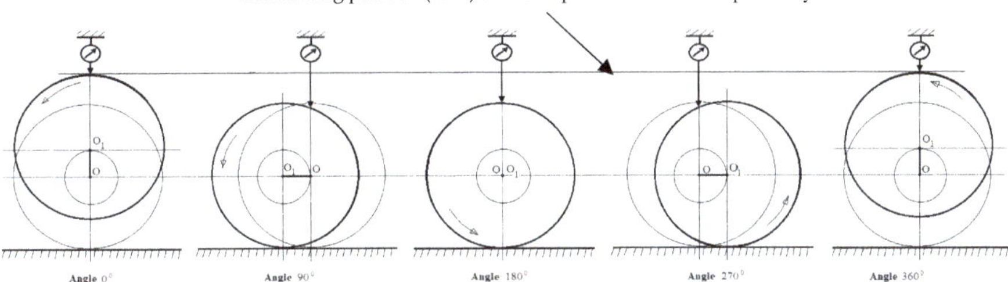

Figure 12. Supplementary diagram of the successive stages of shaft deflection caused by its eccentricity with respect to the axis of rotation of the measuring system when the support retains its fixed height.

If the dial indicator is zeroed in the top dead-center of journal no. five, and if the downward movement of the probe stylus is treated as a negative indication, then in the 90° angular position of the shaft the sensor indicator displays a value of −0.03 mm. For this angle of rotation, journal no. five will contact the support and, as the shaft rotates further (in the angular range of 90° to 180°), the reaction force will gradually increase at the point where journal no. five and the support get into contact. Simultaneously, journals adjacent to journal no. five (i.e., journals no. four, three, six, and seven) will be lifted upwards, losing contact with their supports due to bending of the shaft (caused by increased pressure of journal no. five on the support). Journal no. five is permanently in contact with its support, so the sensor will indicate a constant deflection of 0.03 mm, the same as for the angle of 90°. At shaft angles ranging from 180° to 270°, journal no. five will remain in contact with the support, which will exert gradually decreasing reaction forces on the journal. At the same time, the shaft becomes less bent at the locations of journals no. four, three, six, and seven. At 270° rotation, journal no. five loses contact with its support, and journals no. four, three, six, and seven rest on their supports. For shaft angles ranging from 180° to 270°, the displacement sensor still shows a constant deflection of –0.03 mm. For shaft rotation angles from 270° to 360°, journal no. five is gradually lifted upwards, and the displacement sensor indicates a change from –0.03 mm (at 270°) to 0.00 (at 360°). Figure 13 shows the displacements indicated by the sensor for journal no. five for shaft rotation angles from 0° to 360°.

It can be seen that in general the actual value of eccentricity can be measured by the value of this deviation and the vertical location of the support in relation to the supported main journal. Using the results obtained, a supplementary graph was drawn to show the measurable value w of eccentricity e as a function of the vertical position x of the support (Figure 14).

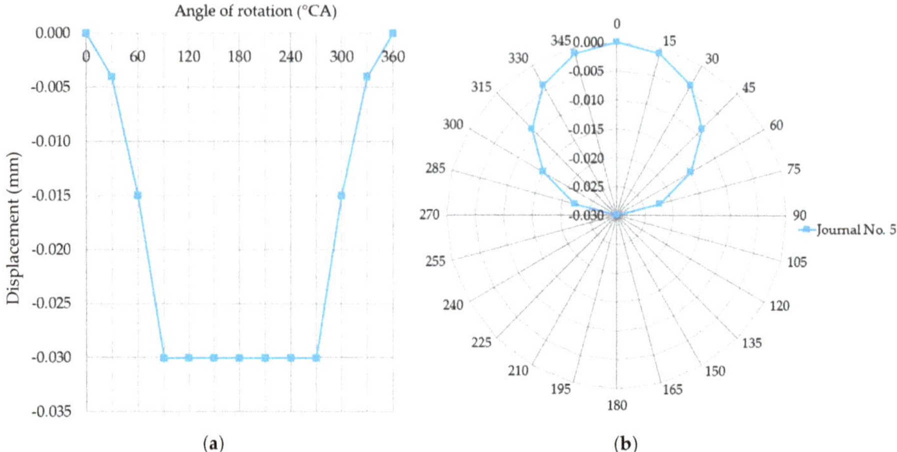

Figure 13. The journal displacements measured by the sensor for a full shaft rotation (0°–360° angle) are recorded when the support maintaining a constant height restricts these displacements, as shown in the (**a**) Cartesian and (**b**) polar coordinate systems.

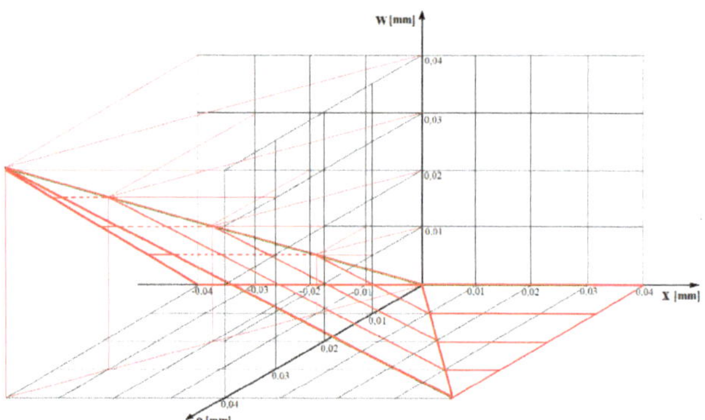

Figure 14. Graph showing the measurable value w of eccentricity e as a function of the vertical position x of the support.

As results from the previous analysis, another issue to be considered is periodic non-overlap between the displacement direction of the sensor's probe stylus and the axis of the journal being measured. This is caused by the eccentric movement of this axis during shaft rotation. As shown in Figure 15, for a given angular position of the shaft (angle φ), the quantity being measured is p, whereas the quantity that should be measured is p'.

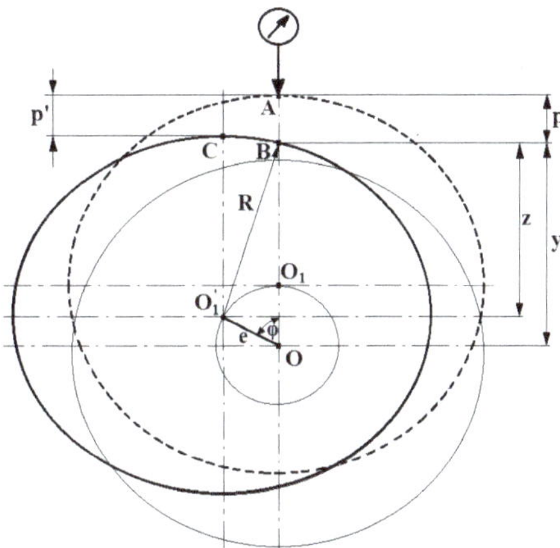

Figure 15. A supplementary diagram to determine the measurement error caused by a journal's eccentric displacement when measuring geometric deviations of the shaft.

By analyzing the geometrical and trigonometric relationships shown in the supplementary diagram (Figure 15), we can find a mathematical relationship that describes the measured value of p and the measurement error, i.e., the difference between p and p'. According to the following diagram:

$$p' = e - e\cos\varphi \tag{1}$$

Since

$$y = z + e\cos\varphi \tag{2}$$

Whereas:

$$z = \sqrt{R^2 - (e\sin\varphi)^2} \tag{3}$$

Thus, the final form is:

$$p = R + e - y = R + e - \left(\sqrt{R^2 - (esin\varphi)^2} + ecos\varphi\right) \tag{4}$$

And

$$\Delta p = p - p' = R - \sqrt{R^2 - (esin\varphi)^2} \tag{5}$$

Due to the deformation, the axis of the object being measured (supported by vee-blocks) takes the angular position ϑ with respect to the probe stylus of the sensor [1,2].

Thus, considering the location of the section to be measured relative to the support points, the location of the center of the section being measured moves with respect to the axis of rotation (determined by the measuring system) by the value of the elastic (f) or permanent (y) deformation (Figure 16). Consequently, the measured profile of roundness is distorted by the so-called apparent eccentricity and ovality. If it is possible to determine the angular deformation ϑ and the arrow of deformations f or y (e.g., from strength calculations or measurements), the resulting measurement errors are systematic errors and should be used as correction factors when evaluating the proper first and second harmonic (after expanding the measurements of the roundness profile into a Fourier series [28]).

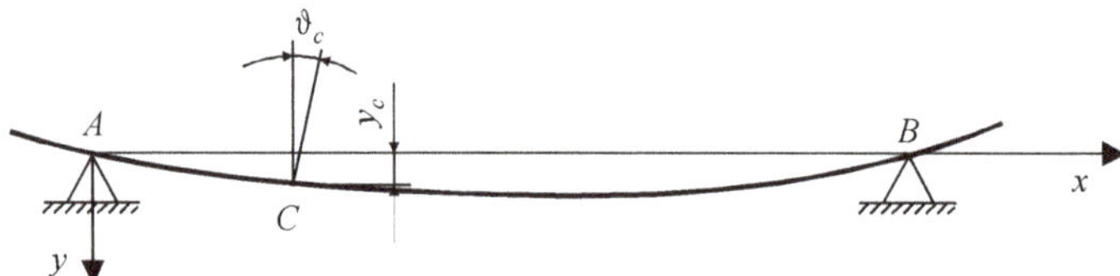

Figure 16. Deformation of the object supported by two vee-blocks and the linear and angular displacements resulting from this deformation.

3.3. *Journal Position Misalignment Taking into Account Eccentricity*

In the considerations presented so far, it has been assumed that journals have an ideal circular profile, but machining processes involve unavoidable errors that give journals irregular roundness profiles. In general, when the shaft is fixed on vee-blocks, the shape and axial position deviations are measured in individual cross-sections of the main journals of the rotating crankshaft. In the case of a misaligned journal position, the center of the measured journal's profile may move relative to the axis of rotation determined by the measuring system. In this case, the measurements describe the shape profile of the given cross-section, as well as the eccentricity that represents the profile center position of the section measured relative to the axis of rotation determined by the measuring system. Taking into account that a rigid support limits the detectability of geometric deviations in the main journal axes of a crankshaft (which has been demonstrated), an analysis was conducted to determine how the limited detectability of axis position deviations affects the evaluation of the main journal's roundness profile. To accomplish this, the deviations and shape profiles of the main journals were measured for the tested crankshaft.

Figure 17 shows an example of a roundness profile measured by the reference method (with a MUK 25-600 sampling cell) corresponding to the roundness contour of pin no. five and a discrete amplitude spectrum obtained from the harmonic analysis. Table 4 shows the values of the individual harmonics.

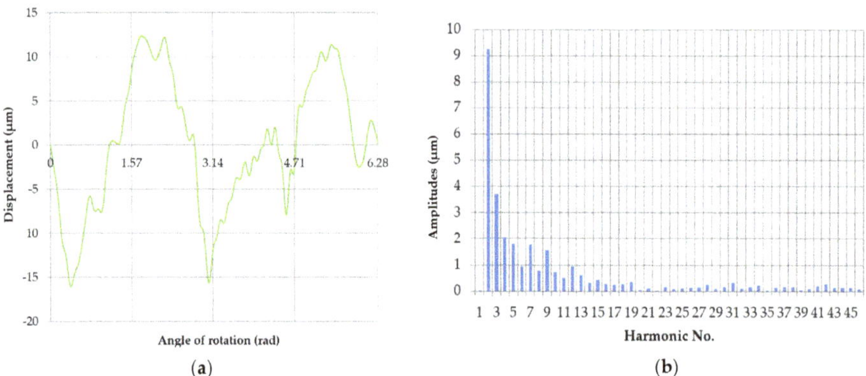

Figure 17. Roundness profile of journal no. 5 measured by the reference method (**a**); the discrete amplitude spectrum (**b**).

Table 4. Amplitudes of the harmonic components for the roundness profile of journal no. 5, measured by the reference method.

Harmonic Amplitudes [μm]					
n	0	10	20	30	40
n + 0		0.733112	0.048821	0.160022	0.078616
n + 1		0.519559	0.100569	0.321846	0.20063
n + 2	9.253684	0.958851	0.030333	0.117518	0.266181
n + 3	3.736699	0.637086	0.153547	0.154698	0.127993
n + 4	2.063435	0.326033	0.084396	0.233796	0.136985
n + 5	1.80228	0.441492	0.116786	0.028151	0.134468
n + 6	0.958385	0.279591	0.139388	0.128674	0.082787
n + 7	1.778264	0.254698	0.142395	0.159856	0.062551
n + 8	0.773096	0.277479	0.244795	0.176508	0.088632
n + 9	1.558816	0.344293	0.073347	0.049974	0.007096

The image of the eccentric profile of the measured roundness contour center at an eccentricity e = 0.03 mm, assuming that the support does not limit the shaft displacement. The corresponding discrete amplitude spectrum is shown in Figure 18, which includes only the first harmonic (Table 5).

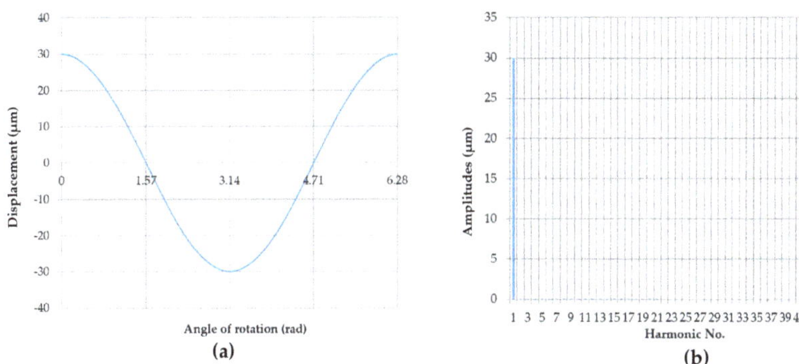

Figure 18. Eccentric movement profile of the center of the roundness profile being measured. The eccentricity was 0.03 mm, and the support did not limit the shaft displacement (**a**); discrete amplitude spectrum (**b**).

Table 5. Amplitudes of the harmonic components for the eccentric movement of the center of the roundness profile being measured for eccentricity $e = 0.03$ mm where the support does not limit the shaft displacement.

Harmonic Amplitudes [μm]					
n	0	10	20	30	40
n + 0		0.009154	0.002274	0.001011	0.00057
n + 1	29.99956	0.007553	0.002062	0.000947	0.000542
n + 2	0.00059	0.006339	0.001879	0.000889	0.000517
n + 3	0.000221	0.005396	0.001719	0.000836	0.000493
n + 4	0.000118	0.004649	0.001579	0.000788	0.000471
n + 5	7.37×10^{-5}	0.004048	0.001455	0.000743	0.000451
n + 6	5.06×10^{-5}	0.003556	0.001345	0.000703	0.000431
n + 7	3.69×10^{-5}	0.003149	0.001248	0.000665	0.000413
n + 8	2.81×10^{-5}	0.002808	0.00116	0.000631	0.000397
n + 9	2.21×10^{-5}	0.00252	0.001082	0.000599	0.000381

The image of the profile corresponding to the eccentric displacement of the center of the measured roundness profile, presented in the Cartesian system for the case when all the supports were situated at the same height (x = 0 mm), with the eccentricity of one of the main pins equal to $e = 0.03$ mm and the discrete amplitude spectrum is shown in Figure 19. As can be seen, the amplitude spectrum contains only even harmonic components, the values of which are summarized in Table 6.

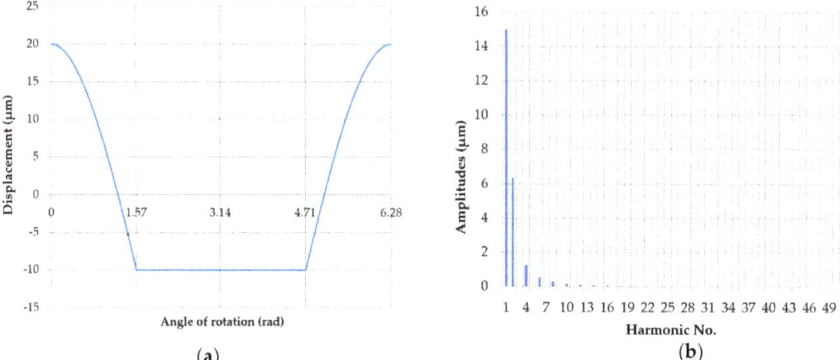

Figure 19. Eccentric movement profile of the center of the roundness profile being measured when the supports were set at the same height ($x = 0$ mm), with an eccentricity of the main journal $e = 0.03$ mm (**a**); discrete amplitude spectrum (**b**).

Table 6. Amplitudes of the harmonic components for the eccentric movement of the center of the roundness profile being measured when the supports were set at the same height ($x = 0$ mm), with an eccentricity of the main journal $e = 0.03$ mm.

Harmonic Amplitudes [µm]					
n	0	10	20	30	40
n + 0		0.1930	0.0479	0.0213	0.0120
n + 1	14.9993	0.0001	0.0000	0.0000	0.0000
n + 2	6.3665	0.1336	0.0396	0.0187	0.0109
n + 3	0.0006	0.0001	0.0000	0.0000	0.0000
n + 4	1.2731	0.0980	0.0333	0.0166	0.0099
n + 5	0.0001	0.0001	0.0000	0.0000	0.0000
n + 6	0.5458	0.0750	0.0284	0.0148	0.0091
n + 7	0.0002	0.0000	0.0000	0.0000	0.0000
n + 8	0.3032	0.0592	0.0245	0.0133	0.0084
n + 9	0.0001	0.0001	0.0000	0.0000	0.0000

When the measured round contour was superimposed on the displacement profile of the pin center without being limited by the support, the total profile and the discrete spectrum shown in Figure 20 was obtained. The corresponding amplitude values are shown in Table 7.

When the measured round contour was superimposed on the displacement profile of the pin center at the support limit, the total profile was obtained, which is shown in Figure 21 for the starting position. The amplitude values are shown in Table 8.

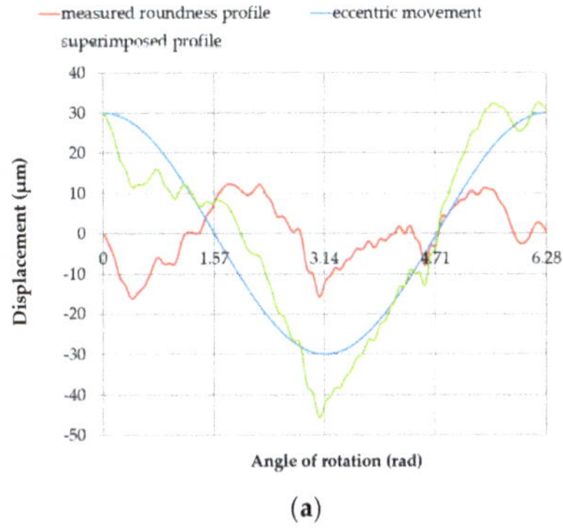

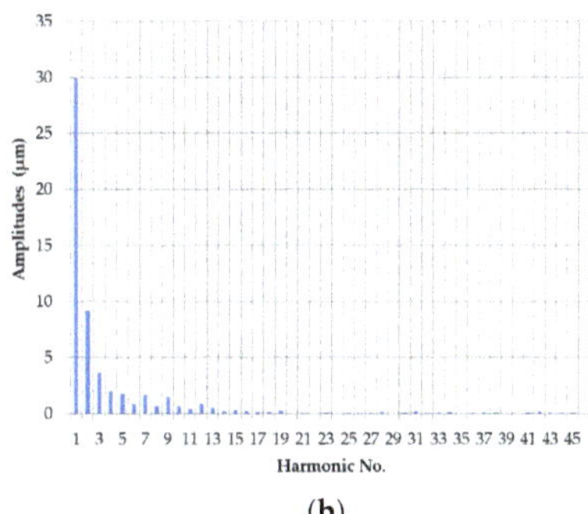

(a) (b)

Figure 20. The profile obtained by superimposing the measured roundness profile of journal no. 5 on the full profile of the eccentric movement of the measured roundness profile center (**a**); discrete amplitude spectrum of the superimposed profile (**b**).

Table 7. Amplitudes of harmonic components for the profile obtained by superimposing the measured roundness profile of journal no. 5 on the full eccentric movement profile of the center of the measured roundness profile.

	Harmonic Amplitudes (μm)				
n	0	10	20	30	40
n + 0		0.73	0.05	0.16	0.08
n + 1	30.00	0.52	0.10	0.32	0.20
n + 2	9.25	0.96	0.03	0.12	0.27
n + 3	3.74	0.64	0.15	0.15	0.13
n + 4	2.06	0.33	0.08	0.23	0.14
n + 5	1.80	0.44	0.12	0.03	0.13
n + 6	0.96	0.28	0.14	0.13	0.08
n + 7	1.78	0.25	0.14	0.16	0.06
n + 8	0.77	0.28	0.24	0.18	0.09
n + 9	1.56	0.34	0.07	0.05	0.01

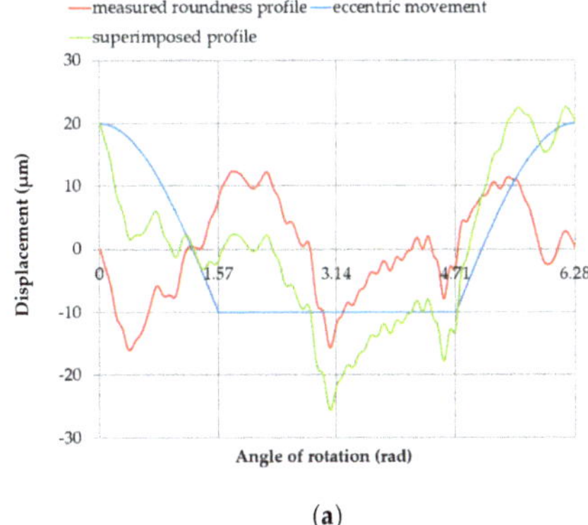

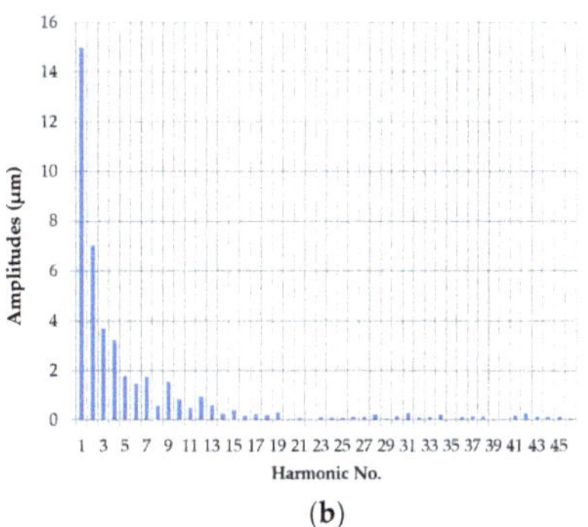

(a) (b)

Figure 21. Profile obtained by superimposing the measured roundness profile of journal no. 5 on the eccentric movement profile of the center of the measured roundness profile with the supports set at the same height ($x = 0$ mm), with an eccentricity of the main journal $e = 0.03$ mm (**a**); discrete amplitude spectrum of the superimposed profile (**b**).

Table 8. Amplitudes of harmonic components for the profile obtained by superimposing the measured roundness profile of journal no. 5 on the profile of the eccentric movement of the center of the measured roundness profile when the supports are set at the same height ($x = 0$ mm); with an eccentricity of the main journal $e = 0.03$ mm.

Harmonic Amplitudes [μm]					
n	0	10	20	30	40
n + 0		0.8665	0.0542	0.1725	0.0686
n + 1	14.9990	0.5196	0.1006	0.3218	0.2006
n + 2	7.0416	0.9778	0.0484	0.1178	0.2767
n + 3	3.7370	0.6371	0.1536	0.1547	0.1280
n + 4	3.2658	0.2978	0.1164	0.2504	0.1456
n + 5	1.8022	0.4414	0.1168	0.0282	0.1345
n + 6	1.4820	0.2104	0.1314	0.1350	0.0857
n + 7	1.7784	0.2547	0.1424	0.1599	0.0625
n + 8	0.6172	0.2376	0.2624	0.1841	0.0867
n + 9	1.5587	0.3442	0.0734	0.0500	0.0071

The image of the profile obtained after superimposition and shifting by 60°, followed by 90° (relative to the assumed starting point) of the measured roundness contour of pin no. 5, on the simultaneously displayed profile of the eccentric movement at the measured roundness contour center limited by the support are shown in Figures 22 and 23, respectively. The amplitudes of the individual harmonics are shown in Tables 9 and 10.

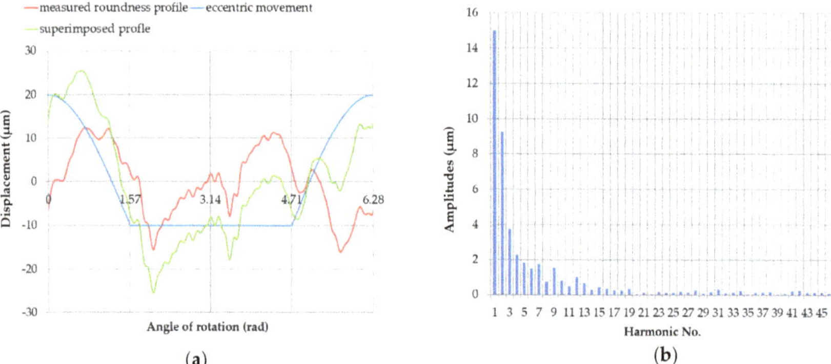

Figure 22. Profile obtained by superimposing the measured roundness profile of journal no. 5—rotated by 60° with respect to the reference profile—onto the profile of the eccentric movement of the center of the measured roundness profile, with the supports set at the same height ($x = 0$ mm); with an eccentricity of the main journal $e = 0.03$ mm (**a**); discrete amplitude spectrum of the superimposed profile (**b**).

According to the accepted interpretation of the measured round contour geometrical features of the analysis based on harmonics, the first term in the Fourier series of the function characterizing the course is the deviation of the axis position, namely the eccentricity. Eliminating this harmonic makes it possible to treat the sum of the remaining harmonics as the theoretically measured roundness contour.

This interpretation was used to qualitatively and quantitatively compare the measured and theoretical (excluding harmonic no. 1) roundness contours, obtained from superimposing the measured roundness contour of pin no. 5 on the full eccentric profile and the eccentricity profile of the measured round outline center at the support limit. A graphical representation of the compared profiles of journal no. 5 (Figures 24 and 25) is helpful to visually assess the profile quality and compare the amplitude spectra.

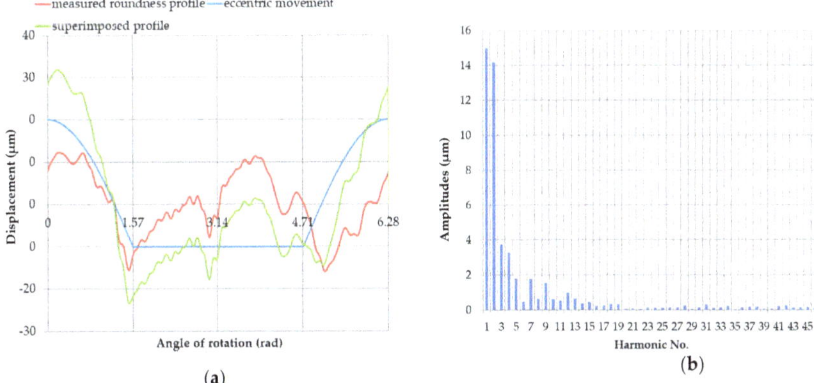

Figure 23. Profile obtained by superimposing the measured roundness profile of journal no. 5 (green)—rotated by 90° with respect to the reference profile—onto the eccentric movement profile of the center of the measured roundness profile (blue), with the supports set at the same height ($x = 0$ mm), with an eccentricity of the main journal $e = 0.03$ mm (**a**); discrete amplitude spectrum of the superimposed profile (**b**).

Table 9. Amplitudes of the harmonic components for the profile obtained by superimposing the measured roundness profile of journal no. 5—and by rotating it by 60° with respect to the reference profile—onto the eccentric movement profile of the center of the measured roundness profile when the supports were set at the same height ($x = 0$ mm), with an eccentricity of the main journal $e = 0.03$ mm.

Harmonic Amplitudes [µm]					
n	0	10	20	30	40
n + 0		0.7893	0.0343	0.1623	0.0844
n + 1	15.0011	0.4982	0.1035	0.3275	0.2169
n + 2	9.2514	0.9881	0.0176	0.0956	0.2657
n + 3	3.7533	0.6462	0.1620	0.1455	0.1292
n + 4	2.2880	0.2782	0.1145	0.2314	0.1355
n + 5	1.8372	0.4334	0.1085	0.0266	0.1317
n + 6	1.5002	0.3518	0.1749	0.1302	0.0869
n + 7	1.7701	0.2455	0.1344	0.1570	0.0000
n + 8	0.7415	0.2474	0.2530	0.1678	0.0000
n + 9	1.5501	0.3552	0.0730	0.0419	0.0000

Table 10. Amplitudes of harmonic components for the profile obtained by superimposing the measured roundness profile of journal no. 5—and rotating it by 90° with respect to the reference profile—onto the profile of the eccentric movement of the center of the measured roundness profile with the supports set at the same height ($x = 0$ mm), with an eccentricity of the main journal $e = 0.03$ mm.

Harmonic Amplitudes [µm]					
n	0	10	20	30	40
n + 0		0.6211	0.0587	0.1537	0.0677
n + 1	14.9993	0.5201	0.1002	0.3213	0.2007
n + 2	14.1993	0.9872	0.0539	0.1215	0.2559
n + 3	3.7376	0.6367	0.1542	0.1540	0.1275
n + 4	3.2729	0.3710	0.1142	0.2185	0.1461
n + 5	1.8030	0.4409	0.1167	0.0283	0.1351
n + 6	0.4735	0.2148	0.1557	0.1315	0.0783
n + 7	1.7790	0.2554	0.1431	0.1595	0.0000
n + 8	0.6317	0.3263	0.2645	0.1673	0.0000
n + 9	1.5596	0.3435	0.0730	0.0496	0.0000

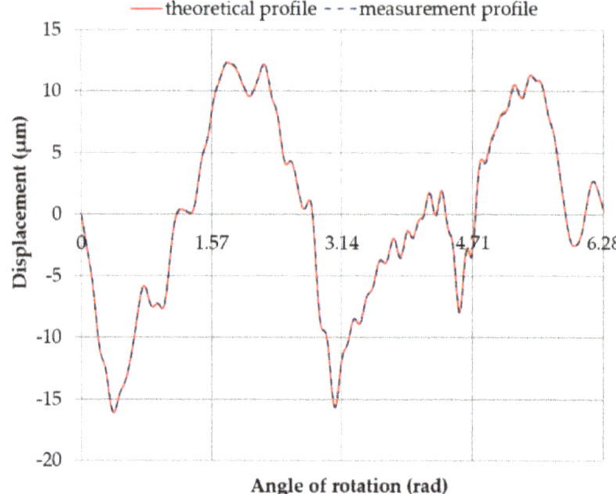

Figure 24. Measured and theoretical profiles (excluding harmonic component no. 1) obtained by superimposing the measured roundness profile of journal no. 5 onto the full eccentric movement profile of the center of the measured roundness profile.

The quantitative evaluation was conducted by determining the roundness deviations between the compared profiles and finding the correlation coefficient between the profiles provided by the formula [1,2,4,41–43]:

$$\rho\left(\gamma_{\phi}\right) = \frac{2\int_{0}^{2\pi} r_1(\phi) r_2\left(\phi + \gamma_{\phi}\right) d\phi}{\int_{0}^{2\pi} r_1(\phi)^2 d\phi + \int_{0}^{2\pi} r_2(\phi)^2 d\phi} \quad (6)$$

where: $r_1(\varphi)$—roundness profile obtained from measurements performed by the reference method. $r_2(\varphi)$—roundness profile obtained from measurements performed by the proposed method. γ_{ϕ}—phase shift between the compared profiles.

The adopted procedure involved repeatedly superimposing the measured roundness profile (with angular rotation) onto the displacement profile using the support to create a limitation. This approach allowed the relative angular position of the compared profiles to be determined. It can also be used to determine the maximum and minimum correlation coefficients between the actual standard profile and the theoretical profile obtained by superimposing the measured profile onto the displacement profile, as well as the roundness deviations resulting from this procedure.

Table 11 presents the correlation coefficients ρ between the compared profiles and the roundness deviations of the evaluated profiles Δ_o. Figure 26 shows a graph of ρ as a function of the angular shift between the compared profiles. The minimum correlation coefficient was $\rho_{min} = 0.7962$, whereas the roundness deviation of the assessed profile was $\Delta_o = 27.03$ µm (Figure 25d). The maximum correlation coefficient was $\rho_{max} = 0.9717$, and the roundness deviation of the assessed profile was $\Delta_o = 41.39$ µm (Figure 25e).

Table 11. Roundness deviations Δ_o for the profiles compared in Figures 21 and 22. The correlation coefficients ρ between the compared profiles.

Profile/Figure	Roundness Deviation Δ_o [µm] Standard (Reference)	Evaluated	Correlation Coefficient ρ
Figure 25	28.40	28.40	0.9999
Figure 25a		26.86	0.8009
Figure 25b		32.95	0.8206
Figure 25c		41.65	0.9330
Figure 25d		27.03	0.7962
Figure 25e		41.39	0.9717

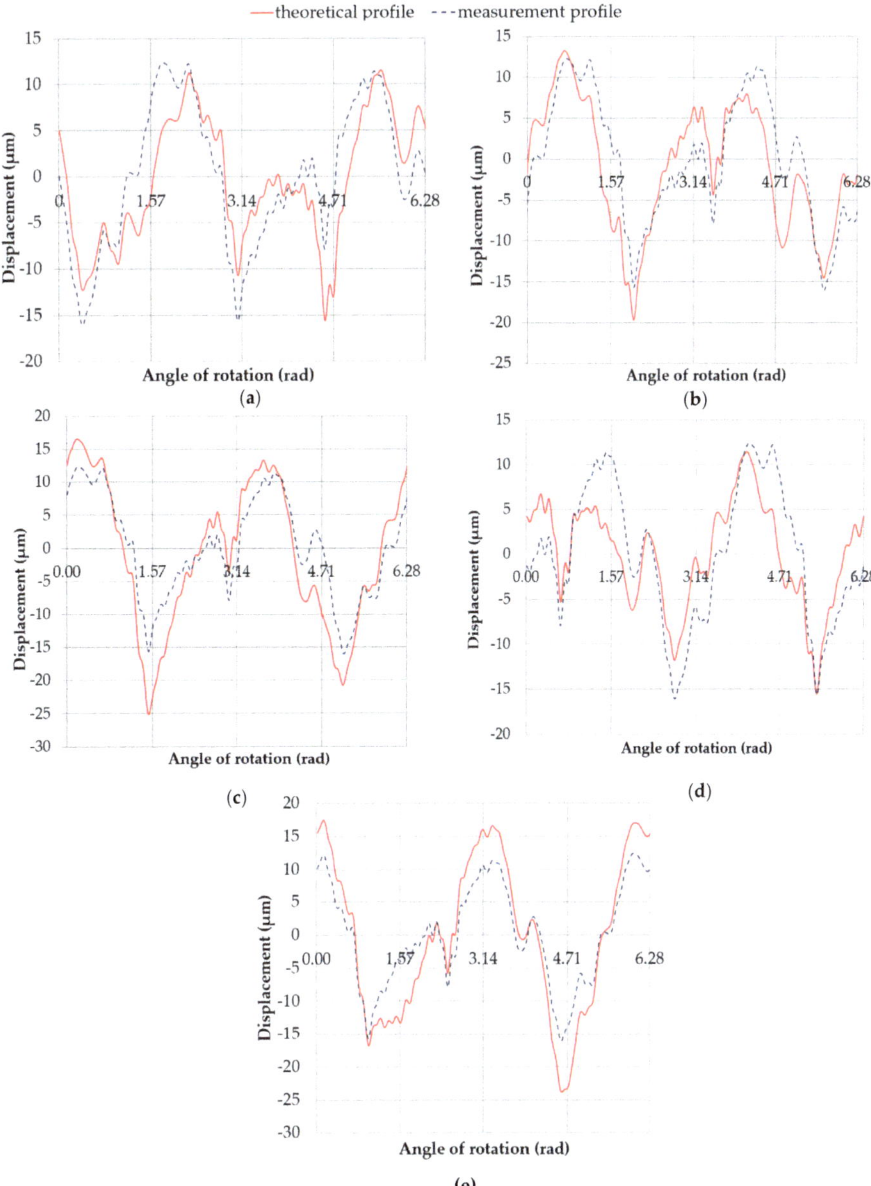

Figure 25. Measured and theoretical profiles (excluding harmonic component no. 1) obtained by superimposing the measured roundness profile of journal no. 5 onto the eccentric movement profile of the center of the measured roundness profile with the supports set at the same height ($x = 0$ mm), with an eccentricity of the main journal $e = 0.03$ mm, (**a**) at a starting position; (**b**) rotated 60°; (**c**) rotated by 90°; (**d**) rotated by 225°; (**e**) rotated by 300°.

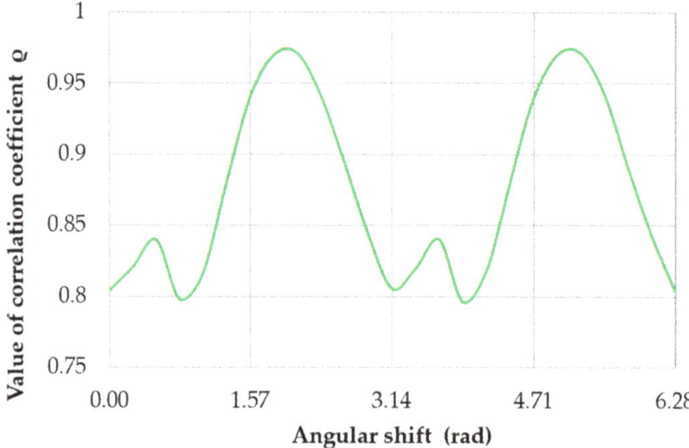

Figure 26. Variation in the correlation coefficient ρ as a function of the angular shift between the measured and evaluated profiles.

4. Conclusions

The study presented in this article confirmed that the detectability of geometric deviations is limited when the shaft is supported in an uncontrolled manner (with a set of rigid vee-block supports), which was especially true for the main journal deviations. However, the limited detectability of large crankshafts due to the support conditions was observed for positional deviations of journal axes and also for shape profile deviations in journals. The shape deviation measurements may vary significantly in terms of their values and profiles relative to the actual shape of an object. Referring to the denotations used in the article, when a journal axis moved eccentrically as the shaft rotated, the investigated parameter values were directly influenced by the eccentricity e, the support location x, and the location of the measured journal profile reflecting the journal axis displacement.

The results of this study show the importance of ensuring appropriate support conditions to eliminate deflections, and thus elastic deformations of the crankshaft under the influence of its own weight, as well as those caused by its geometric deviations. These deflections can only be eliminated if there is constant contact between the supports and the main journals of the shaft. Such conditions cannot be guaranteed by supporting the shaft with a set of rigid vee-block supports maintaining a fixed height. For deviations in the position of main journal axes, unintentional pre-deflections generate elastic deformations when the shaft rotates. This state causes interactions between geometrical deviations and elastic deformations (which are interrelated), and the geometric evaluation of the shaft geometry becomes unreliable.

Therefore, to ensure correct measurement conditions, the main journals of the shaft should be supported with a set of supports that compensate for its deflections and elastic deformations under the influence of its own weight, as well as those caused by geometric deviations of the shaft. The reaction forces at the contact between support heads and main journals vary along the shaft, and also depend on the angle of rotation of the shaft being supported, thereby ensuring zero deflections at the journals.

5. Patents

1. Nozdrzykowski, K. Device for measuring positional deviation of axis of crankshaft pivot set. Polish Patent Office, PL393829-A1; PL218653-B1.

Author Contributions: Conceptualization, K.N.; methodology, K.N., Z.G., S.A.; software, K.N., Z.G.; validation, K.N., Z.G., S.A.; formal analysis, K.N.; investigation, K.N., Z.G., P.D.; resources, K.N., Z.G.; data curation, K.N., Z.G.; writing—original draft preparation, K.N., Z.G., P.D.; writing—review and editing, K.N., P.D.; visualization,

K.N., Z.G., P.D.; supervision, P.D., Z.G. and K.N.,; project administration, K.N.; funding acquisition, K.N. and Z.G. All authors have read and agreed to the published version of the manuscript.

Funding: This research was co-funded by the Ministry of Science and Higher Education of Poland from Grant 1/S/KPBMiM/20.

Conflicts of Interest: The authors declare no conflict of interest.

References

1. Stepien, K.; Janecki, D.; Adamczak, S. Investigating the influence of selected factors on results of V-block cylindricity measurements. *Measurement* **2011**, *44*, 767–777. [CrossRef]
2. Adamczak, S.; Janecki, D.; Stepien, K. Cylindricity measurement by the V-block method–Theoretical and practical problems. *Measurement* **2011**, *44*, 164–173. [CrossRef]
3. Liu, W.; Fu, J.; Wang, B.; Liu, S. Five-point cylindricity error separation technique. *Measurement* **2019**, *145*, 311–322. [CrossRef]
4. Kong, Q.; Yu, Z.; Mao, X.; Zhou, J.; Lia, F.; Li, H.; Tang, J. Rotation error modeling and compensation of spindle based on Clarke transformation in straightness error measurement of regular hexagon section shaft. *Measurement* **2020**, *166*, 108233. [CrossRef]
5. Liu, W.; Zhou, X.; Li, H.; Liu, S.; Fu, J. An algorithm for evaluating cylindricity according to the minimum condition. *Measurement* **2020**, *2020*, 107698. [CrossRef]
6. Sun, C.; Li, C.; Liu, Y.; Wang, H.; Wang, B.; Wang, X.; Tan, J. A cylindricity evaluation approach with multi-systematic error for large rotating components. *Metrologia* **2020**, *57*, 025020. [CrossRef]
7. Uekita, M.; Takaya, Y. On-machine dimensional measurement of large parts by compensating for volumetric errors of machine tools. *Precis. Eng.* **2016**, *43*, 200–210. [CrossRef]
8. Zakharov, O.V.; Kochetkov, A.V. Minimization of the systematic error in centerless measurement of the roundness of parts. *Meas. Tech.* **2016**, *58*, 1317–1321. [CrossRef]
9. Liu, W.; Zeng, H.; Liu, S.; Wang, H.; Chen, W. Four-point error separation technique for cylindricity. *Meas. Sci. Technol.* **2018**, *29*, 075007. [CrossRef]
10. Nozdrzykowski, K.; Chybowski, L. A force-sensor-based method to eliminate deformation of large crankshafts during measurements of their geometric condition. *Sensors* **2019**, *19*, 3507. [CrossRef]
11. Chybowski, L.; Nozdrzykowski, K.; Grządziel, Z.; Jakubowski, A.; Przetakiewicz, W. Method to Increase the Accuracy of Large Crankshaft Geometry Measurements Using Counterweights to Minimize Elastic Deformations. *Appl. Sci.* **2020**, *10*, 4722. [CrossRef]
12. Chybowski, L.; Nozdrzykowski, K.; Grządziel, Z.; Dorobczyński, L. Evaluation of Model-Based Control of Reaction Forces at the Supports of Large-Size Crankshafts. *Sensors* **2020**, *20*, 2654. [CrossRef]
13. Yu, H.; Xu, M.; Zhao, J. In-situ roundness measurement and correction for pin journals in oscillating grinding machines. *Mech. Syst. Sig. Process.* **2015**, *50*, 548–562. [CrossRef]
14. Song, M.-H.; Nam, T.-K.; Lee, J. Self-Excited Torsional Vibration in the Flexible Coupling of a Marine Propulsion Shafting System Employing Cardan Shafts. *J. Mar. Sci. Eng.* **2020**, *8*, 348. [CrossRef]
15. Sun, J.; Wang, J.; Gui, C. Whole crankshaft beam-element finite-element method for calculating crankshaft deformation and bearing load of an engine. *Proc. Instit. Mech. Eng. Part J J. Eng. Tribol.* **2010**, *224*, 299–303. [CrossRef]
16. Nozdrzykowski, K.; Chybowski, L.; Dorobczyński, L. Model-based estimation of the reaction forces in an elastic system supporting large-size crankshafts during measurements of their geometric quantities. *Measurement* **2020**, *155*, 107543. [CrossRef]
17. Shen, N.; Li, J.; Ye, J.; Qian, X.; Huang, H. Precise alignment method of the large-scale crankshaft during non-circular grinding. *Int. J. Adv. Manuf. Technol.* **2015**, *80*, 921–930. [CrossRef]
18. Liu, W.; Zhou, X.; Hu, Y.; Hu, P. A V-block three-probe error separation technique for portable measurement of cylindricity. *Precis. Eng.* **2019**, *59*, 37–46. [CrossRef]
19. Okuyama, E.; Goho, K.; Mitsui, K. New analytical method for V-block three-point method. *Precis. Eng.* **2003**, *27*, 234–244. [CrossRef]
20. Fonte, M.; Reis, L.; Infante, V.; Freitas, M. Failure analysis of cylinder head studs of a four stroke marine diesel engine. *Eng. Fail. Anal.* **2019**, *101*, 298–308. [CrossRef]
21. Sirata, G.G. Fatigue Failure Analysis of Crankshafts-A Review. *IJISET* **2020**, *7*.

22. Król, K.; Wikło, M.; Olejarczyk, K.; Kołodziejczyk, K.; Siemiątkowski, Z.; Żurowski, W.; Rucki, M. Residual stresses assessment in the marine diesel engine crankshaft 12V38 type. *J. KONES* **2017**, *24*, 117–123.
23. Fonseca, L.G.; de Faria, A.R. A deep rolling finite element analysis procedure for automotive crankshafts. *J. Strain Anal. Eng. Des.* **2018**, *53*, 178–188. [CrossRef]
24. Kurbet, S.; Kuppast, V.; Talikoti, B. Material testing and evaluation of crankshafts for structural analysis. *Mater. Today Proc.* **2020**, (in press). [CrossRef]
25. Midas, I.T. *User's Manual of midas NFX*; MIDAS IT: Seongnam, Korea, 2011.
26. Miądlicki, K.; Jasiewicz, M.; Gołaszewski, M.; Królikowski, M.; Powałka, B. Remanufacturing System with Chatter Suppression for CNC Turning. *Sensors* **2020**, *20*, 5070. [CrossRef]
27. Shen, N.Y.; Li, J.; Wang, X.D.; Ye, J.; Yu, Z.X. Analysis and detection of elastic deformation of the large-scale crankshaft in non-circular grinding. In *Proceedings of the Mechatronics and Applied Mechanics III*; Trans Tech Publications Ltd.: Baech, Switzerland, 2014; Volume 532, pp. 285–290.
28. Nozdrzykowski, K. Applying Harmonic Analysis in the Measurements of Geometrical Deviations of Crankshafts–Roundness Shapes Analysis. *Multi. Aspects Prod. Eng.* **2018**, *1*, 185–189. [CrossRef]

Article

Analysis of the Dynamic Height Distribution at the Estuary of the Odra River Based on Gravimetric Measurements Acquired with the Use of a Light Survey Boat—A Case Study

Krzysztof Pyrchla [1], Arkadiusz Tomczak [2,*], Grzegorz Zaniewicz [2], Jerzy Pyrchla [3,*] and Paulina Kowalska [3]

1 Faculty of Electronics, Telecommunications and Informatics, Gdańsk University of Technology, 80-233 Gdańsk, Poland; krzpyrch@student.pg.edu.pl
2 Faculty of Navigation, Maritime University of Szczecin, 70-500 Szczecin, Poland; g.zaniewicz@am.szczecin.pl
3 Faculty of Civil and Environmental Engineering, Gdańsk University of Technology, 80-233 Gdańsk, Poland; paulina.kowalska@pg.edu.pl
* Correspondence: a.tomczak@am.szczecin.pl (A.T.); jerpyrch@pg.edu.pl (J.P.)

Received: 19 September 2020; Accepted: 21 October 2020; Published: 23 October 2020

Abstract: This article presents possible applications of a dynamic gravity meter (MGS-6, Micro-g LaCoste) for determining the dynamic height along the Odra River, in northwest Poland. The gravity measurement campaign described in this article was conducted on a small, hybrid-powered survey vessel (overall length: 9.5 m). We discuss a method for processing the results of gravimetric measurements performed on a mobile platform affected by strong external disturbances. Because measurement noise in most cases consists of signals caused by non-ideal observation conditions, careful attempts were made to analyze and eliminate the noise. Two different data processing strategies were implemented, one for a 20 Hz gravity data stream and another for a 1 Hz data stream. A comparison of the achieved results is presented. A height reference level, consistent for the entire estuary, is critical for the construction of a safe waterway system, including 3D navigation with the dynamic estimation of under-keel clearance on the Odra and other Polish rivers. The campaign was conducted in an area where the accuracy of measurements (levelling and gravimetric) is of key importance for shipping safety. The shores in the presented area of interest are swampy, so watercraft-based measurements are preferred. The method described in the article can be successfully applied to measurements in all near-zero-depth areas.

Keywords: gravity anomalies and earth structure; gravimetric river survey; geophysical river survey; Fourier analysis; numerical modelling; time-series analysis; Europe

1. Introduction

Poland's region of river estuaries is flat; therefore, the river gradients are also small. The vertical reference network in the region of the main Polish rivers, the Odra and the Wisła (Vistula), is necessary for hydrographic, hydrological and hydrodynamic surveys and would have a great impact on developing 3D vessel navigation systems with the dynamic determination of under-keel clearance. The network can also serve for developing a regional digital terrain model. Today, gravity missions in space are an important tool for obtaining global data by providing coverage all over the world [1]. In large estuaries, created vertical reference networks are based on satellite data. With regard to the estuaries of rivers flowing into the Baltic Sea, the resolution of satellite data is insufficient [2] to provide the basis for an analysis of vertical reference networks. Additional data are usually required in such areas to ensure regional geopotential models of acceptable spatial resolution [3]. These data can be obtained from local accurate measurements of gravity [4,5].

An interesting challenge is the estimation of the height in the inland water bodies, such as rivers and lakes. If the whole area of interest is accessible from land, the solution may be a land gravimetric campaign [6]. However, the diversity of environmental conditions in the area of natural inland water bodies can lead to a situation where significant parts of the area of interest are impossible to reach for land measurements. A practical solution is to employ a research vessel, but such a vessel must have a small draft and high maneuverability for operation in a swampy river estuary, which are conditions that reduce the maximum size of such a vessel. During our measurements, gravimetric equipment was installed on a small survey vessel, 9 m in length.

A small vessel engaged in gravimetric measurements has to be prepared for the measurement itself and for retrieving the signal from noise generated by the environment [5,7]. One possible solution is the application of a next-generation strapdown marine gravimeter, which is able to achieve accuracy better than 1 mGal and can perform measurements on curved profiles [8,9]. Another solution could be the GNSS (Global Navigation Satellite Systems) estimation of the sea surface height and the use of the proper corrections to calculate the physical height [3]. In this work, we focused on another solution, namely, the application of a standard marine gravimetry system for measurements within a river estuary. This is not a common operation, because marine gravity sensors are generally sensitive to rapid platform motions [10]. This requires the development of measurement technology in which the maximum possible number of sources interfering with the measurements are eliminated. This objective is adopted in all gravimetric dynamic measurements, where a ship or aircraft provides the platform [11,12]. We can assume that the level of precision in such cases is determined by the manufacturer of the measuring device, and this level should be adopted as a reference value. The manufacturer Micro-g LaCoste indicates a measurement accuracy of 1 mGal [13], but this refers to ships more than 40 m in length. However, after all the necessary dynamic corrections are made, the accuracy of the measured gravimetric signal can be improved even beyond the general limits indicated by the manufacturer [13].

Filtration is one of the key elements of dynamic gravimetric data processing. In this analysis, it was decided to use a low-pass fast Fourier transform (FFT) filter, as described in [14]. Parameters such as the cut-off frequency and transition band were chosen following the spectral analysis of the collected data. The spectral analysis included horizontal accelerations of the vessel to determine in which parts of the band their energy was concentrated. During data analysis, the ship's accelerations were found to be the main source of the noise in the recorded data. Therefore, we focused on selecting frequency filters that would reduce the impact of accelerations as much as possible.

The standard procedure for verifying the consistency of gravimetric measurement data is to determine their internal accuracy [15,16]. Such accuracy is verified by performing part or all measurements on a survey line at least twice and statistically analyzing the obtained differences.

However, such procedures do not provide information on external data accuracy. To determine the value of the absolute error, it is not sufficient to analyze internal data consistency only, but a comparison of the results with reliable data from an independent campaign must also be made. Based on this assumption, a land-based measurement campaign was conducted along the riverbank. A CG-5 gravimeter from Micro-g LaCoste was used for the measurements. A representative group of data gathered from the campaign enabled reliable comparisons. The adopted measurement methodology allows for the determination of the internal and external accuracies of the measurement data so that the filtration effectiveness can be assessed.

This article presents experimental gravimetric research carried out on a river, in northwest Poland. The work consisted of taking high-resolution gravimetric images of increased density and achieving better accuracy for the measurements. We determined the character of the gravitational acceleration field in inland waters of the Odra River mouth, where previously, such an accurate and professional gravimeter had not been used. The results of the measurements were used for modelling variations in the dynamic surface of the river. The system was also used to improve the technology of gravimetric data analysis referring to areas of so-called zero depth.

The rest of the paper proceeds as follows: Section 2 provides a description of the experimental setup and data processing methods, Section 3 presents the results achieved using the described methods, and the Section 4 contains some brief conclusions based on our findings.

2. Materials and Methods

The dynamic marine gravity meter MGS-6 (Micro-g LaCoste, Lafayette, CO 80026, USA) from LaCoste & Romberg—Scintrex Inc., a sixth-generation Marine Gravity System, was used for recording changes in gravity during the campaign. Its basic advantage is the capability of performing gravimetric measurements on a mobile platform. The MGS-6 guarantees very accurate measurements of gravitational acceleration (gravity force). The system, based on a TAGS-6 platform, has a frame supporting a gimbal and a sensor. Vibrations are damped by a gimbal, suspension strings and air-filled shock absorbers. The gimbal holds a gravity meter sensor and keeps it horizontal when the system is moving. The gravity meter sensor contains a gravity-detecting element, an oven and an electronic unit of the platform. The system is operated from a laptop computer, which also records gravity data.

The position of the research vessel was measured using two GNSS antennas and receivers, a Trimble R8 and a Leica GS 15. The SMC IMU-10 was used as an inertial sensor. The data flow in the whole measurement setup is shown in Figure 1. The physical location of each sensor is indicated in Figure 2a.

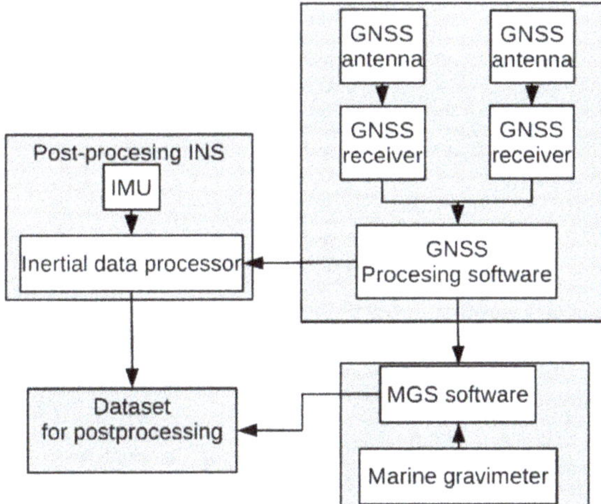

Figure 1. The scheme showing the flow of data during the measurement campaign. Abbreviations used: Global Navigation Satellite Systems (GNSS), Inertial Navigation System (INS), Inertial Measurement Unit (IMU), Marine Gravity System (MGS).

The manufacturer of the MDS-6 marine gravity measurement system recommends installation on vessels more than 40 m in length. We therefore decided to determine what results would be possible to achieve after the installation of this system on a small ship of 9 m in length. The vessel chosen for the measurements was a Hydrograf XXI. Despite its small size, this research boat has one advantage—it has hybrid propulsion (gasoline–electrical). As a result of battery pack installation, the center of gravity (COG) is lowered and the overall mass is increased in comparison to boats without this modification. During measurements, gasoline propulsion was used, but the additional mass allowed rapid hull oscillation to be avoided. Furthermore, it was assumed that the mobilization of the complete navigation sensor setup and filtration of registered signals allowed reliable and interesting results to be achieved.

For the gravimetric sensor installation, thick chipboard plates were used. The installation project used two layers of plates. The first one was fitted to the deck of the ship, so it would not slide along it. The second layer was the base to which the frame of the MGS-6 was screwed. The thickness of the plates, which was 38 mm, allowed us to mill holes for mounting screw heads. As a result, both plates adhered to each other over a large area. The other benefit of this setup is that the sensor frame could be placed in the best location before the top plate was screwed with wood screws to the bottom one. This configuration is rigid and allowed us to precisely set the points of offset measurements using geodetic techniques. The whole structure provided a solid mount for the gravimeter frame on the ship despite the lack of pre-designed mounting places on the deck. Figure 3 shows the relative location of the gravimeter frame on the mounting plates.

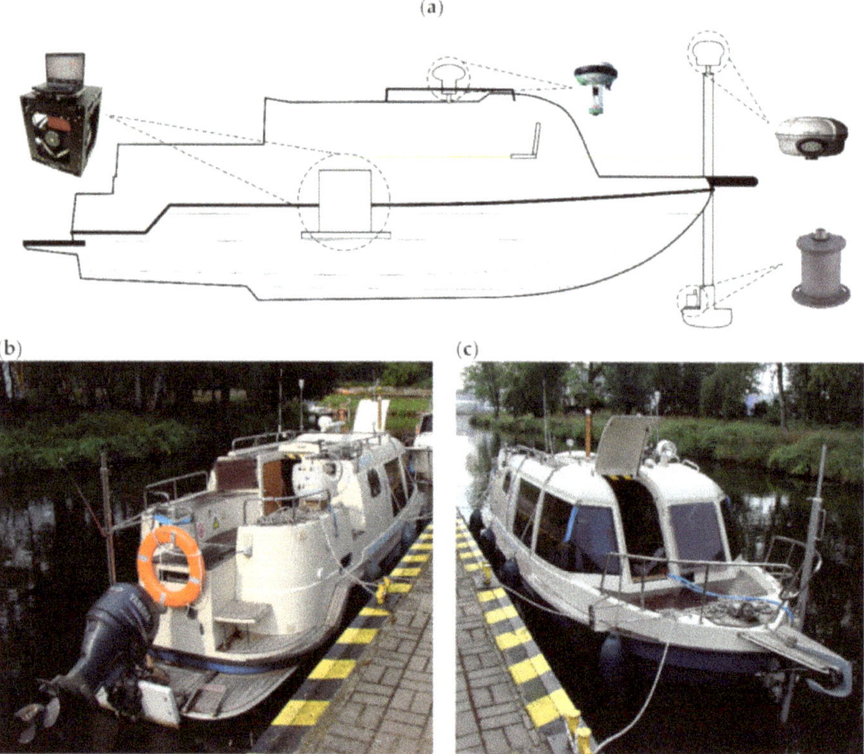

Figure 2. (**a**) Locations of the sensors on the research vessel, from left to right: MGS-6, GS 15, R8 and IMU-10. (**b**) The research vessel during preparation for the campaign: view from stern; (**c**) view from bow.

Dynamic gravimetric measurements are closely related to accurate inertial measurements. Sea measurements are known to have high accuracy and resolution for the acquired data [17] compared with other types of dynamic gravimetric measurements (satellite-, aviation- [18] or altimetry-based methods). The quality of the data obtained is inherently dependent on the stability of the measurement platform, because such measurements involve recording kinematic parameters of the platform transporting the gravity meter. Theoretically, all accelerations of the platform will be recorded by the measurement unit of the gravity meter. In practice, each dynamic gravity meter features a finite passband width, so it can record accelerations in a certain range of frequencies.

The MGS-6 gravimetric system can record gravity data with frequencies of 1 Hz or 20 Hz. For regular marine gravity data collection on large-scale areas of water, 1 Hz sampling is sufficient, so by default, the device software is configured to measure with this frequency. We decided to perform two campaigns on the same area using two different configurations. In the first campaign, the gravimeter was set to measure at a 1 Hz sampling rate, and during the second, it was set to 20 Hz.

Despite the slow speeds used by the survey ships and low variations in heading and speed during measurements (compared to those for airborne platforms), the accelerations onboard a vessel navigating along a survey line have an amplitude a few orders of magnitude greater than the desired anomalies in the gravitational field. A typical solution to that problem is an assumption that these disturbances occupy a band of signals with periods significantly different from those of the expected signals coming from variations in the gravitational field. For instance, the period of vibrations caused by sea waves ranges from 4 to 15 s [19]. This allows us to assume that after using a frequency filter of any type used in marine gravimetry [14,20,21], the impact of the vibrations will be eliminated from the signal.

Figure 3. MGS-6 mounted on the survey vessel. On the left is shown how the frame was fitted to the deck using thick chipboard plates.

Taking for granted such an approach to the filtration of gravimetric data is currently considered to be unjustified. While the accuracy of sea gravimetric systems continues to increase, there is a demand for increasingly higher-quality data with a low margin of uncertainty [22]. In modern dynamic gravimetry, it is crucial to accurately determine the kinematic parameters of the mobile platform. Unlike in airborne gravimetry, these parameters do not have to be processed immediately to obtain corrections for acceleration. To determine the corrections precisely, we need a vessel with an accurate positioning system and knowledge of the gravity meter pulse response. An indirect approach can be taken in which a spectral analysis of the collected kinematic data is performed to determine the band comprising the disturbances from accelerations, and then identify the optimal method of filtration [15].

The authors attempted to perform dynamic gravimetric measurements on the Odra River near its mouth. The depths necessitated a smaller survey vessel, and the measurements had to be conducted on substantially shorter lines. The area of measurements included Szczecin's urban surroundings, introducing additional difficulty due to increased vessel traffic that was not stopped at the time of measurements.

For these reasons, the data analysis started with a detailed review of the possible disturbances of the measurement results. First, we analyzed the influence of a large-scale curvature of the measurement trajectory. The trajectory curvature was calculated from survey vessel positions. Outliers were removed from position data obtained from a differential global positioning system (DGPS), and the gaps were patched using an autoregression model. Then, the data were decimated to a spatial resolution of 35 m. The distance between the measurement points to which the data were decimated was selected by examining the numerical convergence of the curvature determination method. Based on Equation (1), the data were then used to calculate the trajectory curvature "k".

$$k = \frac{\left|y''x' - y'x''\right|}{\left(x'^2 + y'^2\right)^{3/2}} \tag{1}$$

It is assumed in the formula that the curve is described by a parametric system of equations of the co-ordinates $x(t)$ and $y(t)$. In Equation (1), x' and y'' denote the first derivatives of the parameter, and x'' and y'' denote the second derivatives.

With the value of the curvature thus calculated, its impact on the gravimetric measurement results was examined. The value of the centrifugal acceleration on the trajectory was determined directly from the trajectory curvature value. It was assumed that the long-term impact of such acceleration would be similar to that of an unlevelled (tilted) gravity meter. Based on the formula for small roll corrections [13], the value of the correction for the trajectory curvature was calculated. The results are shown in Figures 4 and 5.

This correction exhibited very small values and did not exceed 1 mGal. This was caused by the small curvature of the vessel's trajectory on the survey line and slow linear speed of approximately 2.5 m/s. Despite that, this correction was added at the final stage of data analysis.

The Eotvos correction is extremely important for dynamic measurements. The Eotvos effect may yield very high values—tens of mGal in marine gravimetry and hundreds of mGal in airborne gravimetry. The value of the Eotvos correction g_e can be calculated from Equation (2):

$$g_{e}= 2\,\omega\, V\, \cos(Lat)\sin(A) + \frac{1}{r}\,V^2 \tag{2}$$

In Equation (2), V denotes the velocity of the gravity meter movement relative to the Earth, A is the direction of the velocity vector, Lat is the geographical latitude at the moment of measurement, ω is the rotation angular velocity of the Earth and r denotes the Earth's radius.

Clearly, vessel heading and speed stability are vital. When these requirements are maintained, the correction will not vary much and can be calculated from Equation (2). To ensure high accuracy for the speed and heading measurements, the kinematic parameters of the survey vessel were measured using three devices, namely, two GPS receivers and one inertial measurement unit. The acquired data were cleaned by removing outlier data and merged in postprocessing.

Another important source of disturbance, namely, the horizontal and vertical acceleration of the survey vessel, results from the local curvature of the trajectory and the vessel's yawing and speed variations. An attempt was made to determine in which regions of the band the concentrated energy of these oscillations was present. This was done by examining data from the dilatometers installed in the measurement sensor of the MGS-6. Example values of the acceleration recorded by these systems are shown in Figure 6.

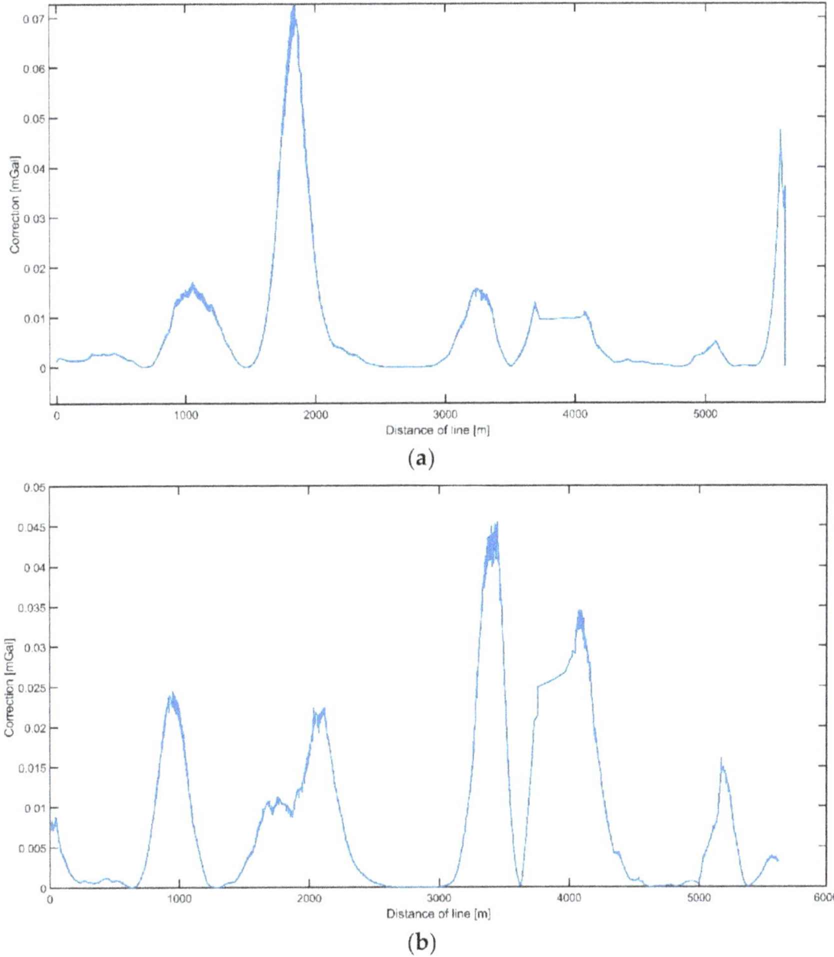

Figure 4. The distribution of the track curvature correction on the measurements of the first line (**a**) and second line (**b**).

The spectral power density was calculated for signals from both dilatometers. The results are shown in Figure 6.

An analysis of the data gathered by the dilatometers shows that the energy of the vibrations on the longitudinal axis was concentrated around the 10 s band. Therefore, it was concentrated in a band distant from the one where gravimetric signals are found. These vibrations are also less important because the values of the accelerations on this axis and the calibration constant provided by the manufacturer allowed us to calculate the vertical cross-coupling (VCC) correction, which was added before filtration and eliminated the cross-coupling effect on that axis.

In the case of vibrations on the cross axis, accelerations were found in a much broader band and reached 500 s. In arm-spring gravimeters, these accelerations, theoretically, are not coupled with the measurement result [19] but can cause the gimbal unit to swing, thus disturbing measurements if they have a sufficiently high amplitude and low frequency.

Therefore, having analyzed the spectral data, we decided to use a low-pass FFT filter with a cut-off period of 500 s. The width of the passband was 200 s. The low-pass FFT filter implementation, as described in the literature [14], was used for frequency filtration. This type of filter enables better separation of the signal, mainly because it is easier, compared to the case of a Butterworth filter, to control the cut-off frequency and the passband width. Consequently, the signal band with noise detected in a spectral analysis can be cut off more precisely.

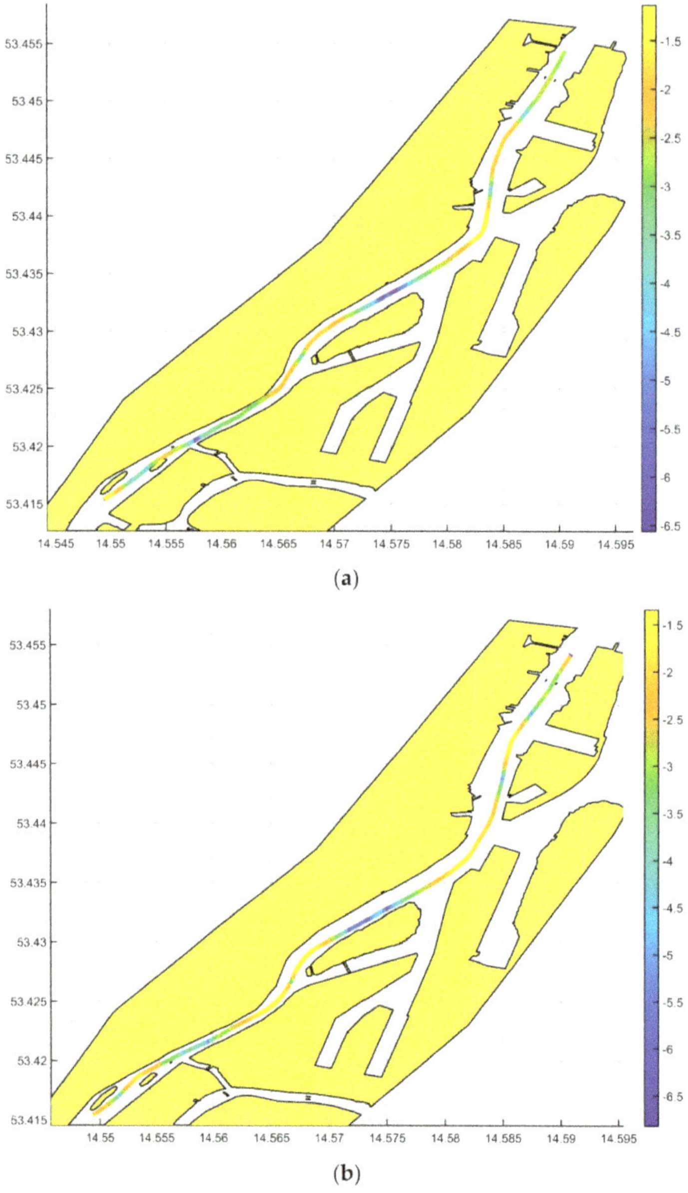

Figure 5. Impact of the curvature of the measurement profile on the recorded gravimetric signal. The spatial distribution of the correction is shown on the maps for the first line (**a**) and for the second line (**b**); a logarithmic scale is used.

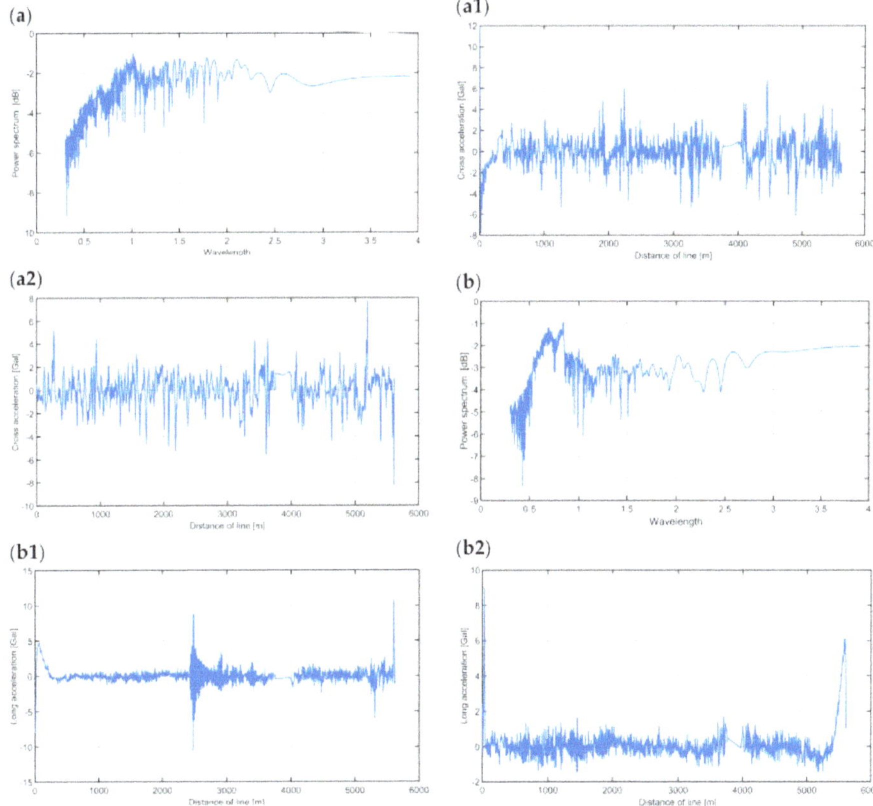

Figure 6. Signals from dilatometers installed in MGS-6. (**a**) Spectral power density for dilatometer signals for cross direction; (**a1**) values of cross accelerations of the first profile; (**a2**) values of cross accelerations of the second profile; (**b**) spectral power density for longitudinal direction; (**b1**) values of longitudinal accelerations for the first profile; (**b2**) values of longitudinal accelerations for the second profile.

In addition, the gravimetric signals recorded on the survey lines were smoothed using a second-order Savitzky–Golay filter with a window length of 500 s. It was required in order to eliminate residual oscillations in the data after frequency filtration and because there was no need for a higher spatial resolution, which was already limited by the selected frequency filtration parameters.

Using the conclusions from the first campaign, the second campaign was designed. The gravimeter was configured to record the gravity at a higher sampling rate equal to 20 Hz. This is the highest sampling frequency available in this model of gravimeter. The kinematic data have to be sampled faster with the increase in gravity data sampling, so the 10 Hz data acquisition frequency was used.

The higher sampling frequency allowed us to utilize a different type of data processing (Figure 7). The raw gravity data were corrected by the application of both the VCC correction and the Eotvos correction calculated from the navigation data. In the second stage, the resulting corrected gravity was carefully examined in order to remove all outliers from the data, after which the data were filled using the autoregressive model (ARMA). The resulting data were filtered using the same one-pass low-pass FFT that was used for processing the data from the first campaign. The cut-off wavelength was set to 600 s, and the stopband, to 400 s. No additional filtration was used.

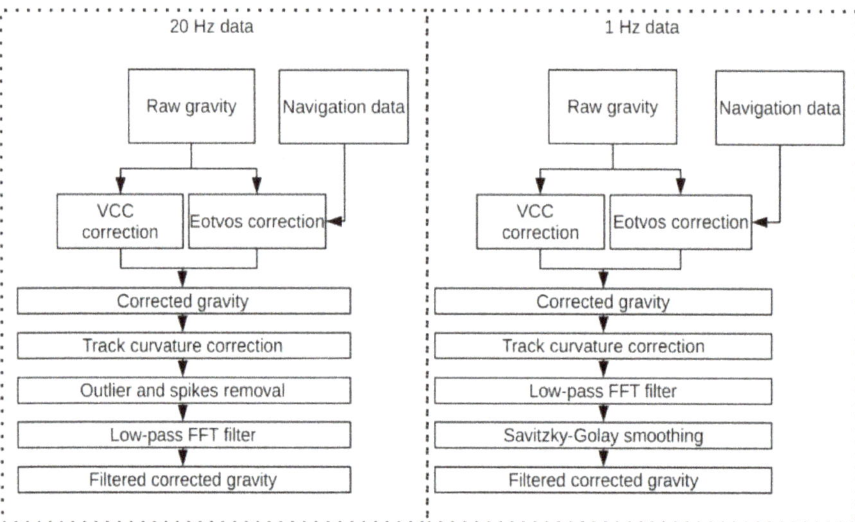

Figure 7. The block scheme showing the data processing routes for 20 Hz and 1 Hz gravimetric data. The FFT refers to Fast Fourier transform.

3. Results

In this work, the data collected in two campaigns performed on the same water area are presented. The results from both of them are presented together on similar types of plots, which allowed for the reliable evaluation of the results and showed the influence of the setup used on the data quality. First, the data from the 1 Hz setup campaign were analyzed.

In the analysis of the results, the internal and external accuracies were verified. The results are shown in Figure 8, including the combined routes in both directions and measurements at land-based stations along the river. The land-based measurements were reduced to the height of the marine gravimeter using free-air reduction [23]. Figure 9 shows the distribution of the differences between the two survey lines. To estimate the internal consistency of the data (internal accuracy), we examined the differences between the results obtained on both survey lines. The internal error was calculated as the mean square from the difference between the two survey lines.

The estimated internal error of the measurement data was 1.13 mGal. The varying differences between the measurement data on the two lines are plotted in Figure 9. The next step was to calculate the external accuracy by comparison with measurements conducted on the riverbank.

Figure 10 depicts the difference between the average measurements on the two lines and on land. An analysis of thew Odra riverbank measurement differences allowed us to estimate the external consistency of the data. The values of the differences between the points acquired by a land-based gravimeter and those from a marine gravimeter are shown in Figure 11. Because gravimetric measurements had not been conducted before on the Odra, no historical data existed for comparison. Therefore, the external accuracy was verified by comparing measurements made on the riverbank. Based on those results, the calculated external accuracy value was 0.61 mGal. The accuracy was calculated as the root-mean-square (RMS) values of the differences, similarly to the internal accuracy.

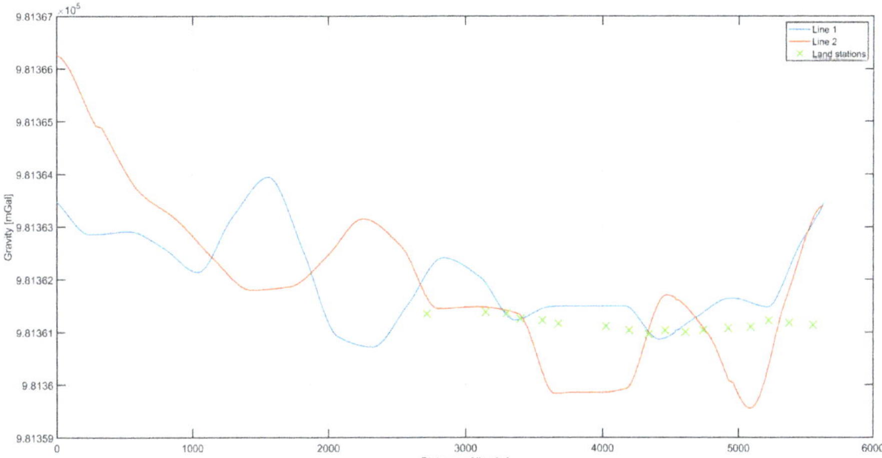

Figure 8. The superimposed values of gravitational acceleration on the two survey lines. In addition, land measurement points are marked, with the obtained values reduced to the height of the dynamic gravimeter.

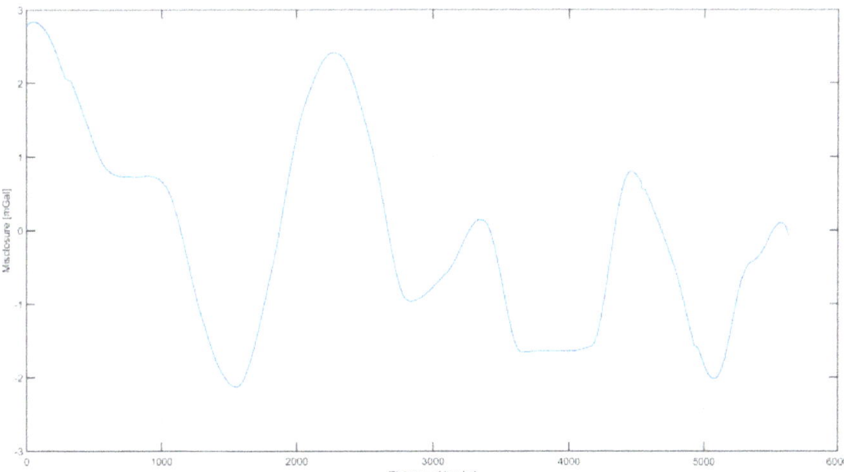

Figure 9. The distribution of differences in the values of gravitational acceleration between the survey lines.

Figure 12 presents data from the second campaign, with the usage of a higher (20 Hz) registration frequency. Once again, gravity was measured on the river during two passages, called lines 1 and 2. These lines overlapped each other, so the differences in values between them could be used as the measure of misclosure (internal accuracy). The green crosses in Figure 12 indicate the points where land gravity data were available. The height difference caused by the pier height was reduced using free-air reduction.

Figure 13a depicts the differences both between the lines and between the land data. The RMS misclosure between the lines was 0.0818 mGal.

The differences between the data collected on the river and the data collected on the land are plotted in Figure 13b. Based on those results, the calculated external accuracy value was 0.2 mGal. The accuracy was calculated as the RMS values of the differences.

After this analysis, the data from all the measurements were used to create the final plot (Figure 14). The data from the second campaign showed significantly fewer fluctuations, which was reflected in the lower internal and external error estimates.

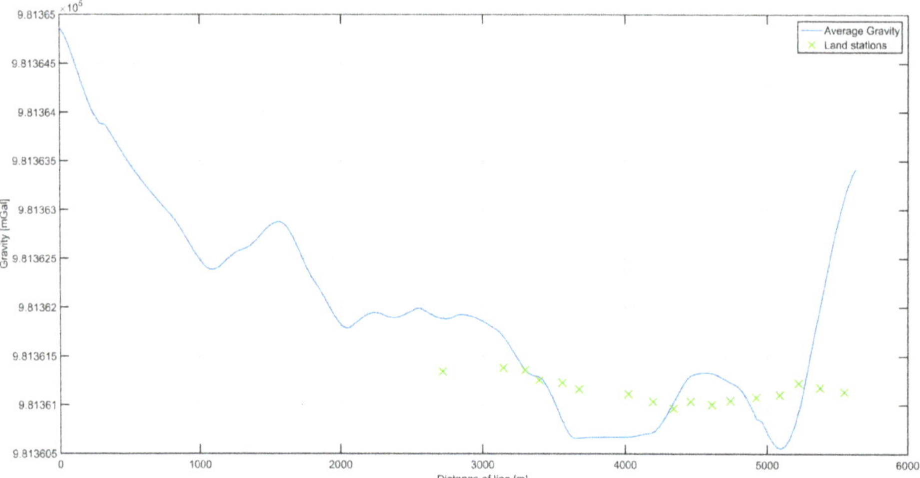

Figure 10. The mean measurements on the river, and data recorded at land-based measuring points.

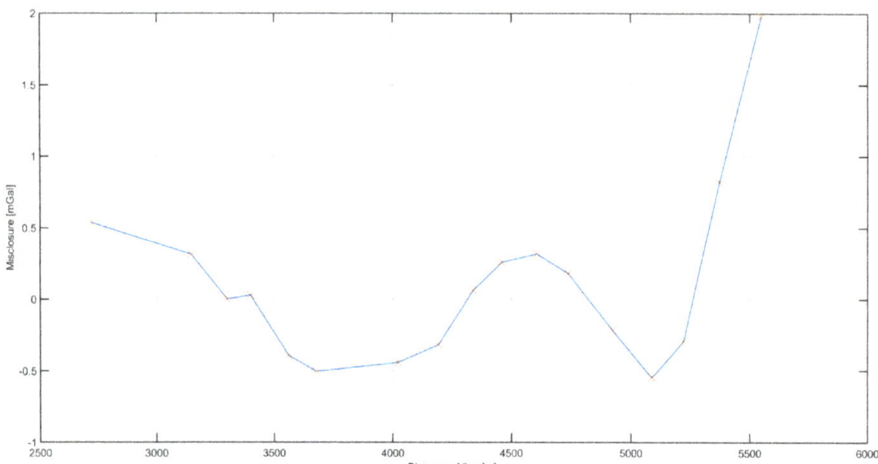

Figure 11. Distribution of differences between the gravity obtained on land and that measured on a survey line. The height differences have been reduced.

The data from the measurement line provided a basis for calculating the dynamic height on the section of the river covered by the measurements. The value of the dynamic height [23] within the area of measurements was calculated from DGPS and gravimetric results. The difference in dynamic height, ΔH_{dyn}, between Points A and B can be calculated from Equation (3).

$$\Delta H_{\mathrm{dyn}} = \int_{\mathrm{A}}^{\mathrm{B}} \mathrm{d}n + \int_{\mathrm{A}}^{\mathrm{B}} \frac{g - \gamma_{45}}{\gamma_{45}} \mathrm{d}n \tag{3}$$

In Equation (3), d_n denotes the height increments between intermediate points, forming a sequence between Points A and B. The second term of the formula is a dynamic correction, depending on the values of the acceleration of gravity g at intermediate points and on the value of the reference acceleration γ_{45}. In this case, the adopted reference acceleration was normal acceleration for the geodetic latitude of 45 degrees.

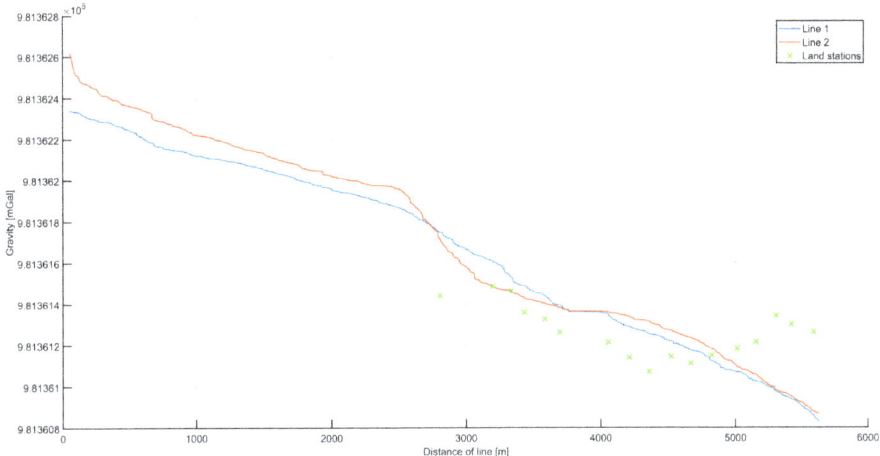

Figure 12. The superimposed values of gravitational acceleration on the two survey lines. In addition, land measurement points are marked, with the obtained values reduced to the height of the dynamic gravimeter. The data presented were processed as a result of the 20 Hz campaign.

The reference height was adopted as the height obtained during the binding of the gravimeter still value to the pier point.

Figure 15 presents the distribution of the dynamic height values in the surveyed area.

The data analysis led us to conclude that it is possible to obtain gravimetric data of the gravity distribution on a river with an accuracy comparable to that characterizing marine campaigns. Consequently, such data can be used for calculating the distribution of the dynamic height on rivers using the methodology adopted in this article.

The obtained variation in dynamic height on the river is not large, as predicted for this area. The surveyed river section is located close to the river mouth, so the topography of the surroundings varies only slightly.

The analysis of the data reveals that the changes in the operation frequency of the gravity registration system can lead to significant changes in data quality. For this reason, we have to sum up the results separately.

During the first campaign, it should be noted that the estimated internal accuracy of the gravimetric measurements was greater than the estimated external accuracy. As mentioned before, vessel traffic on the river caused some disturbances. This observation is confirmed by the graphs of the mean amplitude of the acceleration in Figure 6. The differences between the signals from both measurement lines recorded on the survey vessel reached the greatest value of about 2 mGal in places with strong horizontal acceleration. In the area of land-based measurements, no significant disturbances induced by passing other vessels occurred, which decreased the differences between the values of gravity recorded on land and those recorded on the vessel. In addition, it should be noted that the value of the averaged gravity on both profiles was used in the external accuracy analysis. This allowed us to eliminate the residual errors of the Eotvos correction (the movement direction on the two tracks was opposite, and the speeds, nearly identical). Taking into account the above facts, it should be concluded that the error estimation for external accuracy was more reliable than that for internal accuracy.

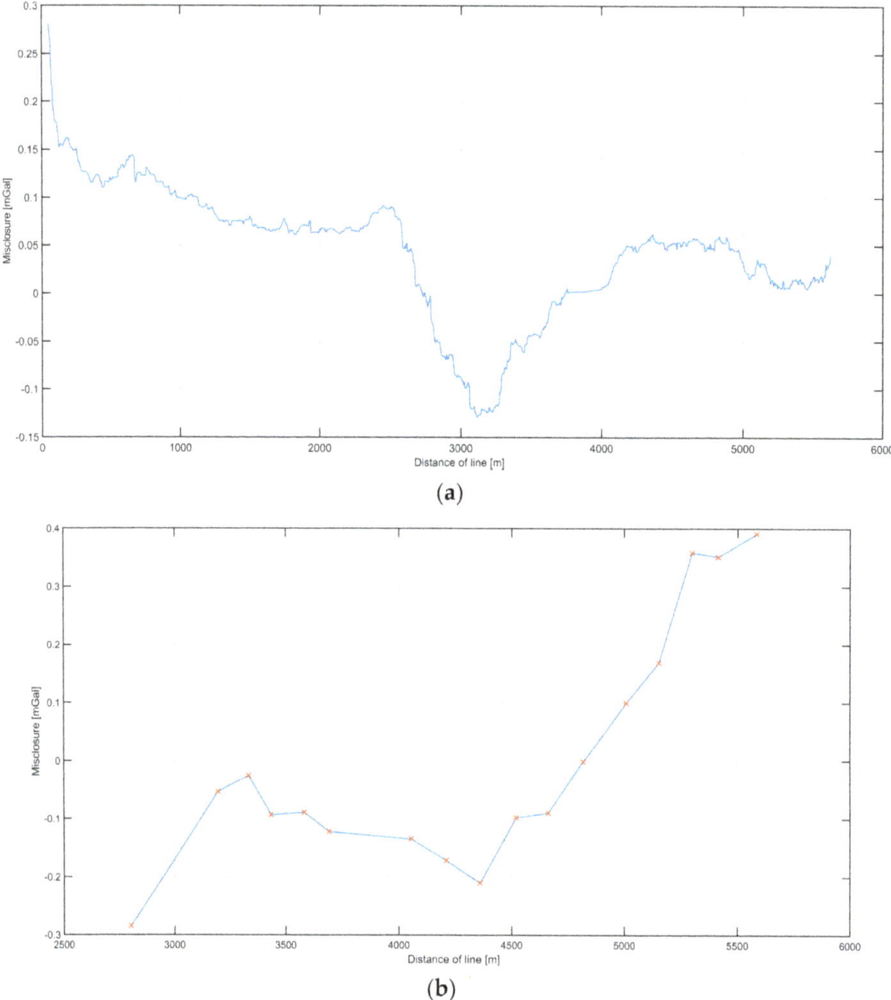

Figure 13. Misclosure analysis of repeated campaign: (**a**) The distribution of differences in the values of gravitational acceleration between the survey lines. (**b**) Distribution of differences between the gravity obtained on land and that measured on a survey line. The height differences have been reduced.

During the second campaign, the environmental conditions were very similar to those during the first campaign. The traffic on the river was also significant. Despite this, the data collected during the second campaign were characterized by much lower errors, both internal and external. The data processing in the second campaign compared to the first was much more resistant to the noise and disturbances caused by the environmental conditions due to the outlier data. The track on which the measurements were taken was only a few kilometers long. In such conditions, the periods of time in which no external disturbances occurred were relatively short. It was crucial to obtain as many valid data as possible in this short period of time, so the sampling frequency of the whole system and proper outlier data removal were the key features.

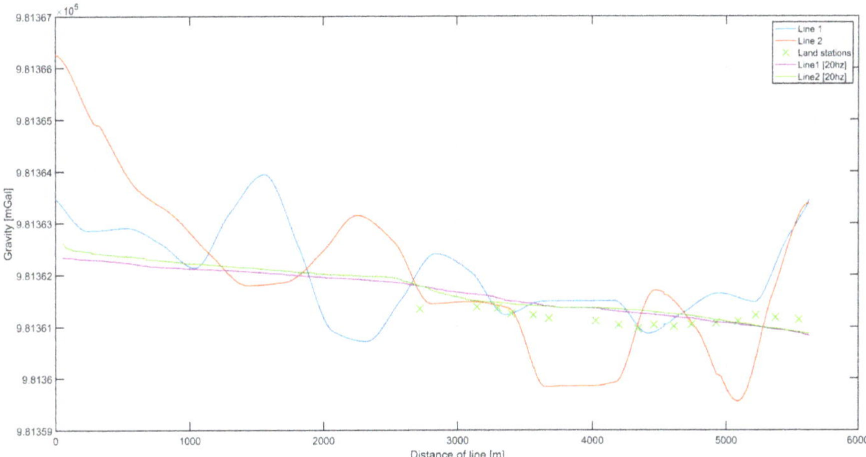

Figure 14. The comparison of results from all campaigns performed on the Odra river.

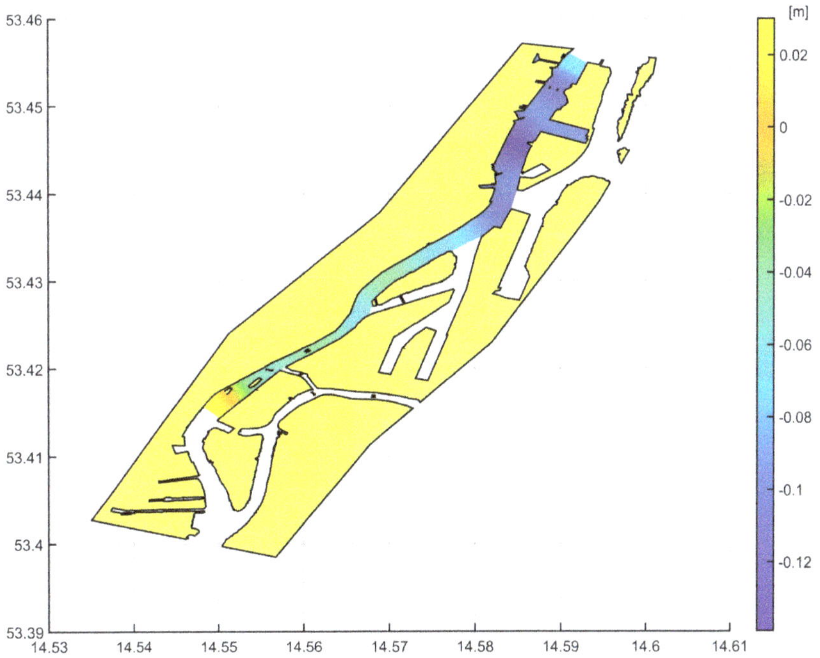

Figure 15. The distribution of dynamic height on the river.

4. Conclusions

This article examines the dynamic height determined from gravimetric data in an area where vertical accuracy is crucial. As in other European countries, the use of physical heights in river navigation has become more important. The determination of dynamic height requires accurate and reliable data on the distribution of the gravity field in the area concerned. Gravimetric surveys conducted on board ships are an effective method of acquiring detailed local data [24]. The accuracy of the recorded data depends largely on the external conditions prevailing during a survey [25].

The proposed measurement technique allows us to make gravimetric measurements on water bodies where access using land techniques is limited or impossible. This technique can be also used in near-zero-depth coastal zones, which is important for filling the gap between marine and land gravimetric measurements.

The results suggest that river estuary mapping can be performed using a marine gravimetric system with a gimbal sensor. The curvatures of European rivers insignificantly affect survey results. The quality of the data obtained does not differ considerably from that achieved during survey campaigns carried out on much bigger vessels at sea.

The study showed that several possible error sources have to be considered in the determination of dynamic height. Most of these errors seem to have a small impact on the results obtained, and almost all can be minimized by the careful planning of measurements, high-speed registration and careful processing. In the case of our campaign, a higher (20 Hz) frequency of the registration allowed us to achieve better results. This is because by increasing the sampling rate, we could collect more valid data during the period of time when external disturbances were smaller and the recording was less distorted.

We analyzed the possibility of calculating the dynamic height of a river, which is the outcome of this study. Carefully applied, the method of dynamic height determination seems to be a promising technique for examining some current circulation phenomena, at least on large rivers. The method seems to be more useful for determining surface current patterns and speed than for sub-surface water flow systems. The use of this method on the Odra River produced logical and reasonable results, internally consistent and in line with data collected by land-based gravimeters.

Small research vessels can be stable platforms for gravimetric measurements if their centers of gravity (COGs) are sufficiently low. A good example of such a vessel is the Hydrograf XXI, which is a boat with hybrid propulsion (electrical and gasoline engines). The batteries lower its COG and give it additional mass, which is important for dampening hull oscillations.

The next logical step would be to extend the scope of the dynamic height surveys with the use of this method to cover the entire navigable part of the Odra estuary. This would allow for better predictions of depth distribution on the river at a specific water level, which are essential for the optimal use of the river water resources.

Author Contributions: Conceptualization: J.P. and A.T.; data curation: K.P., P.K., A.T. and G.Z.; investigation: K.P., A.T. and G.Z.; methodology: J.P. and K.P.; supervision: J.P.; visualization: K.P.; writing—original draft: K.P.; writing—review and editing: A.T. and G.Z. All authors have read and agreed to the published version of the manuscript.

Funding: The study presented was co-financed by the European Union from the European Regional Development Fund under the 2014–2020 Operational Programme Smart Growth. The project titled "Development of technology for acquisition and exploration of gravimetric data of foreshore and seashore of Polish maritime areas" was implemented as part of the National Centre for Research and Development competition: 1/4.1.4/2018 "Application projects".

Acknowledgments: The authors would like to thank the Directorate of the Chair of Geomatics at the Faculty of Navigation, Maritime University of Szczecin, for the opportunity to conduct research from the survey Hydrograf XXI, as well as the vessel crew for their technical support during the measurements.

Conflicts of Interest: The authors declare no conflict of interest.

References

1. Sánchez, L.; Sideris, M.G. Vertical datum unification for the International Height Reference System (IHRS). *Geophys. J. Int.* **2017**, *209*, 570–586. [CrossRef]
2. Breili, K.; Rolstad, C. Ground-based gravimetry for measuring small spatial-scale mass changes on glaciers. *Ann. Glaciol.* **2009**, *50*, 141–147. [CrossRef]
3. Lavrov, D.; Even-Tzur, G.; Reinking, J. Extraction of geoid heights from shipborne GNSS measurements along the Weser River in northern Germany. *J. Geod. Sci.* **2016**, *5*, 148–155. [CrossRef]

4. Pyrchla, K.; Pyrchla, J. The Use of Gravimetric Measurements to Determine the Orthometric Height of the Benchmark in the Port of Gdynia. In Proceedings of the 2018 Baltic Geodetic Congress (BGC-Geomatics), Olsztyn, Poland, 21–23 June 2018; pp. 349–352. [CrossRef]
5. Ayers, J.C. A Dynamic Height Method for the Determination of Currents in Deep Lakes, Limnol. *Oceanography* **1956**, *1*, 150–161. [CrossRef]
6. Koler, B.; Jakljič, S.; Breznikar, A. Analysis of different height systems along the Sava River. *Geod. Cartogr.* **2009**, *35*, 92–98. [CrossRef]
7. Panet, I.; Kuroishi, Y.; Holschneider, M. Wavelet modelling of the gravity field by domain decomposition methods: An example over Japan, Geophys. *J. Int.* **2011**, *184*, 203–219. [CrossRef]
8. Wang, W.; Gao, J.; Li, D.; Zhang, T.; Luo, X.; Wang, J. Measurements and accuracy evaluation of a strapdown marine gravimeter based on inertial navigation. *Sensors* **2018**, *18*, 3902. [CrossRef]
9. Jensen, T.E.; Forsberg, R. Helicopter test of a strapdown airborne gravimetry system. *Sensors* **2018**, *18*, 3121. [CrossRef]
10. Tomoda, Y. Gravity at sea—A memoir of a marine geophysicist—. *Proc. Jpn. Acad. Ser. B. Phys. Biol. Sci.* **2010**, *86*, 769–787. [CrossRef]
11. Forsbergbi, R.; Olesen, A.V.; Alshamsi, A.; Gidskehaug, A.; Ses, S.; Kadir, M.; Peter, B. Airborne Gravimetry Survey for the Marine Area of the United Arab Emirates. *Mar. Geod.* **2012**, *35*, 221–232. [CrossRef]
12. Guo, J.; Liu, X.; Chen, Y.; Wang, J.; Li, C. Local normal height connection across sea with ship-borne gravimetry and GNSS techniques. *Mar. Geophys. Res.* **2014**, *35*, 141–148. [CrossRef]
13. Przyborski, M.; Pyrchla, J.; Pyrchla, K.; Szulwic, J. MicroGal Gravity Measurements with MGS-6 Micro-g LaCoste Gravimeter. *Sensors* **2019**, *19*, 2592. [CrossRef] [PubMed]
14. Childers, V.A.; Bell, R.E.; Brozena, J.M. Airborne gravimetry: An investigation of filtering. *Geophysics* **1999**, *64*, 61–69. [CrossRef]
15. Lu, B.; Barthelmes, F.; Li, M.; Förste, C.; Ince, E.S.; Petrovic, S.; Flechtner, F.; Schwabe, J.; Luo, Z.; Zhong, B. Shipborne gravimetry in the Baltic Sea: Data processing strategies, crucial findings and preliminary geoid determination tests. *J. Geod.* **2019**, *93*, 1059–1071. [CrossRef]
16. Yu, R.; Cai, S.; Wu, M.; Cao, J.; Zhang, K. An SINS/GNSS ground vehicle gravimetry test based on SGA-WZ02. *Sensors* **2015**, *15*, 23477–23495. [CrossRef] [PubMed]
17. Varbla, S.; Ellmann, A.; Märdla, S.; Gruno, A. Assessment of marine geoid models by ship-borne GNSS profiles. *Geod. Cartogr.* **2017**, *43*, 41–49. [CrossRef]
18. Lu, B.; Barthelmes, F.; Petrovic, S.; Förste, C.; Flechtner, F.; Luo, Z.; He, K.; Li, M. Airborne Gravimetry of GEOHALO Mission: Data Processing and Gravity Field Modeling. *J. Geophys. Res. Solid Earth.* **2017**, *122*, 10586–10604. [CrossRef]
19. Dehlinger, P. *Marine Gravity*; Elsevier: Amsterdam, The Netherlands, 1978.
20. Krasnov, A.A.; Sokolov, A.V.; Usov, S.V. Modern equipment and methods for gravity investigation in hard-to-reach regions. *Gyroscopy Navig.* **2011**, *2*, 178–183. [CrossRef]
21. Krasnov, A.A.; Sokolov, A.V.; Elinson, L.S. Operational experience with the Chekan-AM gravimeters. *Gyroscopy Navig.* **2014**, *5*, 181–185. [CrossRef]
22. Hipkin, R. Modelling the geoid and sea-surface topography in coastal areas. *Phys. Chem. Earth, Part A Solid Earth Geod.* **2000**, *25*, 9–16. [CrossRef]
23. Hofmann-Wellenhof, B.; Moritz, H. *Physical Geodesy*; Springer Science & Business Media: Berlin, Germany, 2006.
24. Wang, J.; Guo, J.; Liu, X.; Kong, Q.; Shen, Y.; Sun, Y. Orthometric height connection across sea with ship-borne gravimetry and GNSS measurement along the ship route. *Acta Geod. Geophys.* **2017**, *52*, 357–373. [CrossRef]
25. Rodríguez, G.; Sevilla, M.J. Geoid model in the Western Mediterranean sea. *Phys. Chem. Earth Part A Solid Earth Geod.* **2000**, *25*, 57–62. [CrossRef]

Article

Dimensioning Method of Floating Offshore Objects by Means of Quasi-Similarity Transformation with Reduced Tolerance Errors

Grzegorz Stępień [1,*], Arkadiusz Tomczak [1], Martin Loosaar [2] and Tomasz Ziębka [3]

1 Faculty of Navigation, Maritime University of Szczecin, 70-500 Szczecin, West Pomerania, Poland; a.tomczak@am.szczecin.pl
2 iSurvey, Westhill AB32 6FL, UK; martin.loosaar@gmail.com
3 Geometr Ltd., 71-525 Szczecin, Poland; tziebka@geometr.biz
* Correspondence: g.stepien@am.szczecin.pl; Tel.: +48-91-4877-177

Received: 22 September 2020; Accepted: 11 November 2020; Published: 13 November 2020

Abstract: The human activities in the offshore oil and gas, renewable energy and construction industry require reliable data acquired by different types of hydrographic sensors: DGNSS (Differential Global Navigation Satellite System) positioning, attitude sensors, multibeam sonars, lidars or total stations installed on the offshore vessel, drones or platforms. Each component or sensor that produces information, unique to its position, will have a point that is considered as the reference point of that sensor. The accurate measurement of the offsets is vital to establish the mathematical relation between sensor and vessel common reference point in order to achieve sufficient accuracy of the survey data. If possible, the vessel will be put on a hard stand so that it can be very accurately measured using the standard land survey technique. However, due to the complex environment and sensors being mobilized when the vessel is in service, this may not be possible, and the offsets will have to be measured in sea dynamic conditions by means of a total station from a floating platform. This article presents the method of transformation by similarity with elements of affine transformation, called Q-ST (Quasi-Similarity Transformation). The Q-ST has been designed for measurements on such unstable substrates when it is not possible to level the total station (when the number of adjustment points is small (4–6 points)). Such situation occurs, among others, when measuring before the offshore duties or during the jack up or semi-submersible rig move. The presented calculation model is characterized by zero deviations at the adjustment points (at four common points). The transformation concerns the conversion of points between two orthogonal and inclined reference frames. The method enables the independent calculation of the scale factor, rotation matrix and system translation. Scaling is performed first in real space, and then both systems are shifted to the centroid, which is the center of gravity. The center of gravity is determined for the fit points that meet the criterion of stability of the orthogonal transformation. Then, the rotation matrix is computed, and a translation is performed from the computational (centroid) to real space. In the applied approach, the transformation parameters, scaling, rotation and translation, are determined independently, and the least squares method is applied independently at each stage of the calculations. The method has been verified in laboratory conditions as well as in real conditions. The results were compared to other known methods of coordinate transformation. The proposed approach is a development of the idea of transformation by similarity based on centroids.

Keywords: similarity transformation; affine transformation; rotation matrix; offshore surveying; dimensional control; orthogonal coordinate system; close range photogrammetry; total station

1. Introduction

Similarity transformation is commonly used to convert coordinates between two three-dimensional orthogonal frames of reference [1]. This applies in particular to fields such as hydrography, geodesy, photogrammetry, offshore measurements and navigation [2–4]. Spatial transformations are commonly used in the processing of data from unmanned aerial vehicles [5], point clouds from laser scanning [6,7] and also with the use of computer vision algorithms [8]. Calculations of this type are also performed in engineering geodesy, where measurements are often converted to the coordinate system associated with the measured object, as well as when using, e.g., tunnel boring machines [9,10], and in many others. For big datasets of measurement, algorithms based on machine learning or applying the PCA (principal component analysis) method are also used for data processing [11–13].

Conformal transformation, also called transformation by similarity (or Helmert transformation), includes the change of scale (single accuracy) and isometry, which is related in the orthogonal transformation with translation and rotation [1,14,15]. This transformation concerns three groups of transformations: rotation, translation and scaling, and has seven parameters. The unknowns are: the translation vector, the scale factor and three rotation angles [16,17]. This transformation is usually presented in the form [18]:

$$\begin{bmatrix} X' \\ Y' \\ Z' \end{bmatrix} = \lambda \times R \times \begin{bmatrix} X \\ Y \\ Z \end{bmatrix} + \begin{bmatrix} \overline{X}_0 \\ \overline{Y}_0 \\ \overline{Z}_0 \end{bmatrix} \tag{1}$$

where:

- $\left[\overline{X}_0, \overline{Y}_0, \overline{Z}_0\right]^T$—translation vector,
- $[X, Y, Z]^T$—vector (point) in the original system (subject to transformation),
- $[X', Y', Z']^T$—vector (point) in the secondary system (treated as stationary),
- λ—scale factor,
- R—rotation matrix.

The rotation matrix R in Equation (1) is also commonly used in robotics, mechanics and automation and can be represented using various angle systems: Euler angles Equation (2), directional cosines Equation (3), angles used in photogrammetry Equation (4), angles related to motion, also referred to as Euler (or Tait-Bryan) angles Equation (5), as well as using Hamilton quaternions Equation (6) [18–22]. A rotation matrix using three angles of rotation can be represented by twelve sequences when all rotations are in the same direction (e.g., clockwise). Each sequence can be presented, depending on the adopted direction of the rotation, clockwise or anticlockwise, in six ways. This gives a total of seventy-two possibilities to build the rotation matrix. The in-depth analysis and the relationship between the various systems of angles and sequences of rotations and their relationships with Hamilton quaternions have been discussed in References [2,20,23,24]. In Figure 1 and in Equation (2), one of the possibilities of rotating with the use of the classic Euler angles system is presented, where rotation around one axis occurs twice (the order of rotations around the axis: Z—rotation by the angle ψ, X (W)—rotation by angle θ, Z (3)—rotation by angle φ).

$$\begin{aligned} R &= \begin{bmatrix} \cos\psi & \sin\psi & 0 \\ -\sin\psi & \cos\psi & 0 \\ 0 & 0 & 1 \end{bmatrix} \times \begin{bmatrix} 1 & 0 & 0 \\ 0 & \cos\theta & \sin\theta \\ 0 & -\sin\theta & \cos\theta \end{bmatrix} \times \begin{bmatrix} \cos\varphi & \sin\varphi & 0 \\ -\sin\varphi & \cos\varphi & 0 \\ 0 & 0 & 1 \end{bmatrix} \\ &= \begin{bmatrix} -\cos\theta\times\sin\varphi\times\sin\psi+\cos\varphi\times\cos\psi & \cos\theta\times\cos\varphi\times\sin\psi+\cos\psi\times\sin\varphi & \sin\theta\times\sin\psi \\ -\cos\varphi\times\sin\psi-\cos\theta\times\cos\psi\times\sin\varphi & -\sin\varphi\times\sin\psi+\cos\theta\times\cos\varphi\times\cos\psi & \cos\psi\times\sin\theta \\ \sin\theta\times\sin\varphi & -\cos\varphi\times\sin\theta & \cos\theta \end{bmatrix} \end{aligned} \tag{2}$$

where:

- rotation order: $\psi \rightarrow \theta \rightarrow \varphi$, all rotations were assumed clockwise and with markings as in Figure 1,
- ψ—precession angle, between the OX axis and the OW node line (rotation around the Z axis),

- θ—nutation angle, between the OZ and $O3$ axes (rotation around the OW node line),
- φ—angle of pure rotation (intrinsic rotation), between the line of nodes OW and the axis $O1$ (rotation around axis 3).

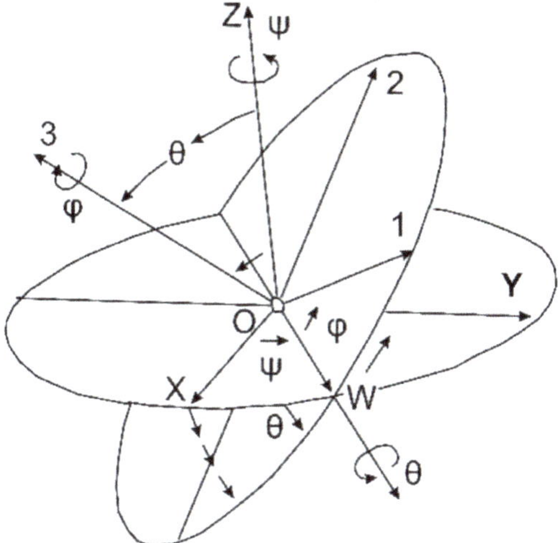

Figure 1. Euler's rotation angles (shown sequence: $\psi \rightarrow \theta \rightarrow \varphi$). The rotation matrix is described by the Equation (2).

The rotation matrix is also represented by directional cosines, where individual columns of the R matrix are represented by the scalar products of the dissipers of individual systems. The target system is also referred to in the literature as a stationary or reference frame [2,20]:

$$R = \begin{bmatrix} cos(x,x') & cos(x,y') & cos(x,z') \\ cos(y,x') & cos(y,y') & cos(y,z') \\ cos(z,x') & cos(z,y') & cos(z,z') \end{bmatrix} \tag{3}$$

The rotation matrix in photogrammetry is determined on the basis of the angles related to the motion in the following order: ω (roll), φ (pitch), κ (yaw), which is written in the form of Equation (4):

$$\begin{aligned} R &= \begin{bmatrix} 1 & 0 & 0 \\ 0 & \cos\omega & \sin\omega \\ 0 & -\sin\omega & \cos\omega \end{bmatrix} \times \begin{bmatrix} \cos\varphi & 0 & -\sin\varphi \\ 0 & 1 & 0 \\ \sin\varphi & 0 & \cos\varphi \end{bmatrix} \times \begin{bmatrix} \cos\kappa & \sin\kappa & 0 \\ -\sin\kappa & \cos\kappa & 0 \\ 0 & 0 & 1 \end{bmatrix} \\ &= \begin{bmatrix} cos\varphi \times cos\kappa & cos\varphi \times sin\kappa & -sin\varphi \\ -cos\omega \times sin\kappa + sin\omega \times sin\varphi \times cos\kappa & cos\omega \times cos\kappa + sin\omega \times sin\varphi \times sin\kappa & sin\omega \times cos\varphi \\ sin\omega \times sin\kappa + cos\omega \times sin\varphi \times cos\kappa & -sin\omega \times cos\kappa + cos\omega \times sin\varphi \times \sin\kappa & cos\omega \times cos\varphi \end{bmatrix} \end{aligned} \tag{4}$$

where:

- rotation order: $\omega \rightarrow \varphi \rightarrow \kappa$,
- ω—the rotation around the X axis (clockwise), the roll angle,
- φ—the rotation around the Y axis (anticlockwise), the pitch angle,
- κ—the rotation around the Z axis (clockwise), the yaw angle.

In navigation, extended notation of the basic Euler system of angles is used and they are often defined in relation to the body frame to an external (often defined by GNSS – Global Navigation Satellite System) reference frame (fixed frame). The rotation sequence is then the opposite to that described with the Equation (4), which for an orthogonal transformation can be written as a transposition of this matrix in the form of the Equation (5):

$$\begin{aligned} R &= \begin{bmatrix} \cos\psi & -\sin\psi & 0 \\ \sin\psi & \cos\psi & 0 \\ 0 & 0 & 1 \end{bmatrix} \times \begin{bmatrix} \cos\theta & 0 & \sin\theta \\ 0 & 1 & 0 \\ -\sin\theta & 0 & \cos\theta \end{bmatrix} \times \begin{bmatrix} 1 & 0 & 0 \\ 0 & \cos\varphi & -\sin\varphi \\ 0 & \sin\varphi & \cos\varphi \end{bmatrix} \\ &= \begin{bmatrix} cos\theta \times cos\psi & -cos\varphi \times sin\psi + sin\varphi \times sin\theta \times cos\psi & sin\varphi \times sin\psi + cos\varphi \times sin\theta \times cos\psi \\ cos\theta \times sin\psi & cos\varphi \times cos\psi + sin\varphi \times sin\theta \times sin\psi & -sin\varphi \times cos\psi + cos\varphi \times sin\theta \times \sin\psi \\ -sin\theta & sin\varphi \times cos\theta & cos\varphi \times cos\theta \end{bmatrix} \end{aligned} \tag{5}$$

where:

- rotation order: $\psi \to \theta \to \varphi$,
- ψ—yaw angle (anticlockwise), rotation around the Z axis,
- θ—pitch angle (clockwise), rotation around the Y axis (after rotation around the Z axis),
- φ—roll angle (anticlockwise), rotation around the X axis (after rotation around the Z and Y axes).

The rotation matrix (5) written with Hamilton quaternions will take the form of the relationship (6). The relation between the Euler angles assumed in Equation (5) and the quaternions described with Equation (6) will then take the form of Equation (7), with the norm in the form of Equation (8):

$$Q = \begin{bmatrix} {q_1}^2 + {q_2}^2 - {q_3}^2 - {q_4}^2 & 2(q_2 \times q_3 - q_1 \times q_4) & 2(q_2 \times q_4 + q_1 \times q_3) \\ 2(q_2 \times q_3 + q_1 \times q_4) & {q_1}^2 - {q_2}^2 + {q_3}^2 - {q_4}^2 & 2(q_3 \times q_4 - q_1 \times q_2) \\ 2(q_2 \times q_4 - q_1 \times q_3) & 2(q_3 \times q_4 + q_1 \times q_2) & {q_1}^2 - {q_2}^2 - {q_3}^2 - {q_4}^2 \end{bmatrix} \tag{6}$$

where:

$$\begin{aligned} q_1 &= \cos\tfrac{\varphi}{2} \times \cos\tfrac{\theta}{2} \times \cos\tfrac{\psi}{2} + \sin\tfrac{\varphi}{2} \times \sin\tfrac{\theta}{2} \times \sin\tfrac{\psi}{2} \\ q_2 &= \sin\tfrac{\varphi}{2} \times \cos\tfrac{\theta}{2} \times \cos\tfrac{\psi}{2} - \cos\tfrac{\varphi}{2} \times \sin\tfrac{\theta}{2} \times \sin\tfrac{\psi}{2} \\ q_3 &= \cos\tfrac{\varphi}{2} \times \sin\tfrac{\theta}{2} \times \cos\tfrac{\psi}{2} + \sin\tfrac{\varphi}{2} \times \cos\tfrac{\theta}{2} \times \sin\tfrac{\psi}{2} \\ q_4 &= \cos\tfrac{\varphi}{2} \times \cos\tfrac{\theta}{2} \times \sin\tfrac{\psi}{2} - \sin\tfrac{\varphi}{2} \times \sin\tfrac{\theta}{2} \times \cos\tfrac{\psi}{2} \end{aligned} \tag{7}$$

$${q_1}^2 + {q_2}^2 + {q_3}^2 + {q_4}^2 = 1 \tag{8}$$

Equations (6)–(8), although not applied in the practical part of the study, are presented in addition to the completeness of the theoretical considerations.

The rotation matrix is also presented in a parametric Equation (9), which is very convenient to use because it can then be used to calculate the rotation matrix in any system of angles:

$$M = \begin{bmatrix} m_{11} & m_{12} & m_{13} \\ m_{21} & m_{22} & m_{23} \\ m_{31} & m_{32} & m_{33} \end{bmatrix} \tag{9}$$

In general, the transformation matrix M has nine unknowns, including six independent ones, but after applying the orthogonal transformation Equation (10), only three of them will be linearly independent, which leads to Equation (11):

$$\begin{aligned} m_{11}^2 + m_{12}^2 + m_{13}^2 &= 1 \\ m_{21}^2 + m_{22}^2 + m_{23}^2 &= 1 \\ m_{31}^2 + m_{32}^2 + m_{33}^2 &= 1 \\ m_{11} \times m_{12} + m_{21} \times m_{22} + m_{31} \times m_{32} &= 0 \\ m_{11} \times m_{13} + m_{21} \times m_{23} + m_{31} \times m_{33} &= 0 \\ m_{12} \times m_{13} + m_{22} \times m_{23} + m_{32} \times m_{33} &= 0 \end{aligned} \tag{10}$$

$$M \times M^T = I \tag{11}$$

where:

- I—identity matrix.

In the case of the affine transformation, in the nine-unknown approach, the scale factor is replaced by the scale factor vector $\lambda = diag[\lambda_X, \lambda_Y, \lambda_Z]^T$, which allows a different scale for each axis. If all the scale factors are the same and after the orthogonality condition in the Equation (10) or Equation (11) is satisfied by the rotation matrix, the affine transformation goes into a transformation by similarity. Various variants of the affine transformation are presented, e.g., in References [25–27], and in this study, they will not be described in more detail.

In practice, in many applications, the rotation matrix is reduced to the infinitesimal (aiming at zero) values of the angles and presented in the form of the Equation (12):

$$m = \begin{bmatrix} 1 & -\psi & \theta \\ \psi & 1 & -\varphi \\ -\theta & \varphi & 1 \end{bmatrix} \tag{12}$$

where:

- ψ, θ, φ—infinitesimal values of angles expressed in radians, the angles are the same as in Equation (5).

The matrix (12) is also called the small rotation angles due to the infinitesimal values of the angles (expressed in radians).

Transformation by similarity based on infinitesimal values of angles (expressed in radians), with a lattice scale close to one, will take the form described as the Buršy–Wolf transformation [28,29]:

$$\begin{bmatrix} X \\ Y \\ Z \end{bmatrix} = (1+ds) \times \begin{bmatrix} 1 & \epsilon_Z & -\epsilon_Y \\ -\epsilon_Z & 1 & \epsilon_X \\ \epsilon_Y & -\epsilon_X & 1 \end{bmatrix} \times \begin{bmatrix} X' \\ Y' \\ Z' \end{bmatrix} + \begin{bmatrix} t_X \\ t_Y \\ t_Z \end{bmatrix} \tag{13}$$

where:

- ϵ_Z—rotation angle around the Z axis,
- ϵ_Y—rotation angle around the Y axis,
- ϵ_X—rotation angle around the X axis,
- $1 + ds$—scale factor, where ds corresponds to linear system distortion,
- $[t_X, t_Y, t_Z]^T$—translation vector.

The rotation matrix expressed in Equation (13) is the inverse of the matrix shown in Equation (12).

On the other hand, Badekas [30] provides a record of the transformation between the average terrestrial system and the geodetic system in the form of the relationship (14):

$$\begin{bmatrix} X \\ Y \\ Z \end{bmatrix} = \begin{bmatrix} d_{x_0} \\ d_{y_0} \\ d_{z_0} \end{bmatrix} + \begin{bmatrix} X_0 \\ Y_0 \\ Z_0 \end{bmatrix} + \begin{bmatrix} 1 & da_3 & -da_2 \\ -da_3 & 1 & da_1 \\ da_2 & -da_1 & 1 \end{bmatrix} \times \begin{bmatrix} x - x_0 \\ y - y_0 \\ z - z_0 \end{bmatrix} + \epsilon \begin{bmatrix} x - x_0 \\ y - y_0 \\ z - z_0 \end{bmatrix} \tag{14}$$

where:

- $[X\ Y\ Z]^T$—coordinates in the average Earth coordinate system,
- $[x\ y\ z]^T$—coordinates in a geodetic (local) coordinate system,
- da_1, da_2, da_3—(infinitesimal) rotation angles expressed in radians,
- ϵ—scale correction,
- $\begin{bmatrix} X_0 & Y_0 & Z_0 \end{bmatrix}^T$—vector of translation in the terrestrial system,

- $\begin{bmatrix} x_0 & y_0 & z_0 \end{bmatrix}^T$—translation vector in the geodetic (local) system,
- $\begin{bmatrix} d_{x_0} & d_{y_0} & d_{z_0} \end{bmatrix}^T$—the coordinates of the origin of the geodetic (local) coordinate system after rotation and shift to the mean terrestrial system.

In the transformation described by Equation (14), both the reference frame and the geodetic (local) system are shifted to a common centroid, which is the center of gravity of the Earth system. The transformation described by Equation (14) written in accordance with the Buršy–Wolf transformation (13) functions in the literature under the name of the Molodensky–Badekas transformation (15), although many authors present it in different variants [29]:

$$\begin{bmatrix} X \\ Y \\ Z \end{bmatrix}_2 = (1+ds)\begin{bmatrix} 1 & \epsilon_Z & -\epsilon_Y \\ -\epsilon_Z & 1 & \epsilon_X \\ \epsilon_Y & -\epsilon_X & 1 \end{bmatrix} \times \begin{bmatrix} \overline{X} \\ \overline{Y} \\ \overline{Z} \end{bmatrix}_1 + \begin{bmatrix} t_X \\ t_Y \\ t_Z \end{bmatrix}_2 \quad (15)$$

where:

- $\begin{bmatrix} \overline{X} & \overline{Y} & \overline{Z} \end{bmatrix}^T_1$—coordinates in the geodetic (local) system after moving to the centroid, the center of gravity of this system,
- indexes 1 and 2 next to the translation and coordinate vectors mean, respectively: 1—geodetic (local) system, 2—Earth system.

Thus, in Equation (15), there is a double translation, first, the geodetic system (local, marked as 1) is shifted to its own center of gravity, and after rotation and scale correction, a re-translation takes place—back to the Earth system, and inversely to the one to which the Earth system was subjected at the Buršy–Wolf and Molodensky–Badekas transformations can also be written using one of the rotation matrices presented in Equations (2)–(6) and (9) for large values of rotation angles. Centroid as the center of gravity and the influence on its location on the accuracy of coordinates calculation is presented in Reference [31].

Infinitesimal values of transformation parameters are used in the adjustment process, where the least squares method is most often used. The discussion on the use of the least squares method in conformal and affine transformations has been undertaken in many studies, including References [6,16,18,32–34]. The adjustment procedure leads to the search for the most probable solutions, which is associated with the minimization of errors [35,36] and directly derived from the maximization of the probability, P:

$$P = \frac{1}{\sigma\sqrt{2\pi}}\int_{-t}^{t} e^{-\frac{(x-\mu)^2}{2\sigma}}dx \quad (16)$$

The solution (calculation of the vector of unknowns) is presented in a matrix form and written in the form:

$$X = \left(A^T \times P \times A\right)^{-1} \times A^T \times P \times L \quad (17)$$

where:

- X—vector of unknown parameters,
- A—matrix of coefficients with unknowns (partial derivatives)—Jacobian transformation,
- L—vector of constants,
- P—vector of weighting factors (statistical weights).

In practice, the adjustment problem comes down to the formulation of the correction Equation (18) or the formulation of the computational problem in the Equation (19), where one of them is calculated, while another is provided by the data [6,32]:

$$V = A \times X - L \quad (18)$$

$$L = A \times X \tag{19}$$

The correction Equation (18) for the transformation by similarity (1) with the use of a small-rotation matrix (12) from the body frame to the stationary frame (reference frame–fixed frame) can be written as Equation (20), and after transformation into the Equation (21) [37]:

$$\begin{bmatrix} X' \\ Y' \\ Z' \end{bmatrix} - \begin{bmatrix} x \\ y \\ z \end{bmatrix}_{transformed} = \begin{bmatrix} d\lambda & -\psi & \theta \\ \psi & d\lambda & -\varphi \\ -\theta & \varphi & d\lambda \end{bmatrix} \times \begin{bmatrix} x \\ y \\ z \end{bmatrix} + \begin{bmatrix} dx_0 \\ dy_0 \\ dz_0 \end{bmatrix} \tag{20}$$

$$\begin{bmatrix} V_X \\ V_Y \\ V_Z \end{bmatrix} = \begin{bmatrix} x & 0 & -y & z & 1 & 0 & 0 \\ y & -z & x & 0 & 0 & 1 & 0 \\ z & y & 0 & -x & 0 & 0 & 1 \end{bmatrix} \times \begin{bmatrix} d\lambda \\ \varphi \\ \psi \\ \theta \\ dx_0 \\ dy_0 \\ dz_0 \end{bmatrix} + \begin{bmatrix} X' \\ Y' \\ Z' \end{bmatrix} - \begin{bmatrix} x \\ y \\ z \end{bmatrix}_{transformed} \tag{21}$$

The reference of compound (21) to the designations of compound (18) is as follows:

$$V = \begin{bmatrix} V_X \\ V_Y \\ V_Z \end{bmatrix}, A = \begin{bmatrix} x & 0 & -y & z & 1 & 0 & 0 \\ y & -z & x & 0 & 0 & 1 & 0 \\ z & y & 0 & -x & 0 & 0 & 1 \end{bmatrix}, X = \begin{bmatrix} d\lambda \\ \varphi \\ \psi \\ \theta \\ dx_0 \\ dy_0 \\ dz_0 \end{bmatrix}, L = \begin{bmatrix} X' \\ Y' \\ Z' \end{bmatrix} - \begin{bmatrix} x \\ y \\ z \end{bmatrix}_{transformed} \tag{22}$$

In this study, the least squares method is used in stages, and not in one equalization process, e.g., expressed by Equation (20). It is used separately at each stage of the transformation, and therefore it is abbreviated as S2-LSM (Stages Sequence–Least Square Method). Such an approach is implemented in the Quasi-Similarity Transformations (Q-ST) algorithm and results in resetting the deviations on the combined points for four adjustment points. This method was designed for measurements in dynamic inclined systems, in particular in the offshore industry and in close-range geodesy and photogrammetry.

The aim of this publication is to develop a coordinate transformation method dedicated to measurements in a dynamic body frame, ensuring simplicity and speed of calculations with a simultaneous reduction to the minimum of errors on the adjustment points, when their number is small (4–6 points). According to the authors, the developed Q-ST method can be used to convert coordinates in the offshore industry and in traditional geodetic and photogrammetric measurements. The method can also be used in navigation, automation, mechanics and robotics. The advantage of the proposed method is especially visible in the case of a small number of adjustment points (4 or 5), and by zeroing the errors at the adjustment points, according to Equation (16)—the calculated coordinates can be treated as maximally probable.

2. Materials and Methods

2.1. Vessel Offsets Measurements

To achieve the highest accuracy standards, vessels' offsets determination is usually performed with total station. Total station, as in other industries, is an essential instrument used in the offshore industry for surveying and building construction due to the level of accuracy it provides. They are used for dimensional control of offshore vessels, survey boats, jack up and semi-submersible platforms, wind turbines, remotely operated vehicles, including primarily the measurement of offsets of navigational or hydrographic sensors and measurements, and other elements of ship's structure, essential for offshore operations.

Precisely measured offsets in the local (vessel) coordinate frame are entered into the hydrographic software integrating signals from sensors, e.g., GNSS, Multi-beam echosounder, gyrocompass, motion reference unit, acoustic underwater positioning systems, lidars and others, and based on the adopted navigation solution, e.g., Kalman filter, acquired data X, Y, Z are logged in the global coordinate frame.

The robotic total station is also used as a navigation sensor for positioning of the jack up and semi-submersible platforms during approach to installation of offshore structures, e.g., during jacket or wind turbine pile installation projects. They are an alternative to GNSS satellite positioning wherever there is a risk of signal obstruction and loss of platform position. Based on both vertical and horizontal angles and the slope distance, the navigation solution is performed. Total Station mounted on the vessel takes measurements to a particular point (usually prism reflector installed on the offshore installation with known coordinates) and the dynamic vessel's position is determined by hydrographic software and can be monitored to mitigate the risk of collision.

During the survey, the tilt compensator is locked. Each survey station is established in a random (local) coordinate system on board the floating vessel. Usually more than one survey station is needed for covering all the observations. Therefore, at least four common control points between survey stations are established and observed. Additionally, for measuring the sensors of interest, the vessel body must be surveyed for center line alignment and the positioning of the new established local coordinate frame of the vessel (see Figure 2).

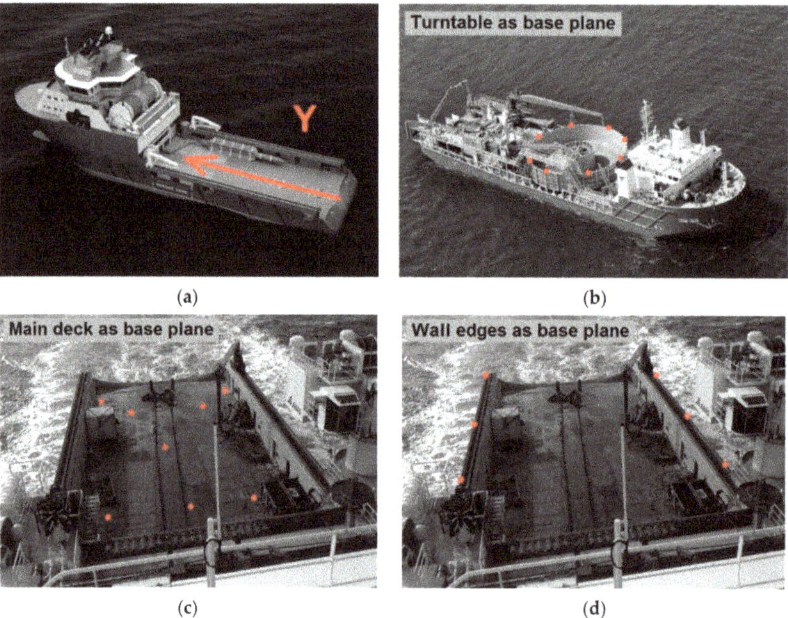

Figure 2. Determination of the ship's center line (Y) and the base plane: (**a**) the ship's center line, define the base plane based on: (**b**) turntable element, (**c**) main deck floor, (**d**) wall edges.

As already mentioned, all the survey stations are independent and established in a random coordinate system. The instrument does not need to be levelled, but for avoiding any bending of the instrument under its own weight, it is good practice to level it approximately. The tilt compensator must be locked, because the instrument must be fixed relatively to the vessel when the vessel is floating. As a precaution for discovering small accidental moves of the instrument during the survey (hitting the tripod with leg, for example), it is wise to start the survey station with measuring the farthest point and to finish the survey station with re-measuring that point. By comparing the initial coordinates against the last ones, a possible instrument move becomes visible. Each survey station must be connected to others through at least four control points. Even though three are minimum, it does not give enough quality control. Good practice is having five or more common points between stations.

Planning and preparing the survey is the most time-consuming part of the whole process. It includes establishing the control points at suitable places all over the vessel (see Figure 3) and planning the survey station locations. Because usually there are multiple operations simultaneously progressing on board the vessel, the surveyor must prepare for sudden unexpected visibility issues raised during the observations. Also, the geometry of the common control points between survey stations must be wide enough for providing wanted precision (due to our practice). A proven approach is: starting the survey from the location with the widest visibility and measuring many control points on all parts of the vessel from there. Other survey stations established later are referenced to the points from the first station. This way, the "chain of stations" is as short as practically possible. In the best-case scenario, all the stations are connected directly to the first survey station control points.

Figure 3. Common control points marked on the vessel.

The meaning of wide geometry of common control points between stations is different from the meaning land surveyors are used to. On terrestrial observations, the Z axis is locked with the help of gravity (by levelling the instrument) and the survey can be considered as two-dimensional regarding the rotations. It is enough to measure 2 points with decent distance between them for establishing the position of the total station. However, on a floating object, at least 3 points are needed for establishing the position of the instrument, and these 3 points cannot be adjusted with LSM. In a perfect case, they form an equilateral triangle. While planning the survey, the surveyor must analyze the situation in three-dimension (3D)—control points should stretch out to various directions.

Usually, some navigation systems on board ships already have a local coordinate system defined, and for simplicity, the surveyor should use the same location for his coordinate frame origin as specified in vessel systems (Ultra Short Baseline (USBL) transducer in Dynamic Positioning system (DP)). That means, during the total station observations, a common element must be surveyed, for knowing where the pre-defined coordinate frame's origin is located. That element is an observable sensor, which coordinates (X, Y, Z values) can be read out from the navigation system, for example, USBL pole or GPS antenna. The origin of the coordinate system of navigation systems can therefore be determined during surveys in the local random coordinate system of the total station.

Below can be found the complete list of items to observe during the total station survey (in addition to the control points for the current survey and for future use—marked in Figure 3):

1. Sensors of interest (the purpose of the survey),
2. Sensors for connecting the survey to a pre-established coordinate frame of vessel system (USBL, DP),
3. Ship elements for connecting the survey to General Arrangement (GA) drawing of a vessel,
4. Prepared pitch, roll and heading calibration points,
5. Ship body for alignment and positioning of the new local coordinate frame:
   - Z-axis direction (base plane),
   - Y-axis direction,
   - Origin X-position (center line of vessel),
   - Origin Y-position,
   - Origin Z-position.

Data processing is done in dimensional control software, like, for example, SC4W. For simplicity, the survey data is imported into the software as (X,Y,Z) coordinates in 1.0 or 0.1 mm precision, instead of raw angles and distances. The next step is merging the survey stations. The first survey station is considered as fixed and other survey stations are rotated and shifted in the best possible way to match common control points with the first station. This is usually done with the help of data processing software's least squares best fit function. After the merging process, all the surveyed points are in the form of one point cloud instead of independent unaligned point clouds of each station. During the data processing, all survey stations are merged, and one point cloud is formed. After that, the point cloud is aligned with the vessel body and as the effect, the origin of the coordinate frame is positioned on the desired location. After the aligning process and coordinate frame positioning related to the vessel is completed, the sensor coordinates are obtained.

For quality control and for positioning the surveyed points relative to the vessel, it is useful to have GA drawings of the ship in CAD format. For being able to relate the GA drawing with total station survey, few well-defined common elements must be found which are easily observable and also presented on the GA drawing. In most cases, vessel attitude sensor calibration follows the offsets survey. Therefore, it is wise to pre-install poles for RTK (Real Time Kinematic) Rovers or prisms, depending on the calibration method. These poles are observed during the total station survey and local coordinates are obtained.

Due to the presented approach, similarity transformation algorithms using the least squares method are used to connect point clouds. For on-board measurements, as mentioned, the number of common points is generally four or five. This publication presents a similarity transformation method that is suited to this small number of common points. This method is characterized by a minimum value of errors of alignment at common points and is based on the Q-ST algorithm.

2.2. Quasi-Similarity Transformations (Q-ST) Algorithm

In the presented Quasi-Similarity Transformations (Q-ST) method, it was assumed that the transformed system associated with the measurement instrument is named as local frame—LF, and the target—stationary system to which the transformation takes place, was named vessel frame—VF (fixed-frame). The composition of the transformation groups: rotations, translations and scaling, is arbitrary in the transformation and results in a mathematical notation appropriate to the adopted model. The functional model of the Q-ST transformation is presented by the formula (23):

$$X_{VF} = R \times (\lambda \cdot X_{LF} - X_{CLF}) + X_{CVF} \tag{23}$$

where:

- X_{VF}—coordinates in the target system—stationary (vessel reference frame—VF),
- R—rotation matrix,

- λ—scale factor,
- X_{LF}—coordinates in the transformed system (local reference frame—LF),
- X_{CLF}—coordinates of the center of gravity in the transformed system (centroid of local reference frame),
- X_{CVF}—coordinates of the center of gravity in the target system—stationary (centroid of vessel reference frame).

The Q-ST (23) is performed in the steps shown in Figure 4.

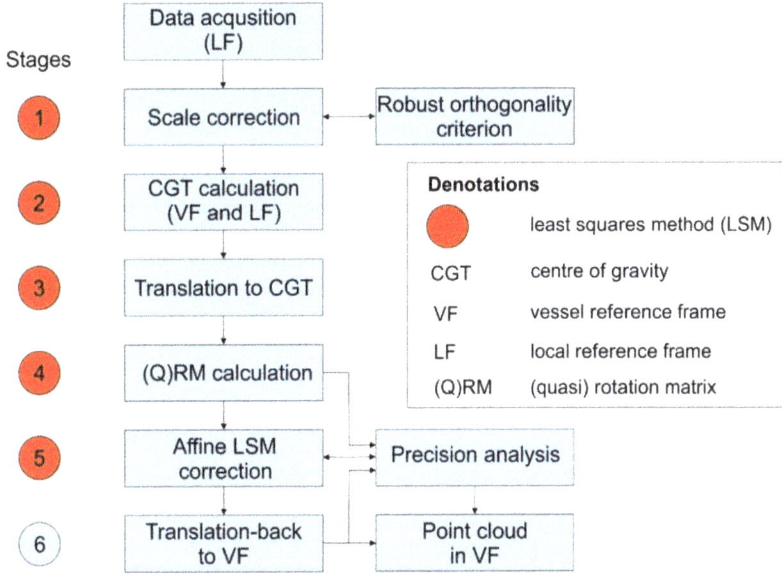

Figure 4. Calculation scheme of the Quasi-Similarity Transformation (Q-ST) method.

After collecting the data by means of measurements in the local system of the LF instrument (local reference frame), they are converted to the vessel's VR coordinate system (vessel reference frame). The calculations are performed in the steps shown in Figure 4. Below are descriptions of each step of the Q-ST.

Stage 1:

The first stage of transformation is related to the scale adjustment between the LF and VF systems. The scale factor λ for a pair of homologous vectors—built from points known in both systems—LF and VF—is determined from the dependence (24):

$$D_{VF} = \lambda \times D_{LF} \tag{24}$$

where:

- D_{VF}—vector length in the ship's coordinate system (VF),
- D_{LF}—the length of the vector in the local coordinate system associated with the instrument (LF).

The scale factor, using Equation (24), can be determined independently for each pair of homologous vectors known in the VF and LF systems. Calculation of the global scale factor between the LF and VF systems for all homologous vectors is performed using the least squares method. The computational problem is reduced to the form of the relationship (19), and the solution written according to the Formula (17) will then take the form of the relationship (25):

$$\lambda = \left(D_{LF}{}^{T} \times D_{LF}\right)^{-1} \times D_{LF}{}^{T} \times D_{VF} \tag{25}$$

where:

- $D_{LF} = [D_1, D_2, \ldots, D_N]^T$—a vector consisting of the length of the homologous vectors in the local LF system,
- $D_{VF} = [D_1, D_2, \ldots, D_N]^T$—a vector consisting of the lengths of the homologous vectors in the VF ship system,
- $D_1, D_2, \ldots, D_N$—lengths of homologous vectors in the LF and VF coordinate systems,
- N—the number of homolog vector pairs known in LF and VR.

The reference of the signs in Equation (25) to the signs in Equations (17) and (19) is as follows:

$$A = D_{LF},\ X = \lambda,\ L = D_{VF} \tag{26}$$

At this stage of transformation, the homogeneity of scale changes is also tested, which is related to the behavior of the orthogonal transformation criterion (robust orthogonal criterion in Figure 4). For each pair of homologous vectors in LF and VF, scale corrections are made, which according to relation (18) and notations (26) and taking into account the Equation (25), can be written as Equation (27):

$$V = D_{LF} \times \left(D_{LF}{}^{T} \times D_{LF}\right)^{-1} \times D_{LF}{}^{T} \times D_{VF} - D_{VF} \tag{27}$$

For the corrections calculated with the use of the Equation (27), the standard deviation σ_D is calculated using Equation (28):

$$\sigma_D = \sqrt{\frac{V^T \times V}{N-1}} \tag{28}$$

Then, for each deviation from the calculation scale, the condition written in the form (29) is checked using the Equation (27):

$$V_i \geq 2\sigma_D \tag{29}$$

The deviations from the scale testify to the heterogeneity of its change between the LF and VF systems. The criterion of exceeding the double value of the standard deviation is statistically justified for twenty lengths of homologous vectors, in the case of normal distribution of their values. In the case of five connecting points known in both systems, ten pairs of homologous vectors can be constructed on their basis. Therefore, exceeding the standard deviation twice can be treated as a disturbance of the scale, which in this case can also be justified using the Chauvenet criterion [36]. Condition (29) is a notation of a robust orthogonality transformation criterion. According to this condition, pairs of homologous vectors for which the condition (29) is satisfied introduce a scale disturbance and are excluded from the determination of this coefficient. After their elimination, the scale factor is determined again using the Formula (25).

Stage 2:

In the second step, the center of gravity of both the LF and VF coordinate systems is determined. The coordinates of the center of gravity have the form of relations (32) and are the result of searching for the extreme (minimum) of the function F (30), for which the sum of the squares of the distance from the searched point is the minimum:

$$\begin{gathered} F(x, y, z) = D_1{}^2 + D_2{}^2 + \ldots + D_n{}^2 \\ D_i^2 = (x_C - x_i)^2 + (y_C - y_i)^2 + (z_C - z_i)^2 = minimum \end{gathered} \tag{30}$$

At the stationary point, the partial derivatives of the function (30) take the extreme (minimum) value, which was written using the compounds (31). Directly from the compounds (31), after ordering, the coordinates of the center of gravity are obtained in the form (32) [38]. The center of gravity is therefore determined by minimizing the squared distances to each of the common points in the LF and VF systems:

$$\begin{array}{l} \frac{\partial F}{\partial x} = 2(x_C - x_1) + 2(x_C - x_2) + \ldots + 2(x_C - x_n) = 2(nx_C - x_1 - x_2 - \ldots - x_n) = 0 \\ \frac{\partial F}{\partial y} = 2(y_C - y_1) + 2(y_C - y_2) + \ldots + 2(y_C - y_n) = 2(ny_C - y_1 - y_2 - \ldots - y_n) = 0 \\ \frac{\partial F}{\partial z} = 2(z_C - z_1) + 2(z_C - z_2) + \ldots + 2(z_C - z_n) = 2(nz_C - z_1 - z_2 - \ldots - z_n) = 0 \end{array} \quad (31)$$

where:

- D_i—length of the homologous vector in each of the LF and VF systems built using two points: the center of gravity and the considered fit point,
- X_C, Y_C, Z_C—coordinates of the center of gravity (center of mass) in LF and VF respectively,
- X_i, Y_i, Z_i—the coordinates of the fitting point, i ∈ (1,2, ..., n),
- n—number of homologous points, adjustments (common points) in LF and RF systems.

$$\begin{array}{l} X_{CLF} = \frac{\sum_{i=1}^{N}(X_i)}{n},\ Y_{CLF} = \frac{\sum_{i=1}^{N}(Y_i)}{n},\ Z_{CLF} = \frac{\sum_{i=1}^{N}(Z_i)}{n}, \\ X_{CVF} = \frac{\sum_{i=1}^{N}(X_i)}{n},\ Y_{CVF} = \frac{\sum_{i=1}^{N}(Y_i)}{n},\ Z_{CVF} = \frac{\sum_{i=1}^{N}(Z_i)}{n}, \end{array} \quad (32)$$

Stage 3:

In the third stage, auxiliary coordinates are determined in the LF and VF systems. These coordinates are calculated for the common points after translating both systems into the centroid—the center of gravity of both systems. The idea of translation is presented in Figure 5.

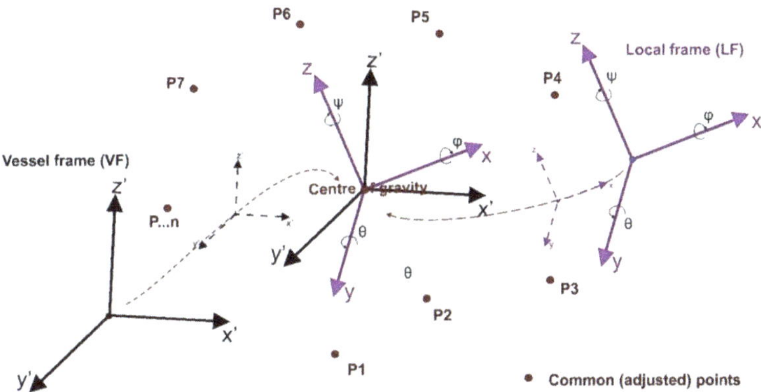

Figure 5. Shift of the two inclined frames of reference (VF and LF) to the centroid—center of gravity.

The systems (Figure 5) are translated along the axis of each of the LF and VF coordinate systems. In this way, the mutual angular orientation of the two systems does not change. As a result, new centroid coordinates are obtained, in the auxiliary computational space, and related to the center of gravity. These coordinates for points in LF and VF systems, after taking into account relations (25) and (32), can be written using Formula (33):

$$\begin{array}{c} X_{iLF} = (\lambda \times X_{LF} - X_{CLF}) = \left[\left(D_{LF}{}^T \times D_{LF} \right)^{-1} \times D_{LF}{}^T \times D_{VF} \right] \times X_{LF} - \frac{\sum_{i=1}^{N}(X_i)}{n} \\ X_{iVF} = (X_{VF} - X_{CVF}) = X_{VF} - \frac{\sum_{i=1}^{N}(X_i)}{n} \end{array} \quad (33)$$

where:

- X_{iLF}—centroid coordinates of points in the LF and VF systems related to the center of gravity,
- X_{iVF}—centroid coordinates of points in the VF system.

Stage 4:

In the fourth step, the rotation matrix of the system LF with respect to VF is determined. The matrix of rotation is calculated using the coordinates defined by Equation (33). The designations of the angles of the LF system with respect to VF were adopted in accordance with the relationship (5) and Figure 5. For the simplicity and convenience of calculations, a parametric matrix (9) is used. For three common points, the rotation matrix is determined from the dependence (34):

$$M = X_{iVF} \times {X_{iLF}}^{-1} \tag{34}$$

where:

- M—a rotation matrix defined by Equation (9),
- X_{iLF}, X_{iVF}—designations as in Equation (33).

In the case when the number of adjustment points is greater than three, the rotation matrix M is determined by the least squares method according to the relationship (17), and the Equation (34) is redefined to the form (19), which can be written in the form of the relationship (35):

$$\begin{bmatrix} X_{iVF1} \\ Y_{iVF1} \\ Z_{iVF1} \\ X_{iVF2} \\ Y_{iVF2} \\ Z_{iVF2} \\ \cdots \\ X_{iVFN} \\ X_{iVFN} \\ X_{iVFN} \end{bmatrix} = \begin{bmatrix} X_{iLF1} & Y_{iLF1} & Z_{iLF1} & 0 & 0 & 0 & 0 & 0 & 0 \\ 0 & 0 & 0 & X_{iLF1} & Y_{iLF1} & Z_{iLF1} & 0 & 0 & 0 \\ 0 & 0 & 0 & 0 & 0 & 0 & X_{iLF1} & Y_{iLF1} & Z_{iLF1} \\ X_{iLF2} & Y_{iLF2} & Z_{iLF2} & 0 & 0 & 0 & 0 & 0 & 0 \\ 0 & 0 & 0 & X_{iLF2} & Y_{iLF2} & Z_{iLF2} & 0 & 0 & 0 \\ 0 & 0 & 0 & 0 & 0 & 0 & X_{iLF2} & Y_{iLF2} & Z_{iLF2} \\ X_{iLFN} & Y_{iLFN} & Z_{iLFN} & 0 & 0 & 0 & 0 & 0 & 0 \\ 0 & 0 & 0 & X_{iLFN} & Y_{iLFN} & Z_{iLFN} & 0 & 0 & 0 \\ 0 & 0 & 0 & 0 & 0 & 0 & X_{iLFN} & Y_{iLFN} & Z_{iLFN} \end{bmatrix} \times \begin{bmatrix} m_{11} \\ m_{12} \\ m_{13} \\ m_{21} \\ m_{22} \\ m_{23} \\ m_{31} \\ m_{32} \\ m_{33} \end{bmatrix} \tag{35}$$

where

$$X = (A^T \times A)^{-1} \times A^T \times L$$

The vector of unknowns, X, calculated using the Equation (35), contains elements of the rotation matrix M. The matrix M describes the relation between the centroid coordinates in the instrument (LF) and ship (VF) systems, which are given by the Equation (33). This relationship can be written in a parametric form in Equation (36):

$$X_{iVF} = M \times X_{iLF} \tag{36}$$

Equation (36) has a simple form, but it does not guarantee that the elements of the matrix M are related only to rotations, which can be stated after checking the orthogonality condition in the form (11). In the case of an orthogonal transformation, the condition (11) should be strictly met, which for a coordinate transformation containing six significant digits in the discussed case means compliance with the unit matrix at the level of 1×10^{-6}. Such a match confirms the condition of orthogonality and transformation by similarity. Otherwise, we should look at the M matrix as a transformation matrix—a quasi-rotation matrix—which we denote here as QR. The QR matrix is a 3×3 matrix and it can also be obtained by the affine transformation of coordinates by Equation (36), which can be written in the form of Equations (37) and (38):

$$X_{iVF} = \mathrm{QR} \times X_{iLF} \tag{37}$$

where:

$$QR = \begin{bmatrix} \lambda_X & 0 & 0 \\ 0 & \lambda_Y & 0 \\ 0 & 0 & \lambda_Z \end{bmatrix} \times \begin{bmatrix} m_{11} & m_{12} & m_{13} \\ m_{21} & m_{22} & m_{23} \\ m_{31} & m_{32} & m_{33} \end{bmatrix} = \begin{bmatrix} q_{11} & q_{12} & q_{13} \\ q_{21} & q_{22} & q_{23} \\ q_{31} & q_{32} & q_{33} \end{bmatrix} \tag{38}$$

If the scale coefficients λ_X, λ_Y, and λ_Z are equal unity, then condition (11) is satisfied, the QR matrix is equal to the M matrix, the transformation is orthogonal and the transformation is a transformation by similarity. In another case, the relationship between the coordinates is described by the Equation (37), and the QR transforming matrix can be described as the quasi-rotation matrix, then, the transformation in question is also defined as quasi-similarity.

Stage 5:

In the fifth step, after calculating the matrix M (QR), the condition for zeroing the deviations between the coordinates of the adjustment points in the VF system (related to the centroid) and the coordinates converted to this system from the LF system using the Equations (36)/(37) is checked. The fulfillment of the condition of zero deviations L, within 1×10^{-4} m (ten-fold more precise than the total station measurement), can be written using the relation (39):

$$L = \begin{bmatrix} L_X \\ L_Y \\ L_Z \end{bmatrix} = \begin{bmatrix} X_{iVF} \\ Y_{iVF} \\ Z_{iVF} \end{bmatrix} - \begin{bmatrix} q_{11} & q_{12} & q_{13} \\ q_{21} & q_{22} & q_{23} \\ q_{31} & q_{32} & q_{33} \end{bmatrix} \times \begin{bmatrix} X_{iLF} \\ Y_{iLF} \\ Z_{iLF} \end{bmatrix} = \begin{bmatrix} 0.0000 \\ 0.0000 \\ 0.0000 \end{bmatrix} \tag{39}$$

If the condition (39) is not met, corrections to the (quasi) QR rotation matrix are sought, which also includes the scale correction for each axis. Additionally, corrections to the position of the centroid of the LF system with respect to VF are sought, which can be generally written in the form of Equation (40), and after transformation in the form of Equation (41):

$$V = dQR \times X_{iLF} + dX_{iLF} - L \tag{40}$$

$$\begin{bmatrix} V_X \\ V_Y \\ V_Z \end{bmatrix} = \begin{bmatrix} 1 & 0 & 0 & X_{iLF} & Y_{iLF} & Z_{iLF} & 0 & 0 & 0 & 0 & 0 & 0 \\ 0 & 1 & 0 & 0 & 0 & 0 & X_{iLF} & Y_{iLF} & Z_{iLF} & 0 & 0 & 0 \\ 0 & 0 & 1 & 0 & 0 & 0 & 0 & 0 & 0 & X_{iLF} & Y_{iLF} & Z_{iLF} \end{bmatrix} \times \begin{bmatrix} dX_{iLF} \\ dY_{iLF} \\ dZ_{iLF} \\ dq_{11} \\ dq_{12} \\ dq_{13} \\ dq_{21} \\ dq_{22} \\ dq_{23} \\ dq_{31} \\ dq_{32} \\ dq_{33} \end{bmatrix} - \begin{bmatrix} L_X \\ L_Y \\ L_Z \end{bmatrix} \tag{41}$$

where

$$V = A \times X - L.$$

Equation (41) therefore contains the affine scale corrections with corrections to the value of revolutions (elements) and corrections to the LF system translation to the centroid—corrections of the LF system center of gravity (elements dX_{iLF}, dY_{iLF}, dZ_{iLF}). The determined corrections are added to the coordinates, and the control is the zeroing of the intercepts L in Equation (39), after taking into account corrections to the coordinates calculated by Equation (41).

Stage 6:

In the next step of transformation, back translation is performed—a shift opposite to the one that was subjected to the VF (vessel frame–fixed frame) system in the third transformation step. Reverse translation is written as the last component of the sum in Equation (23). Thus, the coordinates of the points in the LF system undergo a double change in translation. First, in the third step, it is a translation to the LF center of gravity, and in the sixth step, it is a back translation from the VF center of gravity. Thus, the transformation is completed, and its result is a point cloud in the VF system. Transformation results are assessed on common points and on check points that were common points (known in both LF and VF systems) but did not participate in determining transformation coefficients using Equations (36)/(37) and (40). In the proposed transformation model, it was assumed that the variables are uncorrelated, errors are random, and their values are small. Due to these assumptions, the normal distribution of measurements is used. As a result of the precision analysis, the mean observation error m_0, is calculated as an estimate of the variance coefficient (42), the covariance matrix Q (43) and finally, the mean errors of the coordinates on the connecting points (44):

$$m_0 = \pm\sqrt{\frac{V^T \times V}{n-k}} \tag{42}$$

$$Q = \left(A^T \times A\right)^{-1} \tag{43}$$

$$m_X = m_0 \times \sqrt{Q_{XX}},\ m_Y = m_0 \times \sqrt{Q_{YY}},\ m_Z = m_0 \times \sqrt{Q_{ZZ}} \tag{44}$$

where:

- n—number of matching points,
- k—number of observations (adjustment points) necessary to carry out the transformation (in the considered transformation model, $k = 3$

2.3. Software

For the research, the SC4W, Geonet DC (Dimensional Control) and Mathcad Professional software were used. The SC4W and Geonet DC software are designed for dimensional control, including the processing of data in the object's own system (local coordinate system), as well as for the coordinate transformation in oblique local coordinate systems. In the SC4W and Geonet DC software, transformation by similarity with an unknown engine of calculations is used. In Mathcad Professional software, the Q-ST algorithm was implemented, and calculations were carried out. The calculations were performed with the accuracy of 10^{-4} meters, which for measurements in the range up to 100 m gives six significant digits.

3. Results

3.1. Description of the Experiment

Tests on the accuracy of the Q-ST transformation were carried out based on laboratory data and on the basis of real data—collected during the measurements. Laboratory studies were based on ideal (no disturbance) and disturbed data. The disturbance to the data was generated randomly according to the normal distribution of the data. In the first step, the algorithm's response and accuracy was verified on the test data without disturbances, then on the disturbed data, and finally, on the real measurement data. The real data were collected during the measurements on a floating ship. All data were processed using the software described in Section 2.3. Laboratory tests were performed using 4 and 5 common (adjusted) points, and field studies with 4, 5 and 6 common points. The number of control points (check points) in laboratory tests was 10. Accuracy analysis was performed independently on common and check points.

3.2. Experiment

A set of ideal test data has been generated and then shifted and twisted according to the transformation (1). The rotation matrix R^{-1} was adopted for the rotation (the matrix R is described by the compound (5)), and the transformation scale was 1. The transformation parameters are presented in Table 1.

Table 1. Parameters for the transformation of the VF test dataset to the LF system.

	φ (°)	θ (°)	ψ (°)	X_0(m)	Y_0(m)	Z_0(m)	λ
Value	38.57893	33.30716	227.35478	10.000	10.000	4.000	1.000000

In this way, a set of laboratory data was created in two systems: VF—vessel frame (fixed frame) and LF—local frame (see Table 2 and Figure 6).

Table 2. Laboratory data test set in VF and LF systems.

Point	Vessel Frame—VF			Local Frame—LF		
	X (m)	*Y* (m)	*Z* (m)	*X* (m)	*Y* (m)	*Z* (m)
1.	0.000	0.000	0.000	−14.006	−2.300	−3.815
2.	5.000	−8.000	1.000	−12.468	6.188	−7.763
3.	15.000	−14.000	2.000	−14.990	14.829	−15.244
4.	25.000	−15.000	2.000	−20.037	19.041	−22.846
5.	35.000	−14.000	3.500	−27.137	22.472	−29.255
6.	15.000	14.000	2.000	−32.203	−7.053	−12.257
7.	5.000	8.000	1.000	−22.304	−6.315	−6.055
8.	25.000	0.000	0.000	−28.160	6.276	−22.553
9.	40.000	0.000	15.000	−44.890	19.239	−23.995
10.	38.000	7.000	15.000	−48.060	13.083	−21.749
11.	16.000	10.000	2.000	−30.310	−3.584	−13.433
12.	23.000	−11.000	2.000	−21.364	15.229	−20.920
13.	39.000	−6.000	12.000	−38.988	22.022	−25.846
14.	42.000	6.000	12.000	−48.063	13.673	−26.814
15.	31.000	11.000	3.000	−39.967	1.302	−23.916

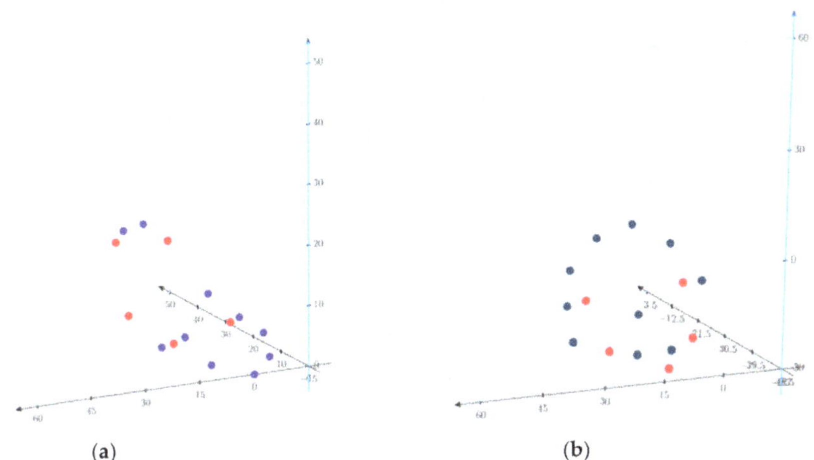

Figure 6. Ideal data distribution in the VF frame (**a**) and in the LF frame (**b**). Common points are marked in red color, check points are blue.

Then, the Q-ST was performed, and its results were compared with the calculations made in the software dedicated to Dimensional Control: SC4W and Geonet DC (Dimensional Control). The calculation results are summarized for two types of common points: for adjustment points and for check points (see Table 3).

Table 3. Mean errors of transformations (standard deviation) for laboratory dataset without distortion by random errors.

Transformation	No. of Common (Adjusted) Points	Mean Errors on Common (Adjusted) Points (mm)				Mean Errors on Check Points (mm)			
		m_X	m_Y	m_Z	m_P	m_X	m_Y	m_Z	m_P
Q-ST	5	0.00	0.00	0.00	0.00 0.00	0.00	0.00	0.00	0.00
SC4W		0.33	0.10	0.18	0.38	0.42	0.49	0.70	0.96
Geonet DC		0.33	0.10	0.18	0.39	0.42	0.49	0.71	0.96
Q-ST	4	0.00	0.00	0.00	0.00 0.00	0.00	0.00	0.00	0.00
SC4W		0.33	0.14	0.20	0.41	0.44	0.52	0.72	1.00
Geonet DC		0.33	0.09	0.28	0.44	0.49	0.49	0.96	1.18

The analysis of the data collected in Table 3 shows the resetting of the deviations on the combined points of adjustment and check in the Q-ST. The deviations are zero for the Q-ST, regardless of whether five fitting points (11–15 in Table 2) or four fitting points (12–15 in Table 2) were used for the transformation. In the programs for dimensional control (SC4W, Geonet DC), the deviations are not zeroed, and the errors in the position of the checkpoints after the transformation are rounded to 1 mm, despite the fact that the data used for the calculations were without disturbance.

The next step was to study the Q-ST on disturbed data. For this purpose, a new set of points was generated, which were located on a sphere with a radius of 50 m. These data were then shifted and twisted (see Table 4), and in the last step, random disturbances were added to them (see Table 5).

Table 4. Parameters for the transformation of the VF test dataset to the LF system before their disturbance.

	φ (°)	θ (°)	ψ (°)	X_0(m)	Y_0(m)	Z_0(m)	λ
Value	27.35478	1.30716	1.57894	15.000	1.000	2.000	1.000000

Table 5. Pseudo-random disturbance for laboratory data (see Table 6).

Point	dX (m)	dY (m)	dZ (m)
1.	−0.0009	−0.0014	−0.0008
2.	−0.0014	−0.0010	−0.0013
3.	−0.0009	0.0011	−0.0003
4.	−0.0019	−0.0005	0.0020
5.	−0.0034	0.0002	0.0004
6.	0.0020	−0.0058	0.0009
7.	0.0017	−0.0043	−0.0010
8.	0.0018	0.0004	−0.0014
9.	0.0013	−0.0012	0.0061
10.	−0.0021	−0.0024	−0.0006
11.	0.0001	0.0002	−0.0026
12.	−0.0015	0.0015	0.0015
13.	0.0014	0.0006	−0.0008
14.	−0.0004	0.0000	−0.0024
15.	−0.0013	−0.0015	−0.0021
Mean error—Calculated (m_P=0.0034 m)	0.0017	0.0022	0.0021

Table 6. Laboratory data test set with disturbances.

Point	Vessel Frame—VF			Local Frame—LF		
	X (m)	X (m)	X (m)	X (m)	Y (m)	Z (m)
1.	28.8670	28.8670	28.8680	51.967	7.250	31.873
2.	28.8670	−28.8670	28.8680	25.445	−44.025	32.681
3.	−28.8670	28.8670	28.8680	0.702	33.739	29.974
4.	−28.8670	−28.8670	28.8680	−25.821	−17.538	30.784
5.	28.8670	28.8670	−28.8680	53.281	5.661	−25.825
6.	0.0000	35.3550	35.3550	29.170	26.430	37.317
7.	0.0000	−35.3550	−35.3550	−1.700	−38.316	−32.360
8.	35.3550	35.3550	0.0000	61.370	9.242	3.146
9.	−35.3550	−35.3550	0.0000	−33.900	−21.120	1.816
10.	0.0000	−50.0000	0.0000	−9.238	−50.347	3.178
11.	50.0000	0.0000	0.0000	58.130	−28.876	4.122
12.	−50.0000	0.0000	0.0000	−30.666	17.002	0.834
13.	0.0000	0.0000	50.0000	12.594	−4.560	52.446
14.	0.0000	50.0000	0.0000	36.702	38.468	1.776
15.	0.0000	0.0000	−50.0000	14.872	−7.317	−47.492

The following errors were adopted to generate, according to the normal distribution, pseudo-random disturbances: $m_X = m_Y = m_z = 0.002$ m, which gives an error in the position of the point $m_P = 0.0035$ m. After generating pseudo-random disturbances for a set of twenty points, on a sphere with a radius of 50 m, the mean error of the point location was $m_P = 0.0034$ m, which slightly differs from the assumed value. The generated laboratory data were therefore shifted and twisted, and then randomly generated disturbances were added to them. In this way, data was obtained in the VF and LF systems (see Table 6).

The computations on disturbed data were performed in two variants: for four adjustment points (points 12–15 in Table 2) and for five adjustment points (points 11–15 in Table 2). In the next step, for the dataset in the VF and LF systems (Table 6), the scale factor was calculated according to the formula (25), which amounted to: $\lambda = 0.99999103$ for four adjustment points and $\lambda = 0.99997818$ for five adjustment points. In all cases, the robust orthogonality criterion for calculating the scale factor was met. During the calculations with the use of the Q-ST algorithm, the condition of orthogonality of the transformation (11) for the QR matrix (38) was checked. The results of this verification were related to the unit matrix, which was presented for four adjustment points in Equation (45) and for five adjustment points in Equation (46):

$$I - QR^T \cdot QR = \begin{bmatrix} 0.0000154 & -0.0000203 & -0.0000114 \\ -0.0000203 & -0.000036 & -0.0000203 \\ -0.0000114 & -0.0000203 & -0.0000302 \end{bmatrix} \tag{45}$$

$$I - QR^T \cdot QR = \begin{bmatrix} -0.0000247 & -0.0000374 & 0.0000104 \\ -0.0000374 & -0.0000624 & 0.0001047 \\ 0.0000104 & 0.0001047 & -0.0000071 \end{bmatrix} \tag{46}$$

The results of the Equations (45) and (46) indicate that the transformation is not orthogonal for common (adjusted) control points with a precision of six decimal places. The transformation shows elements of an affine transformation and is considered a quasi-similarity, as shown in Equation (38).

Then, the Q-ST transformation was performed, and its results were compared with the calculations made in software dedicated to Dimensional Control. The calculation results are summarized in Table 7.

Table 7. Mean errors of transformations (standard deviation) for laboratory dataset with distortion by random errors.

Transformation	No. of Common (Adjusted) Points	Mean Errors on Common (Adjusted) Points (mm)				Mean Errors on Check Points (mm)			
		m_X	m_Y	m_Z	m_P	m_X	m_Y	m_Z	m_P
Q-ST	5	0.57	0.03	0.50	0.76	2.27	3.03	2.89	4.77
SC4W		1.06	0.94	1.35	1.95	2.70	2.34	2.80	4.54
Geonet DC		1.16	0.47	1.03	1.62	2.81	2.54	2.40	4.48
Q-ST	4	0.00	0.00	0.00	0.00 0.00	2.84	2.75	2.07	4.46
SC4W		1.25	0.38	0.97	1.62	2.84	2.49	2.07	4.31
Geonet DC		1.25	0.38	0.97	1.63	2.84	2.49	2.07	4.31

The next and last stage of verification of the Q-ST was field research on real data. Measurements were made with a tilted reference frame using the total station (see Figure 7).

Figure 7. Total station measurements on a tilted reference frame (on the ship in the water).

Measurements were made on the ship in the water, from four measuring stations. The ST1 station was the central one, which was the reference for the other measurement stations. The remaining positions were selected so that they had common points with the first position. Common points (adjustments) to the ST1 station are summarized in Table 8.

Table 8. Points common between positions—in relation to the first position—ST1.

Point Number ST1	Common Points		
	ST2	ST3	ST4
1		+	+
2		+	+
3		+	
5	+		
6	+		
7	+		
M1		+	+
M2	+	+	+
M3	+		
M4	+		
M11			+
Suma	6	5	5

Tables 9–12 summarize the coordinates of the points measured on individual measuring stations.

Table 9. Points measured at the ST1 station.

Station ST1 Point Number	*X* (m) (North)	*Y* (m) (East)	*Z* (m) (Up)
1	234.752	116.514	29.169
2	235.198	107.448	29.096
3	260.012	100.302	32.768
5	333.558	93.845	49.364
6	337.931	104.895	49.292
7	334.844	108.814	49.302
M1	259.083	117.185	29.061
M2	299.949	101.61	49.456
M3	319.777	96.212	49.505
M4	307.113	109.849	48.809
M11	257.723	102.046	32.763

Table 10. Points measured at the ST2 station.

Station ST2 Point Number	*X* (m) (North)	*Y* (m) (East)	*Z* (m) (Up)	Common Point (Yes—Y/No—N)
ST2	300.000	100.000	50.000	N
M2	264.588	99.054	48.506	Y
5	298.653	93.677	48.469	Y
6	302.242	105.017	48.403	Y
7	298.884	108.703	48.410	Y
M4	271.151	107.780	47.873	Y

Table 10. *Cont.*

Station ST2 Point Number	X (m) (North)	Y (m) (East)	Z (m) (Up)	Common Point (Yes—Y/No—N)
FUGRO_STBD_C	273.766	102.844	56.104	N
FUGRO_PORT_C	273.770	100.120	56.141	N
FUGRO_PORT_1	273.822	100.169	56.188	N
FUGRO_PORT_2	273.844	100.131	56.189	N
FUGRO_PORT_3	273.837	100.072	56.196	N
FUGRO_PORT_4	273.813	100.049	56.187	N
FUGRO_STB_1	273.810	102.910	56.147	N
FUGRO_STB_2	273.832	102.874	56.143	N
FUGRO_STB_3	273.831	102.834	56.141	N
FUGRO_STB_4	273.810	102.797	56.144	N
GPS_PORT_1	296.980	91.978	48.604	N
GPS_PORT_2	297.054	91.978	48.602	N
GPS_PORT_3	297.119	91.914	48.599	N
GPS_PORT_4	296.945	91.956	48.604	N
GPS_PORT_5	297.022	91.984	48.603	N
GPS_PORT_6	297.125	91.882	48.599	N
GPS_STBD_1	297.020	110.743	48.519	N
GPS_STBD_2	296.966	110.711	48.519	N
GPS_STBD_3	296.888	110.722	48.518	N
GPS_STBD_4	296.925	110.709	48.518	N
GPS_STBD_5	296.995	110.722	48.519	N
GPS_STBD_6	297.032	110.762	48.518	N
M3	284.740	95.070	48.588	Y

Table 11. Points measured at the ST3 station.

Station ST3 Point Number	X (m) (North)	Y (m) (East)	Z (m) (Up)	Common Point (Yes—Y/No—N)
ST3	300.000	100.000	50.000	N
M2	325.598	92.445	69.835	Y
M1	283.660	104.852	49.435	Y
1	259.446	102.306	49.537	Y
2	260.589	93.302	49.451	Y
3	285.881	88.085	53.128	Y
USBL_1	307.939	86.696	48.413	N
USBL_2	308.280	86.765	48.408	N
USBL_3	307.707	86.259	48.421	N
USBL_4	307.968	85.836	48.419	N
USBL_5	308.458	85.837	48.411	N
USBL_6	308.697	86.172	48.404	N

Table 12. Points measured at the ST4 station.

Station ST4 Point Number	X (m) (North)	Y (m) (East)	Z (m) (Up)	Common Point (Yes—Y/No—N)
ST4	300.000	100.000	50.000	N
M2	356.008	90.648	69.726	Y
M1	314.558	104.564	49.317	Y
M11	313.816	89.377	53.002	Y
1	290.266	102.901	49.412	Y
2	291.082	93.860	49.329	Y
PRISM_SF	328.338	104.061	49.454	N
PRISM_SA	300.156	104.953	49.644	N
PRISM_PA	299.529	84.558	49.710	N

Then, the measurements from the stations ST2, ST3 and ST4 were transformed to the coordinate system of the ST1 station. The transformation was performed with the use of common points. The calculations were made in dimensional control software: SC4W, Geonet DC and with the use of the Q-ST algorithm. As a result of the calculations, the scale factor was determined from the dependence (25), which for the transformation of individual sites in relation to station 1 is presented using Equation (47):

$$\begin{aligned}\lambda_{21} &= 1.0000433\\ \lambda_{31} &= 0.9998868\\ \lambda_{41} &= 0.9999414\end{aligned} \tag{47}$$

In all cases, the robust orthogonality criterion for scale factor calculation, checked by formula (29), was met. The orthogonality condition (11) for the QR matrix was also checked. The results for individual stations are presented by means of (48)–(50)—in the order ST2, ST3 and ST4 against ST1:

$$I - QR^T \cdot QR = \begin{bmatrix} -0.0001745 & 0.0004735 & 0.0065694 \\ 0.0004735 & 0.0006779 & 0.0021945 \\ 0.0065694 & 0.0021945 & -0.0057780 \end{bmatrix} \tag{48}$$

$$I - QR^T \cdot QR = \begin{bmatrix} 0.0002384 & 0.0002747 & -0.0004385 \\ 0.0002747 & -0.0002029 & -0.0003900 \\ -0.0004385 & -0.0003900 & 0.0005775 \end{bmatrix} \tag{49}$$

$$I - QR^T \cdot QR = \begin{bmatrix} 0.0007725 & -0.0003493 & -0.0015617 \\ -0.0003493 & 0.0000938 & 0.0002925 \\ -0.0015617 & 0.0002925 & 0.0011551 \end{bmatrix} \tag{50}$$

The calculated scale factor (47) and the (quasi) rotation matrix indicate that the transformation is not strictly orthogonal with a six-digit precision.

The transformation errors for common (adjusted) points from individual measurement stations, in relation to the measurements from the ST1 station, are presented in Table 13. Due to the small number of common control points between the individual measuring stations, all common points were used only as the adjusted control points.

The data in Table 13 show that the smallest transformation errors are achieved using the Q-ST transformation. With five fitting points, the transformation mean errors are approximately 1.5 mm. In the other transformations—SC4W and Geonet DC—these errors are three times higher. With six adjustment points, the results achieved with the Q-ST are already close to the transformation made in the programs for dimensional control, but still smaller, this time by about 30%. In both calculation variants, with five and six points in common, SC4W and Geonet DC software give similar results and it is difficult to indicate which of them is more accurate.

Table 13. Mean error (standard deviation) for field surveying data set.

Transformation	No. of Common (Adjusted) Points	Mean Errors on Common (Adjusted) Points (mm)				Average Errors on Common (Adjusted) Points (mm)			
		m_X	m_Y	m_Z	m_P	m_X	m_Y	m_Z	m_P
Q-ST	6 (ST2 to ST1)	1.86	2.34	0.75	3.09	1.44	1.82	0.55	2.39
SC4W		3.19	3.00	0.89	4.47	2.50	1.83	0.67	3.17
Geonet DC		3.24	2.88	0.92	4.43	2.54	1.92	0.77	3.28
Q-ST	5 (ST3 to ST1)	0.55	0.20	1.48	1.59	0.43	0.16	1.16	1.25
SC4W		3.81	1.32	2.29	4.64	3.20	1.00	1.80	4.54
Geonet DC		3.61	1.29	2.13	4.39	2.99	1.00	1.78	3.62
Q-ST	5 (ST4 to ST1)	1.31	0.12	0.32	1.350.00	1.03	0.10	0.26	1.07
SC4W		4.30	0.71	2.06	4.82	3.60	0.40	1.80	4.04
Geonet DC		4.56	0.80	1.93	5.02	3.80	0.52	1.68	4.19

The final step in verifying the Q-ST was to compare the transformation results of the measurement points (see Tables 10–12), which were not fitted ("N" designation) to transformation by similarity. The transformation of these points was performed in the following programs: PCMS (Leica), Spatial Analyzer (SA), SC4W, Geonet Dimensional Control and using the Q-ST algorithm. The discrepancy in the position of the points between individual software in relation to the Q-ST algorithm was 4.79 mm on average. This difference is about 70% of the value of the sum of errors on the common points (see Table 13) of the Q-ST compared to other methods. A sparse point cloud formed from all transformed points is shown in Figure 8.

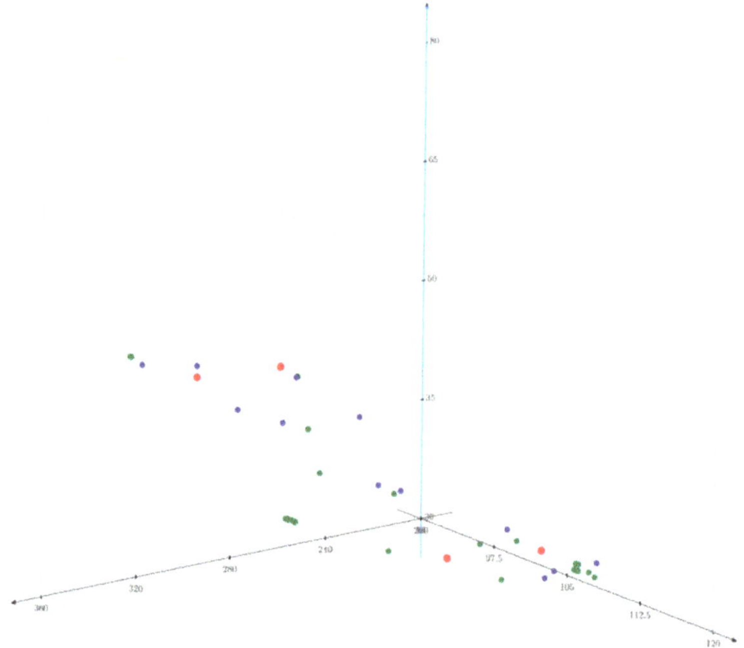

Figure 8. Sparse point cloud. The measuring stations are marked in red, common points are marked in blue and the measuring points are marked in green.

4. Discussion and Conclusions

The presented method of Quasi-Similarity Transformation allows to obtain zero deviations on common points, in the case when the number of these points is four, that is, with one redundant point. The transformation of the coordinates of the connecting points was also performed in the SC4W and Geonet DC programs, which are dedicated to dimensional control, in particular in the offshore industry. Geonet DC and SC4W do not produce zero deviations on combined points when the number of match points is greater than three. Even with ideal, laboratory-generated data, errors within 1 mm are generated using these programs. In the case of five connection points, the Q-ST transformation gave three times greater precision than the software used for dimensional control purposes, which was verified, among others, on real data by transforming points measured on the ship in the inclined frame of reference. For the six connectivity points, the Q-ST gave 30% more accurate results than software used in the offshore industry. The goal of zeroing deviations on total points, set in the paper, when their number is four, has been achieved. When the number of common points is five, the deviations are no longer zero, but are still close to zero (within the range of 0.5–1.5 mm in the conducted experiments). When the number of total points is six, the precision of the Q-ST algorithm is still better, but already close to that obtained in coordinate transformation programs.

The tested Geonet DC, SC4W (for connecting points) and PCMS (Leica), Spatial Analyzer (SA), SC4W and Geonet DC software (for the remaining points) give very similar coordinate transformation results. This allows the conclusion that a similar transformation by similarity algorithm works in them, and that the subtle differences in coordinates result from the compensatory least squares approach. The Q-ST, on the other hand, has a non-standard compensatory approach—the least squares method is used separately at each stage of the calculations. The presented computational approach—the Q-ST—was verified on the following data: laboratory, disturbed and real laboratory. In each calculation variant, the Q-ST gave the smallest errors at the adjustment points, in relation to the compared methods. The assumption of the normal (Gaussian) distribution of the data and the least squares method allows us to conclude that the Q-ST method gives the most probable results of the transformation by similarity, which results directly from smaller errors in relation to other methods. Therefore, the authors believe that the Q-ST is the optimal computational method for coordinate transformations with a small number of common points (4–6 points) and can find application, in particular in the offshore industry, where it is very often necessary to perform geodetic measurements in inclined frames of reference.

Author Contributions: G.S. proposed the original idea for the Q-ST; A.T. proposed the implementation of the Q-ST algorithm in offshore surveying; A.T., T.Z., G.S. and M.L. implemented the experiments; G.S., M.L. and A.T. wrote the paper, with the revision by T.Z. All authors have read and agreed to the published version of the manuscript.

Funding: This research was funded by the grants No. 1/S/IG/17 and 3/SKG/20, financed from a subsidy of the Ministry of Science and Higher Education for statutory activities.

Acknowledgments: G.S. gives special thanks to the professors: Edward Nowak, Andrzej Borkowski and Mieczysław Kwaśniak, for their critical comments on the calculation methods based on the centroids. The authors would like to especially thank Krzysztof Beczkowski for his involvement in the data acquisition. The authors pay tribute with this research to the late Marta Kowalska.

Conflicts of Interest: The authors declare no conflict of interest.

References

1. Deakin, R.E. 3-D coordinate transformations. *Surv. L. Inf. Syst.* **1998**, *58*, 223–234.
2. Grewal, M.S.; Weill, L.R.; Andrews, I.A.P. *Global Positioning Systems, Inertial Navigation, and Integration*, 2nd ed.; John Wiley & Sons: Hoboken, NJ, USA, 2006.
3. Soler, T. A compendium of transformation formulas useful in GPS work. *J. Geod.* **1998**, *72*, 482–490. [CrossRef]
4. Luhmann, T.; Robson, S.; Kyle, S.; Boehm, I.J. *Close-Range Photogrammetry and 3D Imaging*, 2nd ed.; Walter de Gruyter: Berlin/Boston, Germany, 2014.

5. Colomina, I.; Molina, P. Unmanned aerial systems for photogrammetry and remote sensing: A review. *ISPRS J. Photogramm. Remote Sens.* **2014**, *92*, 79–97. [CrossRef]
6. Brazeal, R. Three dimensional coordinate transformations for registering terrestrial laser scanning datasets based on tie points. *Point Cloud Anal.* **2013**, SUR 6905. [CrossRef]
7. Huang, J.; You, S. Point cloud matching based on 3D self-similarity. In Proceedings of the 2012 IEEE Computer Society Conference on Computer Vision and Pattern Recognition Workshops, Providence, RI, USA, 16–21 June 2012.
8. Brazetti, L.; Scaioni, M. Automatic orientation of image sequences for 3D object reconstruction: first results of a method integrating photogrammetric and computer vision algorithms. In Proceedings of the 3D-ARCH 2009: 3D Virtual Reconstruction and Visualization Workshop, Trento, Italy, 25–28 February 2009.
9. Schofield, W.; Breach, M. *Engineering Surveying*, 6th ed.; CRC Press: Boca Raton, FL, USA, 2007.
10. Kilford, W.K. Surveying for Engineers. *Surv. Rev.* **1979**, *25*, 94–96. [CrossRef]
11. Tu, Y. Machine learning. In *EEG Signal Processing and Feature Extraction*; Springer: Berlin/Heidelberg, Germany, 2019; pp. 301–323.
12. Zuur, A.F.; Ieno, E.N.; Smith, G.M. Principal coordinate analysis and non-metric multidimensional scaling. In *Analysing Ecological Data*; Springer: Berlin/Heidelberg, Germany, 2007; pp. 259–264.
13. Graf, M. Coordinate transformations in object recognition. *Psychol. Bull.* **2006**, *132*, 920–945. [CrossRef] [PubMed]
14. Baetslé, P.-L. Conformal transformations in three dimensions. *Photogramm Eng. Remote Sens.* **1966**, *32*, 816–824.
15. Stark, M. *Geometra Analityczna z Wstępem do Geometrii Wielowymiarowej*, 6th ed.; Państwowe Wydawnictwo Naukowe: Warszawa, Poland, 1974. (In Polish)
16. Kutoglu, H.S.; Mekik, C.; Akcin, H. A comparison of two well known models for 7-parameter transformation. *Aust. Surv.* **2002**, *47*, 24–30. [CrossRef]
17. Stępień, G. *Transformacje Symetryczne w Nachylonych Układach Odniesienia z Wykorzystaniem Metod Analizy Funkcjonalnej*; Wydawnictwo Naukowe Akademii Morskiej w Szczecinie: Szczecin, Poland, 2018. (In Polish)
18. Schut, G.H. Similarity Transformation and least Squares. *Photogramm. Eng.* **1973**, *39*, 621–627.
19. Diebel, J. Representing attitude: Euler angles, unit quaternions, and rotation vectors. *Matrix* **2006**, *58*, 1–35.
20. Titterton, D.; Weston, J. *Strapdown Inertial Navigation Technology*, 2nd ed.; IET: London, UK, 2004.
21. Baranowski, L. Equations of Motion of a Spin-Stabilized Projectile for Flight Stability Testing. *J. Theor. Appl. Mech.* **2013**, *51*, 235–246.
22. Allgeuer, P.; Behnke, S. Fused Angles and the Deficiencies of Euler Angles. In Proceedings of the 2018 IEEE/RSJ International Conference on Intelligent Robots and Systems (IROS), Madrid, Spain, 1–5 October 2018.
23. Henderson, D. *Euler Angles, Quaternions, and Transformation Matrices*; Technical Memorandum for NASA, Lyndon B. Johnson Space Center: Houston, TX, USA, 1997.
24. Kielich, S. *Molekularna Optyka Nieliniowa*; Państwowe Wydawnictwo Naukowe: Warszawa-Poznań, Poland, 1977. (In Polish)
25. Liu, H.; Fang, Y. Direct 3D coordinate transformation based on the affine invariance of barycentric coordinates. *J. Spat. Sci.* **2019**. [CrossRef]
26. Andrei, C. 3D Affine Coordinate Transformations. Master's Thesis, School of Architecture and the Built Environment Royal Institute of Technology (KTH), Stockholm, Sweden, March 2006.
27. Späth, H. A Numerical Method for Determining the Spatial HELMERT Transformation in the Case of Different Scale Factors. *Fachbeiträge* **2004**, *6*, 255–257.
28. Wolf, H. Geometric connection and re-orientation of three-dimensional triangulation nets. *J. Geod.* **1963**, *68*, 165–169. [CrossRef]
29. Deakin, R. *A note on the Bursa-Wolf and Molodensky-Badekas transformations*; Bulletin for the School of Mathematican and Geospatial Sciences; RMIT University: Merlbourne, Australia, 2006.
30. Badekas, J. Establishment of an Ideal World Geodetic System. Ph.D. Thesis, The Ohio State University, Columbus, OH, USA, 1969.
31. Dare, O.P.; Okiemute, E.S.; Eteje, S.O. Impact of different centroid means on the accuracy of orthometric height modelling by geometric geoid method. *Int. J. Sci. Rep.* **2020**, *6*, 124–130. [CrossRef]
32. Jue, L. Research on close-range photogrammetry with big rotation angle. *Int. Arch. Photogramm. Remote Sens. Spat. Inf. Sci.* **2008**, *1*, 11–14.

33. Acar, M.; Özlüdemir, M.T.; Akyilmaz, O.; Çelik, R.N.; Ayan, T. Deformation analysis with Total Least Squares. *Nat. Hazards Earth Syst. Sci.* **2006**, *6*, 663–669. [CrossRef]
34. Qin, Y.; Fang, X.; Zeng, W.; Wang, B. General Total Least Squares Theory for Geodetic Coordinate Transformations. *Appl. Sci.* **2020**, *10*, 2598. [CrossRef]
35. Brandt, S. *Data Analysis: Statistical and Computational Methods for Scientists and Engineers*, 4th ed.; North-Holland Publishing Company: Amsterdam, The Netherlands, 2014.
36. Taylor, J.R.; Thompson, W. An Introduction to Error Analysis: The Study of Uncertainties in Physical Measurements. *Phys. Today* **1998**, *51*, 57–58. [CrossRef]
37. Czarnecki, K. *Geodezja Współczesna w Zarysie*; Wydawnictwo Wiedza i Życie: Warszawa, Poland, 1995. (In Polish)
38. Wrona, W. *Matematyka*; Państwowe Wydawnictwo Naukowe: Warszawa, Poland, 1964. (In Polish)

Article

Marine Diesel Engine Exhaust Emissions Measured in Ship's Dynamic Operating Conditions

Artur Bogdanowicz * and Tomasz Kniaziewicz

Faculty of Mechanical and Electrical Engineering, Polish Naval Academy, 81-103 Gdynia, Poland; t.kniaziewicz@amw.gdynia.pl

* Correspondence: a.bogdanowicz@amw.gdynia.pl

Received: 26 October 2020; Accepted: 16 November 2020; Published: 18 November 2020

Abstract: The paper presents the results of research on measuring the emissions from marine diesel engines in dynamic states. The problem is as follows: How to measure emissions of the composition of exhaust gases on board a ship, without direct measurement of fuel consumption and an air flow to marine diesel engine, during maneuvering the ship in the port area. The authors proposed a measurement methodology using an exhaust gas analyzer with simultaneous recording of the load indicator, engine speed, inclinometer, and Global Positioning System (GPS) data. Fuel consumption was calculated based on mean indicated pressure (MIP) tests. Recorded data were processed in the LabView systems engineering software. A simple neural network algorithm was used to model the concentrations of ingredients contained in engine exhaust gases during dynamic states. Using the recorded data, it is possible to calculate the emissions of the composition of exhaust gases from the marine diesel engine and calculate the route emissions of the tested vessel.

Keywords: ship emissions; artificial neural network; dynamic states; ports; hydrographic survey vessel

---

1. Introduction

A typical operation of a ship contains three basic states of operation: Stay at quay, maneuvering, and cruise. During this states, ships perform standard unberthing and mooring maneuvers, in addition they make frequent changes of direction and speed. In ports and offshore areas we can find special units such as tugs and dredgers, in which main propulsions are exposed to variable loads during most of their operation. Load changes of main propulsions affect variable emissions of ingredients contained in engine exhaust gases into the atmosphere. Due to the fact that port areas are close to human agglomerations, maneuvering vessels affect the health of people living there [1,2]. Therefore, the problem of the emissions from marine diesel engine in dynamic states taking place in ports and harbors areas is necessary to be investigated.

In reference [3], the authors comprehensively reviewed the status of the air quality measured in harbor areas. Measured concentrations of the main air pollutants were compiled and intercompared, for different countries worldwide allowing a large spatial representativeness. However, the published studies showed a limited number of available air quality monitoring data in harbor areas, mostly located in Europe.

In Reference [4], a detailed exhaust emission inventory of ships by using Automatic Identification System (AIS) data were developed for Tianjin Port, one of the top 10 world container ports and the largest port in North China. Pollutants were mainly emitted during cruise and hoteling modes, and the highest intensities of emissions located in the vicinity of fairways, berth and anchorage areas in Tianjin Port.

The author of the paper [5] confirmed that the emissions in ports were substantial. In the paper, the ports in Asia and Europe were examined and concluded that most of emissions came from container ships and tankers. It was estimated that approximately 230 million people were directly exposed to the emissions in the top 100 world ports in terms of shipping emissions.

The estimation and analysis of ship emissions in popular tourist ports in Dubrovnik (Croatia) and Kotor (Montenegro) and along the eastern coast of the Adriatic Sea were presented in the paper [6]. There was also a record (2012–2014 years) of cruise ships calling at these ports, which was used to model and estimated exhaust emission factors and their impact on the area of bays and ports.

There is an ongoing debate regarding the measurement of emissions from ships operating at sea. Currently, there are no guidelines or legal requirements defining methods and rules for minimizing the emission of exhaust gases during sea crossing by vessels. In the paper [7], the existing methods for calculating energy consumption and emissions were presented. The authors conceived an attempt to propose the most appropriate method of obtaining the data needed for optimal energy management, which could be applied to any vessel. The considerations mainly concerned the sea crossing.

In Reference [8], the authors presented a method for estimating the emissions of analyzed exhaust gases in ports on the basis of vessel traffic data refer to one year. The research focused mainly on cruise and passenger ships. The analysis concerned only the movement of ships without taking into account auxiliary activities such as maneuvering the vessel or loading at the quay.

In Reference [9], the authors presented the possibility of estimating emissions using the chemical model of transport in the North Sea area. The data for the model were provided by the Dutch research institute MARIN, which is responsible for creating emission reports in the Baltic Sea areas and selected Dutch ports [10,11].

The conducted research on emissions from marine diesel engines is mainly focused on averaging emissions. In the field of aircraft engines, in the paper [12], the model for determining the exhaust emissions of aviation piston engines during the flight of an aircraft was presented. The assessment was carried out in accordance with the guidelines contained in Annex 16 to the Convention on International Civil Aviation Organization (ICAO), while the test covered the operational parameters of the engine corresponding to the approach, landing and operations at the airport.

The civilian ships realizing sea crossing are obliged to comply with provisions on the protection of the marine environment. One of such documents is the International Convention for the Prevention of Pollution from Ships, known as the "MARPOL Convention" with subsequent annexes [13–15].

In Reference [16], the authors examined the potential effects of the emerging international maritime emission regulations on the competition between seaports and the potential underlying economic motivations fostering the discussion of introducing Emission Control Areas (ECA). The legal analysis; however, showed that the current enforcement regime of MARPOL Annex VI should be improved in order to rule out the possibility of a low degree of compliance and to protect the competitiveness of complying ships.

In Reference [17], the International Maritime Organization (IMO) is working on development of new short-term measures to implement greenhouse gas (GHG) strategy. Draft new mandatory measures to cut the carbon intensity of existing ships have been agreed by an IMO working group. It is assumed to reduce carbon intensity of international shipping by 40% by 2030, compared to 2008. The amendments were developed by the seventh session of the Intersessional Working Group on Reduction of GHG Emissions from Ships (ISWG-GHG 7), held as a remote meeting 19–23 October 2020.

In the case of warships, they are exempt from compliance with emission standards. However, naval fleets, including the Polish Navy, implement the regulations on their ships as far as possible.

This article is a continuation of authors previous works on the emissions from marine diesel engines. Due to the fact that research on emissions from marine diesel engines is mainly aimed on averaging emissions, the authors decided to measure the emissions more precisely, focusing on their waveforms. The authors decided to use neural networks for modeling emission from marine diesel engines in dynamic states. The neural networks were used in emission modeling [18], but in the field of marine diesel engines there is a gap. The research questions that this article tries to answer are as follows:

- Is it possible to measure and calculate the emissions of the composition of exhaust gases in dynamic operating conditions of marine diesel engine, taking into account the distance travelled?
- Is it possible to use a low-calculation method for modeling the emissions of the composition of exhaust gases in the dynamic operating conditions of the ship's propulsion system?

2. Materials and Methods

2.1. Research Object

The research was carried out on board the hydrographic ship project 874 (Figure 1) while carrying out survey work at sea. The ship's data are presented in Table 1.

Figure 1. The silhouette of the hydrographic ship [19].

Table 1. Project 874 ship data [19].

Specification	
Standard displacement	1145 t
Full displacement	1214 t
Length overall	61.6 m
Breadth	10.8 m
Draught	3.3 m
Speed	13.7 w
Main engines	2 × SULZER type 6AL25/30
Generators	3 × WOLA39H12

The main purpose of the ship is to perform hydrographic measurements and research, to put up navigation signs, and to mark the shallows in the sounding areas. The tasks are carried out as part of the navigational and hydrographic security of the Polish Navy and for the maritime economy and navigation safety.

The hydrographic ship has two engine compartments. In the main engine room, there are two SULZER 6AL25/30 marine diesel engines and one generator set WOLA 39H12 marine diesel engine. In the auxiliary engine room there are the other two generating sets WOLA 39H12 marine diesel engines. The visualization of the main propulsion system is shown in Figure 2.

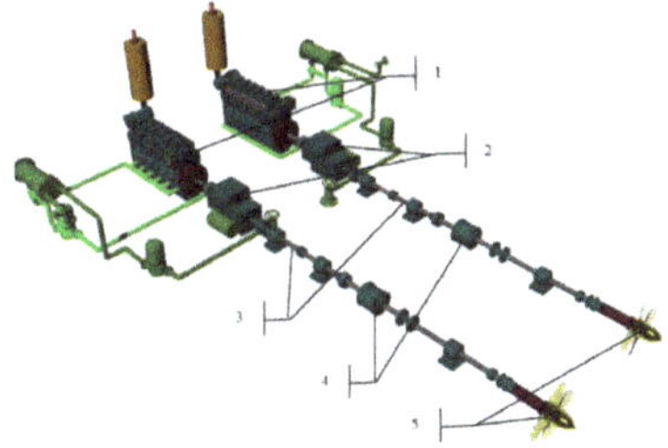

Figure 2. Visualization of the two-shaft main propulsion system with two controllable pitch propellers (CPP): 1—Main diesel engines, 2—reduction gears with auxiliary electric motors, 3—shaft lines, 4—CPP hydraulic boxes, 5—CPP's.

The research was aimed at measuring the concentration of exhaust gases components from the main propulsion marine diesel engine. The SULZER engine data type 6AL25 / 30 are presented in Table 2.

Table 2. Marine diesel engine SULZER type 6AL25/30 [19].

Specification	
Piston arrangement	Inline
Cylinder diameter	250 mm
Piston stroke	300 mm
Displacement volume	1 cyl.—14.726 dm^3
Nominal power	706.08 kW
Starter	pressure compressed air—3 MPa
Number of cylinders	6
Number of valves per cylinder	4

The measurements were realized using portable exhaust gas analyzer TESTO MARITIME 350 [20]. It can be used to measure the gaseous flue gas concentrations of oxygen (O_2), carbon monoxide (CO), carbon dioxide (CO_2), nitrogen oxides (NO_x), and sulfur dioxide (SO_2). The technical data of portable analyzer are presented in Table 3.

Table 3. Parameters and measuring ranges of the TESTO 350 analyzer [20].

Parameter	Measuring Range	Tolerance
Temperature range	from −40 to + 1000 °C	max. ± 5 °C
Oxygen range	from 0 to 25%	According to MARPOL, Annex VI or NO_x Technical Code
Carbon monoxide range	from 0 to 3000 ppm	
Nitrogen monoxide range	from 0 to 3000 ppm	
Nitrogen dioxide range	from 0 to 500 ppm	
Sulphur dioxide range	from 0 to 3000 ppm	
Carbon dioxide range	from 0 to 40%	
Absolute pressure range	from 600 to 1150 hPa	±5 hPa w 22 °C ±10 hPa w −5 do +45 °C

The GPS receiver type BU-353S4 was used to record the position of the ship. The technical data are presented in Table 4.

Table 4. Antenna GPS type BU-353S4 [21].

Specification	
Chipset	SiRF STAR IV GSD4e
Frequency	L1; 1575.42 MHz
C/A code	1.023 MHz chip rate
Channels	48
Sensitivity	−163 dBm
Accuracy position	<2.5 m 2D RMS SBAS Enable
Velocity	0.1 m/s
Time	1 µs synchronized to GPS time

The indicator MA2018 was used to measure the indicated pressure. The device was developed at the Polish Naval Academy and allows measurements with a sampling frequency of 20 kHz and with resolution of 12 bits. This device works with a piezoelectric pressure transducer type 7613B KISTLER. Technical data are presented in Table 5. In addition, the device uses a GIG PDS-1 vibration acceleration sensor, which is mounted on the bolts securing the cylinder head cover.

Table 5. Pressure transducer type 7613B KISTLER [22].

Specification	
Pressure range	0–25 MPa
Maximum indicated pressure	30 MPa
Sensitivity	2 mV/MPa
Resonant frequency	60 kHz
Temperature range	223–623 K
Temperature drift 473, ... , 500 K	3.5%

The ADVANTECH USB-4711A unit is equipped with an onboard terminal block, 16-ch analog input, 2-ch analog output, 16-ch digital I/O, and a counter channel capable of outputting a constant frequency square wave [23]. The technical data are presented in Table 6.

Table 6. ADVANTECH USB-4711A unit [23].

Specification	
Channels	16 analog input, 2 analog output
Resolution	12 bits
Max. sampling rate	150 kS/s max
FIFO size	1024 samples
Sampling mode	Software, onboard programmable pacer, and external
Bipolar	±10 ± 5 ± 2.5 ± 1.25 ± 0.625
Absolute Accuracy (% of FSR)	0.1 0.1 0.2 0.2 0.4

The USB-4711A unit was connected with:

- Rotating speed sensor,
- fuel rack sensor, and
- inclinometer.

Rotating speed was measured using reflection sensor Optom OCOE 02581 [24]. The linear potentiometer LPF 150 [25] was used as fuel rack sensor. The Kubler 8.IS40.23321 inclinometer [26] was used to measure the angles of pitch and roll. Measurement for this type of device is carried out biaxially in the ranges of ±60°. The output signal is a voltage ranging from 0 to 5 V for both axes. It is supplied with a voltage ranging from 10 to 30 V. The device housing provides IP68 protection.

A logging program was written in the Labview [27] system engineering software to record the measurement data. All parameters were recorded with the frequency of 2Hz. Due to TESTO MARITIME 350 analyzer recorded data at a frequency of 1 Hz, there was no need to increase the logging frequency. The portable transducers were mounted on the SULZER marine diesel engine, type 6AL25/30 (Figure 3).

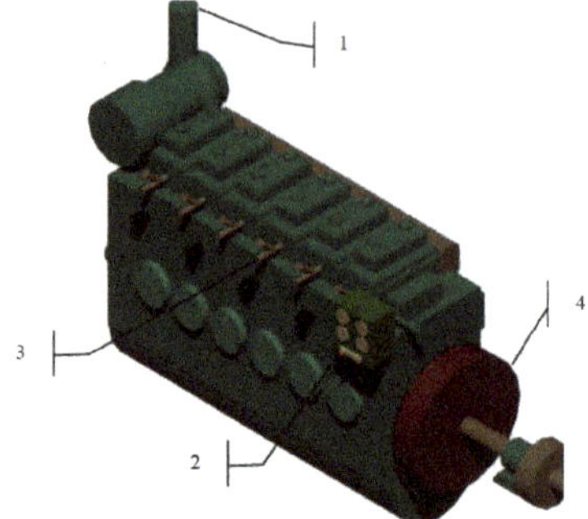

Figure 3. The arrangement of portable measuring transducers on the SULZER type 6AL25/30, 1 –exhaust gas analyzer probe, 2—fuel rack sensors, 3—indicated pressure sensor, 4—rotating speed sensor.

2.2. Measurement Site

The measurements were performed in Gulf of Gdansk and the area north of Władysławowo (Figure 4).

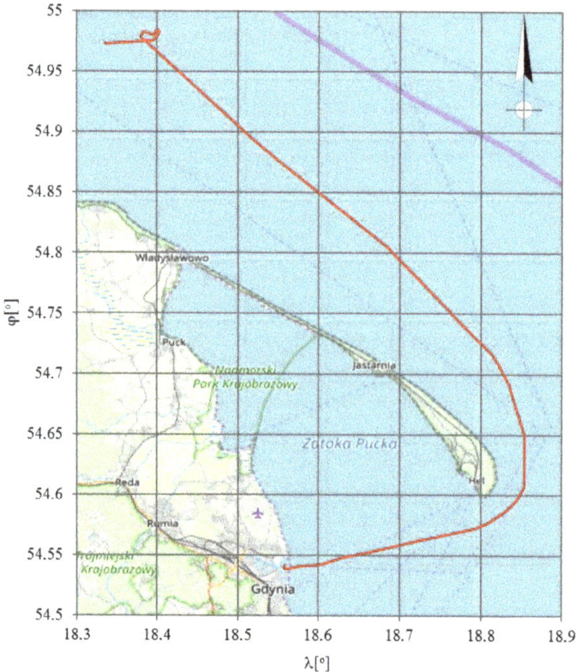

Figure 4. Registered route of the hydrographic vessel.

The measurement was carried out from the maneuvers in the Navy Port Gdynia to sounding area. The hydrometeorological conditions in the area were as follows:

- Wind direction and strength: NW—4,
- sea state: 3,
- atmospheric pressure: 1017 hPa, and
- air temperature: +5 °C.

Due to the large amount of recorded measurement data, the authors focused only on the ship's departure from the Navy Port Gdynia.

2.3. A Neural Network

The study of the nervous systems is an important factor in the advancement of systems theory and its practical applications. As early as 1943, McCulloch and Pitts developed a model of the nerve cell, the idea of which has survived over the years and is still the basic of most models in use. An important element of this model is the sum of the input signals with an appropriate weight and subjecting the obtained sum to the non-linear activation function [28].

The neural network receives information in the form of numerical variables, which are then sent, taking into account the weighting factors, to the individual neurons. Typically, the activation function is a linear, sigmoidal unipolar or bipolar. In modeling emissions of the composition of exhaust gases from marine diesel engine, it was decided to use the unipolar sigmoid function (Equations (2) and (4)). The products of the weight variables are added up and sent to the next layers. The number of operations depends on the number of neurons in the network. Figure 5 shows a diagram of the operation of a neural network with mathematical relationships used in the approximation.

$$h_j = \sum_t w_{t,j} \times x_t \tag{1}$$

$$v_j = \frac{1}{1 + exp\left(-h_j\right)} \tag{2}$$

$$s_k = \sum_j w_{t,j} \times v_j \tag{3}$$

$$y_k = \frac{1}{1 + exp(-s_k)} \tag{4}$$

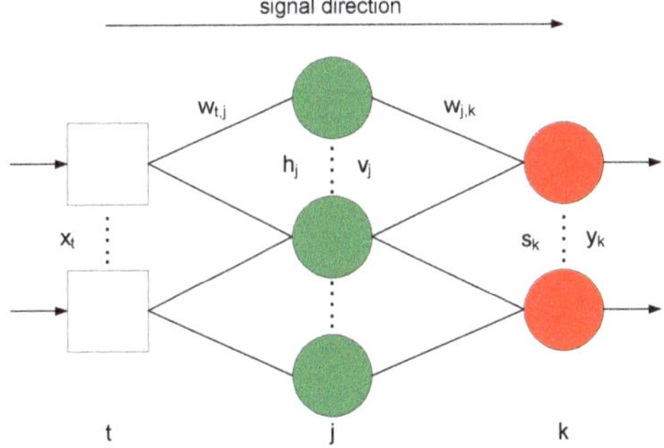

Figure 5. The signal direction in the neural network.

Backpropagation was used as the learning algorithm, which changed the weights, taking into account the minimization of errors:

$$\sigma_k = y_k \times (1 - y_k) \times \left(y_k^n - y_k\right) \tag{5}$$

$$\sigma_j = v_j \times \left(1 - v_j\right) \times \sum_k \sigma_k \times w_{j,k} \tag{6}$$

$$w_{j,k}(t+1) = w_{t,j}(t) + \mu \times \sigma_k \times v_j \tag{7}$$

$$w_{t,j}(t+1) = w_{j,k}(t) + \mu \times \sigma_j \times w_{j,k}. \quad (8)$$

A model of a neural network in the 3-5-1 configuration was built to model the emission concentrations. The choice of this configuration was dictated by the use of the simplest possible network to model emissions in dynamic states. More neurons in the hidden layer could result in the so-called "Overfitting the network" and receiving unreliable results, which would lead to more attempts to learn the neural network. The diagram of the neural network is shown in Figure 6. The network has three inputs, five neurons in the hidden layer, and one output neuron. The sigmoid function was used as the activation function in hidden layer and output neurons.

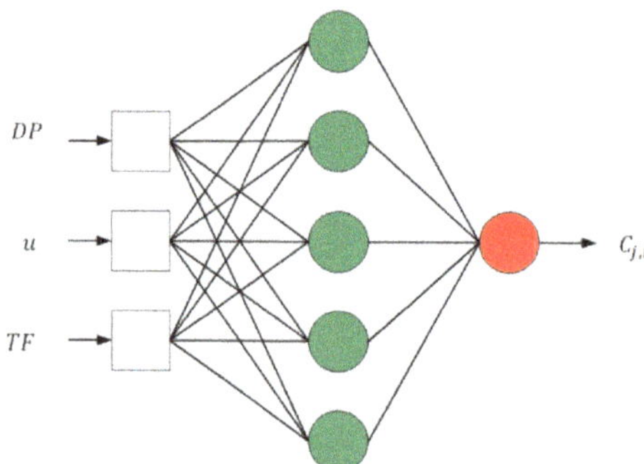

Figure 6. The neural network in configuration 3-5-1. *DP*—the fuel dose $\left[\frac{g}{rot.}\right]$, *u*—the speed of a ship [knots], *TF*—the exhaust gases temperature [°C], $C_{j,i}$—the concentration of exhaust gases components [ppm, %].

2.4. The Durbin–Watson Test

The Durbin–Watson test was used to check the fit of the model and empirical data [29] which uses the residual differences in values between the data obtained from the model and the empirical. The following relationship was used for the calculations:

$$DW = \frac{\sum_{i=1}^{n-1}\left(R_{reg\ C_j(i+1)} - R_{reg\ C_j(i)}\right)^2}{\sum_{i=1}^{n} R^2_{reg\ C_j(i)}} \quad (9)$$

where:

n—the number of data items,
$R_{reg\ C_j(i)}$—the regression residuum value,
$R_{reg\ C_j(i+1)}$—next step value of the regression residuum.

To determine the value of the Durbin–Watson (DW) test it is necessary to use the DW distribution tables [30]. Low (d_l) and high (d_g) limits are specified for the number of predictors and the number of data items in the model. These values determine the range of the residual correlation test. The DW statistic ranges from 0 to 4. The following range of the residual component correlation is assumed:

- if $DW = 2$, there is no correlation,
- if $DW > 2$, then
    - if $DW > 4 - d_l$, there is a negative correlation,
    - if $4—d_g < DW < 4 - d_l$, no conclusion/decision,
    - if $DW < 4 - d_g$, no correlation,
- if $DW < 2$, then
    - if $DW < d_l$, there is a positive correlation,

- if $d_l < DW < d_g$, no conclusion/decision,
- if $DW < 4 - d_g$, no correlation,

2.5. The Emission of the Composition of Exhaust Gases in Dynamic States

Due to the impossibility of installing a fuel flow meter on the ship's engine, authors decided to calculate the fuel consumption on the basis of MIP measurements. For this purpose, the thermal efficiency was determined for SULZER type A engines. It has been done on the basis of previous ship's engine tests and preliminary tests carried out on a laboratory engine SULZER, type 6AL20/24. The indicated power was calculated using the indicated pressure:

$$N_{icyl} = \frac{V_s \cdot n \cdot p_i \cdot z}{60} \tag{10}$$

where:

z—number of ignitions (for 4-s engines z = 0.5),
V_s—displacement volume $\left[m^3\right]$,
p_i—mean indicated pressure $[Pa]$, and
n—rotation speed $\left[min^{-1}\right]$.
The fuel mass flow rate consumed by a diesel engine was calculated:

$$\dot{m}_{fuel} = \frac{\sum_{i=1}^{k} N_{icyl}}{W_d \cdot \eta_i} \left[\frac{kg}{s}\right] \tag{11}$$

where:

η_i—thermal efficiency,
W_d—the calorific value of a fuel (for the NATO F–75 fuel is W_d = 42,700,000 $\frac{J}{kg}$), and
k—number of cylinders.
The fuel dose for one cylinder during one work cycle was calculated:

$$DP = \frac{\dot{m}_{fuel} \cdot 60,000}{n \cdot k} \left[\frac{kg}{rot}\right] \tag{12}$$

where:

$n$rotation speed $\left[min^{-1}\right]$.
The fuel dose was linearly dependent on the load indicator:

$$DP = f(WO) \tag{13}$$

The air mass flow rate and exhaust mass flow rate were calculated from the actual air demand, taking into account the excess air factor λ:

$$L_R = \lambda \times [11.84 \times c + 34.214 \times h] \left[\frac{kg_{air}}{kg_{fuel}}\right] \tag{14}$$

$$\lambda = \frac{20.95}{20.95 - C_{O_2}} \tag{15}$$

$$\dot{m}_{air} = \dot{m}_{fuel} \times L_R \left[\frac{kg}{s}\right] \tag{16}$$

$$\dot{m}_{ex} = \dot{m}_{fuel} + \dot{m}_{air} \left[\frac{kg}{s}\right] \tag{17}$$

where:

C_{O_2}—oxygen concentration in exhaust gases [%].

The composition of the NATO F–75 fuel used in the Polish Navy is c = 0.87 i h = 0.13.

The intensity of mass emissions of individual components were calculated on the basis of the equation:

$$E_{i,j} = u_j \times C_{i,j} \times \dot{m}_{ex} \tag{18}$$

where:

$\dot{m}_{ex}$ exhaust mass flow rate $\left[\frac{\text{kg}}{\text{s}}\right]$,

$C_{i,j}$the concentration of exhaust gas components [ppm, %],

u_j—factor characteristic for a given compound j:

u_{CO} = 0.000966, u_{CO_2} = 15.19, u_{NO_x} = 0.001587

The emissions of components were calculated by integrating the intensity of mass emissions over time:

$$m_{i,j} = \int_{t_{pp}}^{t_{kp}} E_{i,j} dt \text{ [kg]} \tag{19}$$

where:

t_{pp}—the beginning of the dynamic state [s],

t_{kp}—the end of dynamic state [s].

Based on the measurement data from the GPS system, the distance traveled was calculated from the relationship (20) using the law of cosines:

$$d = \text{acos}(\sin(\varphi_1) \times \sin(\varphi_2) + \cos(\varphi_1) \times \cos(\varphi_2) \times \cos(\lambda_2 - \lambda_1) \cdot R \tag{20}$$

where:

φ_1—the latitude of the first point,

φ_2—the latitude of the second point,

λ_1—the longitude of the first point,

λ_2—the longitude of the second point, and

R—radius of the Earth.

The route emission is used to assess the ecological properties of ships in terms of the emission of exhaust gases components. It is used as a reference quantity for the distance traveled by the ship, in emission models and emission inventories. The ship's route emissions were calculated from the following relationship:

$$b_s = \frac{m_{i,j}}{d} \left[\frac{\text{kg}}{\text{NM}}\right] \tag{21}$$

The presented calculation algorithm was implemented in the LabView development environment. By this way, it was possible to analyze any interval of the recorded experiment. The next part of the paper presents the results of the analysis of maneuvers carried out by the hydrographic ship during departure the Navy Port Gdynia.

3. Results

The departure of the hydrographic ship was divided into four stages:

Stage 1—unberthing and hauling off maneuver,
stage 2—moving off maneuver,
stage 3—change of direction maneuver, and
stage 4—acceleration maneuver.

The ship way in port is shown in Figure 7. In the first stage, the main propulsion engines were coupled to a line of shafts driving propellers. The ship's propulsion operated at a constant rotation speed of 600 min^{-1}. The CPP's were set to 0. At this stage, the engines of the generator sets were mainly loaded, which provided electricity for the working bow thruster. After unberthing and hauling off maneuver, the ship was changed course to heads of breakwater and moved off. By the CPP's changing, the main propulsion engines were loaded and the ship reached an ahead speed of 4 knots which kept until change of direction maneuver. The ship turned to port, causing the vessel roll to starboard. At the time of turning, the ahead speed of the ship was slightly reduced. After establishing a course towards the entrance, the ship began accelerating to the ahead speed of 11 knots.

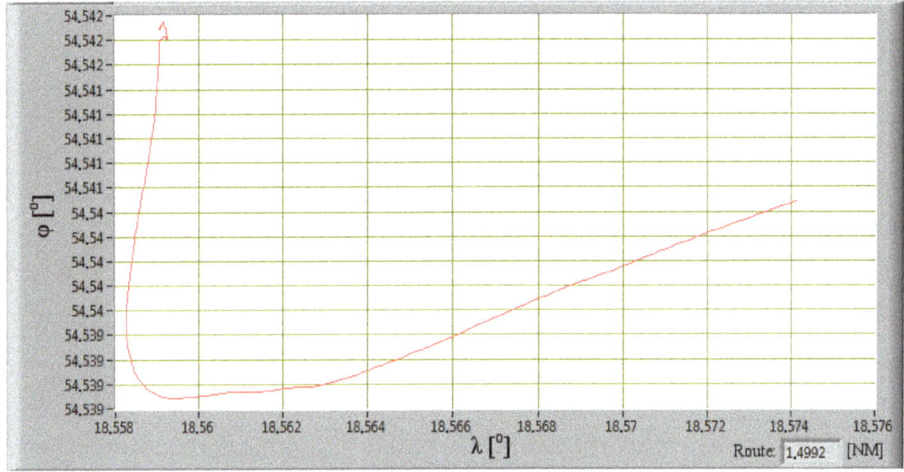

Figure 7. Recorded way during the ship departure from the port.

Table 7 shows values of the distance traveled which were calculated on the basis of data obtained from the GPS system. The ship's speed changes during the departure from the port is presented in Figure 8.

Table 7. Ship's distance in the port.

The Stage	Distance [NM]
No. 1	0.039
No. 2	0.15
No. 3	0.21
No. 4	1.1
Sum:	1.499

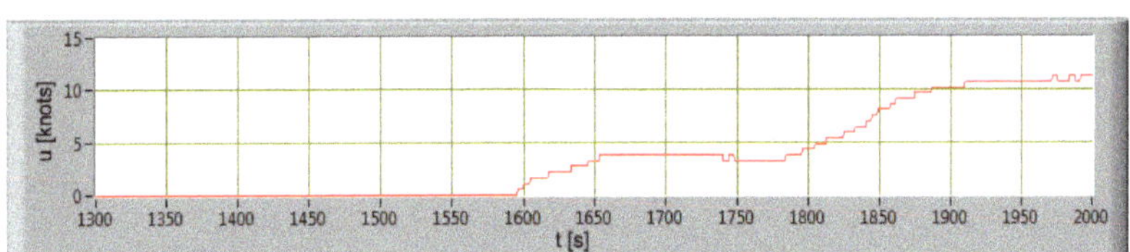

Figure 8. The ship's speed change during the departure from the port as a function of time.

The pitch and roll of the ships were recorded using an inclinometer, which was mounted on the main propulsion engine (port side). Admit mark of direction of pitch and roll are shown in Figure 9.

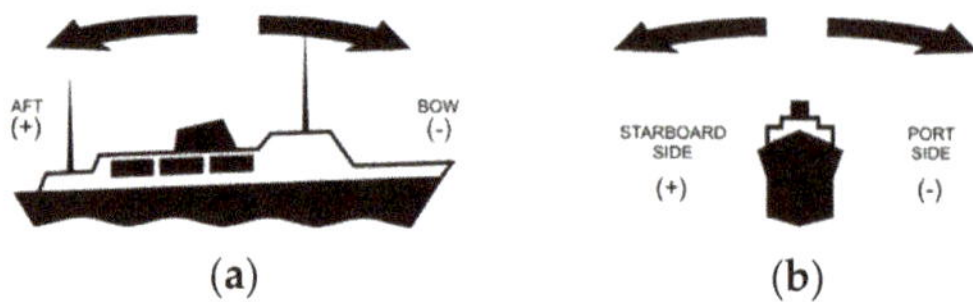

Figure 9. Marks of pitch (**a**) and roll (**b**) changes.

The changes of the ship's pitch and roll during departure from the port are shown in Figure 10. The ship alongside the berth had starboard side list and small trim by the stern.

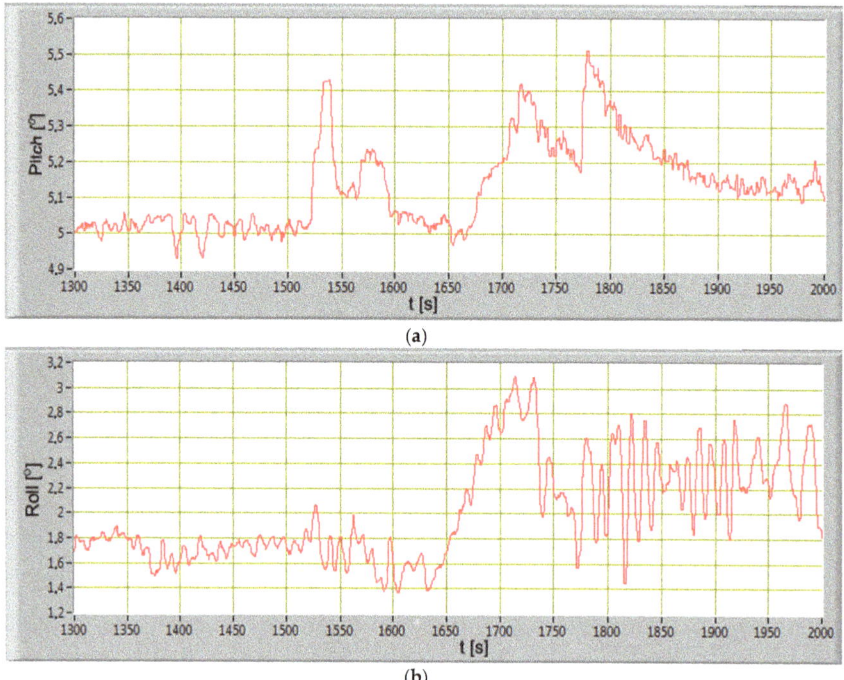

Figure 10. The ship's pitch (**a**) and roll (**b**) change during the departure from the port as a function of time.

The change of main engine load indicator is shown in Figure 11. In stage 1, the engine load was constant because the ship was maneuvering only with the bow thruster. The first change in the main engine load (increase) occurred while accelerating the vessel to the speed of 4 knots. After reaching the set speed, the main engine load stabilized at the load indicator of 32%. The change of direction forced a temporary, slight decrease in the value of the ship's speed and an increase in the load to 48%. This was caused by the helm angle to the port side, causing an increase in the hydrodynamic resistance of the left side propeller and the hull. At the same time, the ship heeled to starboard. In order to maintain a constant main engine rotational speed, the governor forced an increase the fuel dose injected into the cylinders and was recorded on the waveform. The last increase in the load to the value of 81% was caused by the acceleration of the ship to a cruising speed of 11 knots. At the same time, a temporary change in trim towards the stern (immersion of the propeller) and a slight stabilization were recorded.

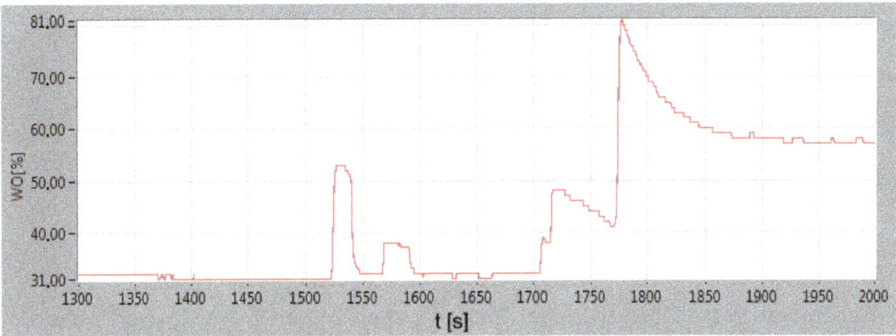

Figure 11. The main engine (port side) load indicator change during the departure from the port as a function of time.

Changing the CPP's (connected with the engine load) caused a change concentrations of analyzed substances in exhaust gases. The courses of changes in individual recorded and calculated on the basis of the neural net's concentration model are presented in Figure 12.

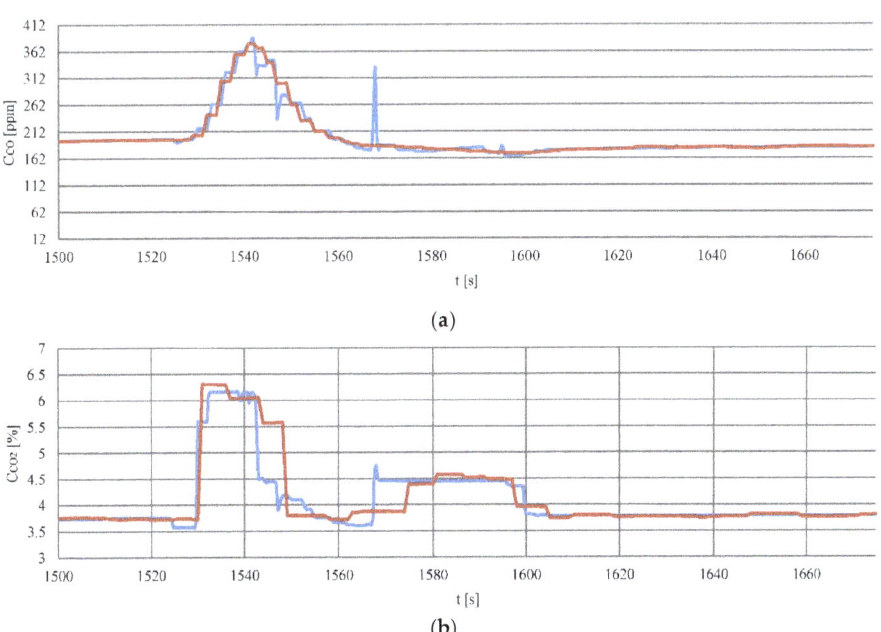

Figure 12. *Cont.*

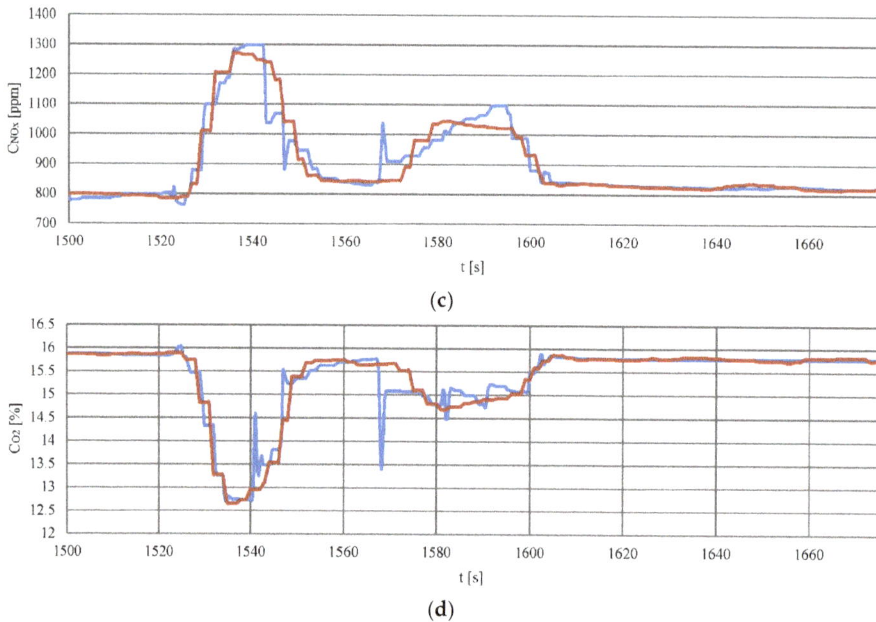

Figure 12. The courses of changes concentrations: (**a**) Carbon monoxide, (**b**) carbon dioxide, (**c**) nitrogen oxides, (**d**) oxygen during moving off maneuver as a function of time, red line—recorded data, blue line—data from the model.

The recorded and calculated data regression differences were used for comparison (Figure 13). The Regression residuum were subjected to the Durbin–Watson test, described in the previous section.

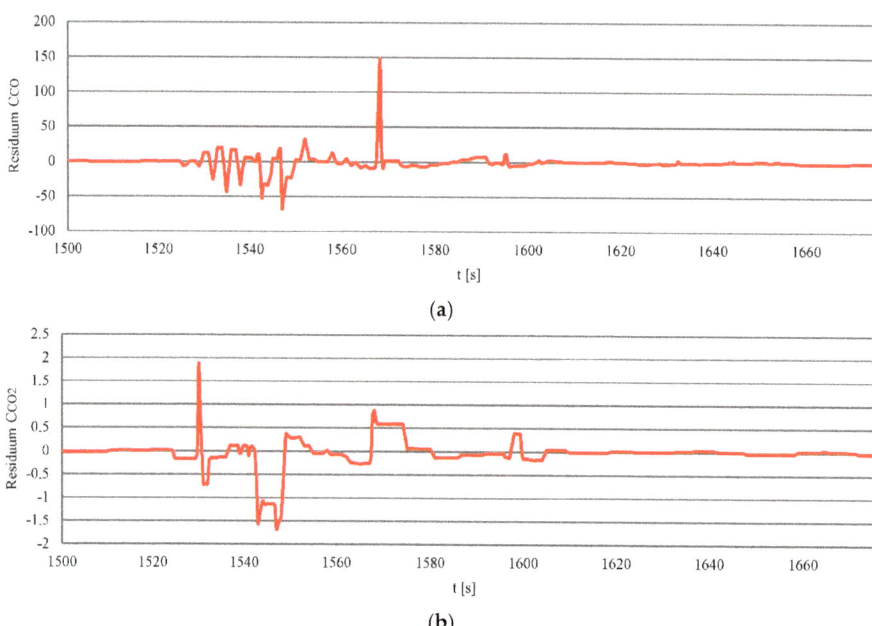

Figure 13. *Cont.*

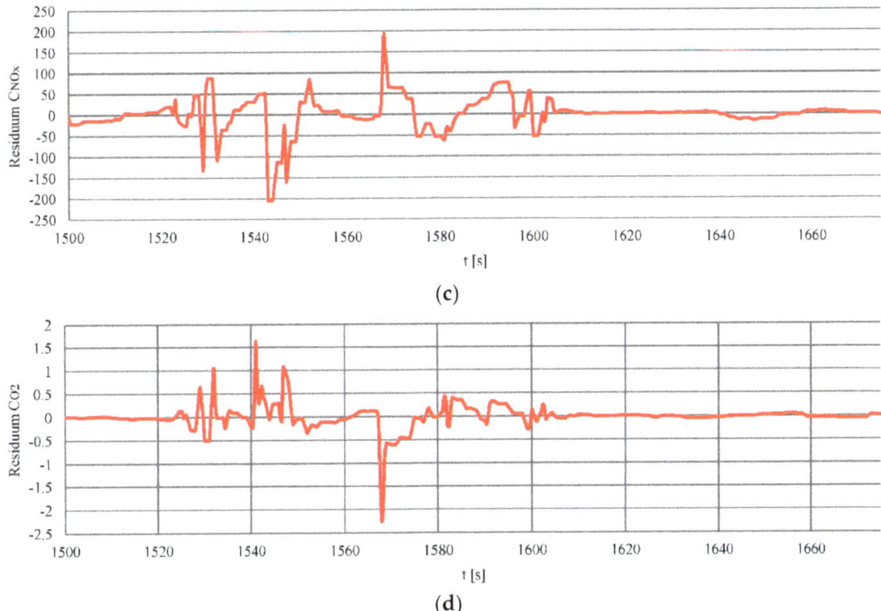

Figure 13. The courses of changes regression residuum: (**a**) Carbon monoxide, (**b**) carbon dioxide, (**c**) nitrogen oxides, (**d**) oxygen during moving off maneuver as a function of time.

The modelling of the concentrations of the composition of exhaust gases were performed only for stages 2 to 4. In stage 1 the ship was maneuvering with the bow thruster and only the generator sets were loaded. This fact was considered irrelevant in terms of modelling dynamic states.

The regression residuum of the composition of exhaust gases was subjected to the DW test. The calculated values show that the model and empirical data show a strong positive correlation at the studied stages. Table 8 shows the values of the regression residuum of the model and empirical data during the departure maneuver from the port.

Table 8. The Durbin–Watson test values for the concentration of exhaust gases components regression residuum.

The Stage	The Concentration	The Durbin–Watson Test
No. 2	C_{CO}	0.694121033
	C_{CO_2}	0.215164508
	C_{NO_x}	0.203601472
	C_{O_2}	0.390373856
No. 3	C_{CO}	0.467848069
	C_{CO_2}	0.857018339
	C_{NO_x}	0.463965328
	C_{O_2}	0.626851209
No. 4	C_{CO}	0.471728494
	C_{CO_2}	0.868353646
	C_{NO_x}	0.149970084
	C_{O_2}	0.778670739

On the basis of the concentrations' courses of carbon monoxide, carbon dioxide, and nitrogen oxides, the intensity of mass emissions of the exhaust gases components were calculated. Numerical integration of the intensity of mass emissions courses made it possible to obtain a curve showing the waveform in emissions during the maneuvers of the hydrographic ship in the port. In Figure 14 the waveforms of the emissions intensity and the emissions of the exhaust gases components during the moving off maneuver of the ship in the port (stage 2) are presented. Analyzing the waveforms, it can be concluded that the fastest reaction to diesel engine load changes can be seen in the change of the nitrogen oxides concentration. In order to maintain the set engine speed during load changes, the governor increased fuel doses of all cylinders, which resulted in an increase in the fuel-air mixture combustion temperature. The sudden engine load change caused a delay in the operation of the turbocharger as a result of incomplete combustion. There was an increase in the carbon monoxide emission intensity in the exhaust gases. The slowest reaction to a load change is seen in carbon dioxide concentration change. By adjusting the ship's speed, the crew forced a smaller change in the engine load and propulsion system. The intensity of mass emissions gradually stabilized after the ship reached the cruising speed set. The stabilization of the engine load and the intensity of mass emissions resulted from the cooperation between the hull and the propulsion system of the ship. The stabilization of the ahead speed of the ship was resulted of equal resistances of the hull components and the thrust force of the propellers. The ship, performing the maneuvers in the port, was sheltered from the hydrometeorological conditions in Gulf of Gdansk. In stage 4, the ship was leaving the port and began to be affected by the waves coming from the waters of the gulf. It was recorded (t = 1750 s to t = 2000 s) in the waveforms of the vessel's pitch and roll angles (Figure 10), the engine load indicator (Figure 11).

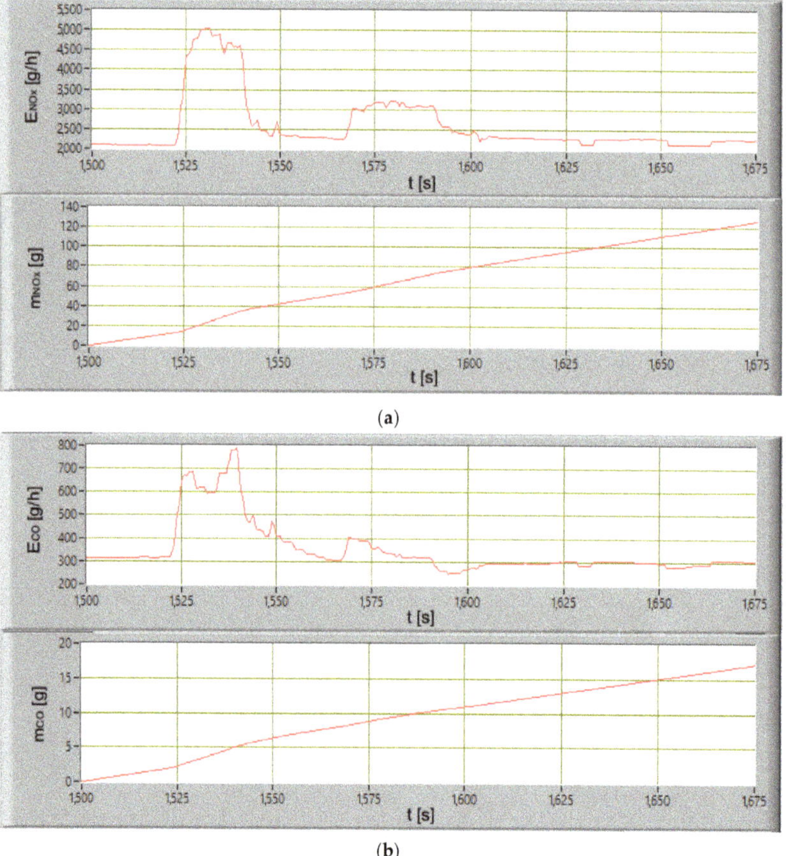

Figure 14. *Cont.*

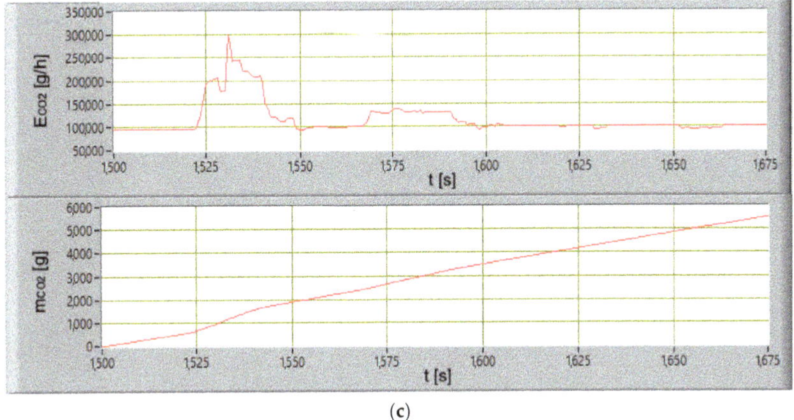

(c)

Figure 14. The courses of changes emissions intensity and emissions of exhaust gases components: (**a**) Nitrogen oxides, (**b**) carbon monoxide, (**c**) carbon dioxide during moving off maneuver as a function of time.

The next step was to calculate the emissions in stages 2 to 4. The calculations were made on the basis of model and empirical data (Figures 15–17). The performed calculations show that the highest values of the exhaust gases components were emitted during the last stage of departure from the port. The lowest values were recorded during changing the direction maneuver. The moving off maneuver to 4 knots (stage 2) slightly increased the values compared to the change of direction maneuver.

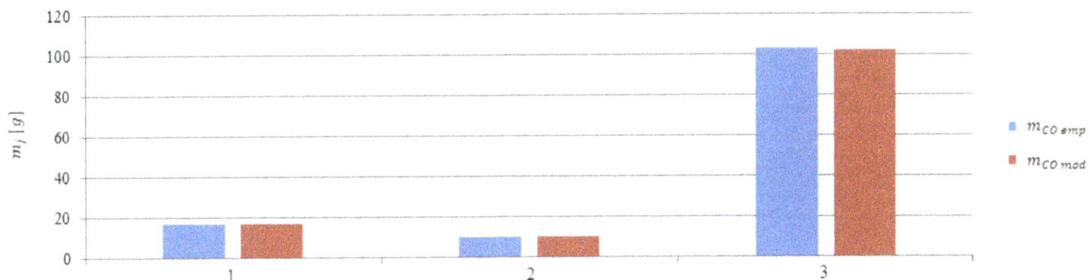

Figure 15. Carbon monoxide emissions from model and empirical data during the maneuvers: 1—Moving off, 2—change of direction, 3—acceleration.

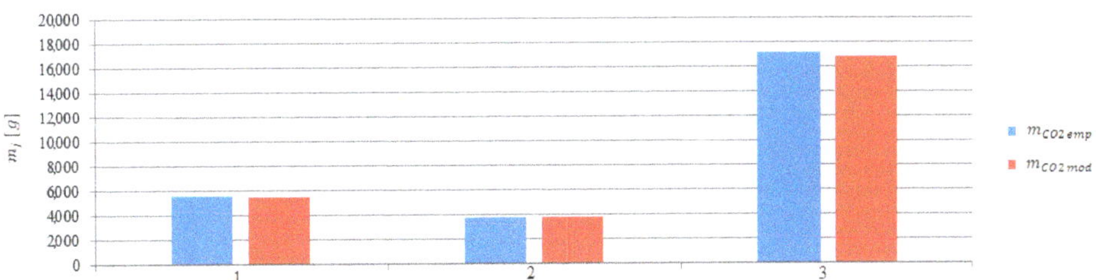

Figure 16. Carbon dioxide emissions from model and empirical data during the maneuvers: 1—Moving off, 2—change of direction, 3—acceleration.

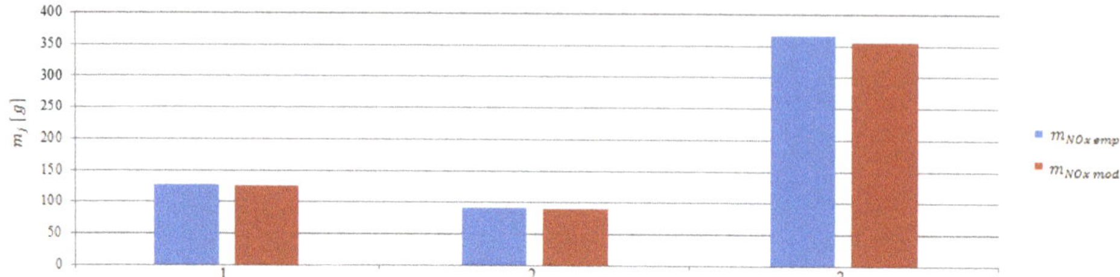

Figure 17. Nitrogen oxides emissions from model and empirical tests during the maneuvers: 1—Moving off, 2—change of direction, 3—acceleration.

The determination of the emissions and the ship's route made it possible to calculate road emissions. In these calculations, the stage 1 was taken into account. The emission of the main engine operated at a constant load was designated. The ship's route emissions are presented in Figures 18–20. The red bar shows the calculated mean value of the ship's route emission for the entire range connecting all maneuvers performed by the ship in the port. In stage 1 and 2 emission reached values greater than the average, while the stage 3 and 4 emission were lower than the average value.

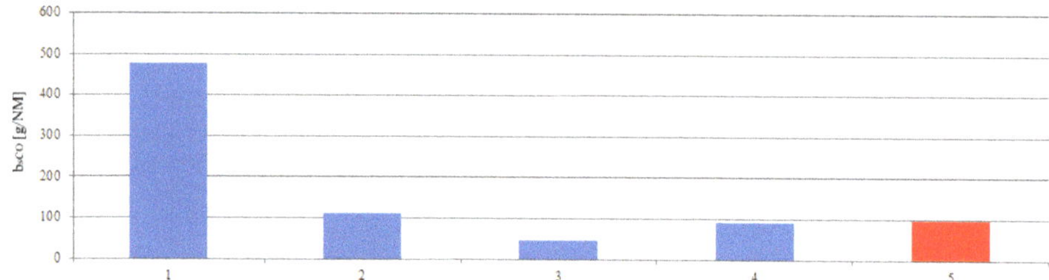

Figure 18. The ship's route carbon monoxide emissions during the stages: 1—Unberthing and hauling off maneuver, 2—moving off maneuver, 3—change of direction maneuver, 4—acceleration maneuver, 5—mean value of the ship's route emission.

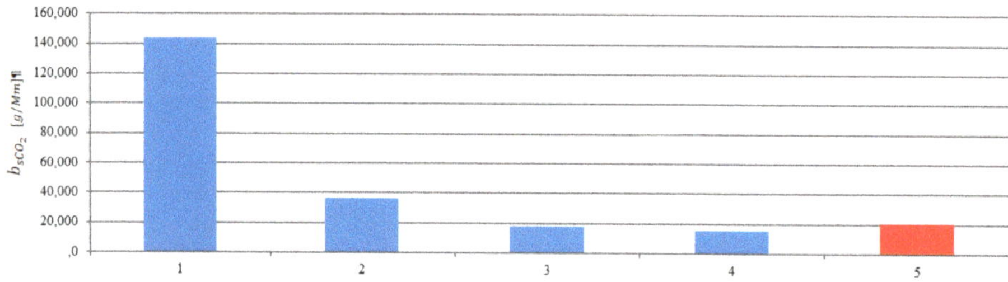

Figure 19. The ship's route carbon dioxide emissions during the stages: 1—Unberthing and hauling off maneuver, 2—moving off maneuver, 3—change of direction maneuver, 4—acceleration maneuver, 5—mean value of the ship's route emission.

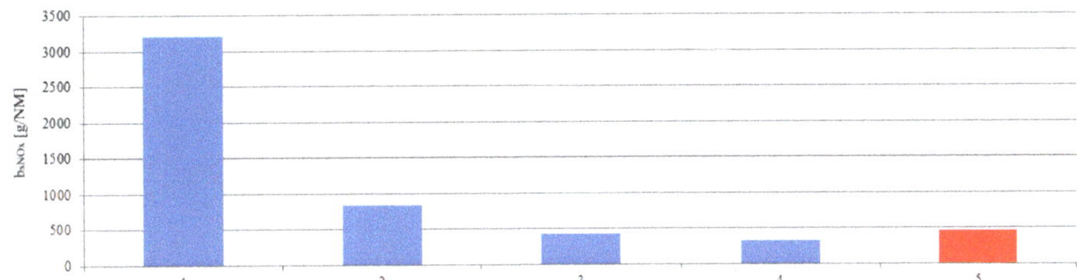

Figure 20. The ship's route nitrogen oxides emissions during the stages: 1—Unberthing and hauling off maneuver, 2—moving off maneuver, 3—change of direction maneuver, 4—acceleration maneuver, 5—mean value of the ship's route emission.

4. Conclusions

The results of the experiment confirmed that it is possible to calculate the emission of exhaust gases components from the main propulsion engine in the dynamic states. If it is not possible to measure fuel consumption directly, indirect methods can be used to obtain these values from MIP measurements. The use of simple neural networks make it possible to model the concentrations of the exhaust gases components in dynamic operating states. They are the input data for the calculation of the emissions of exhaust gases components from the marine diesel engine. The presented experiment took place during a routine task of the hydrographic survey vessel. Due to large amount of registered data, only the ship's departure maneuvers from the Navy Port Gdynia is presented. The results of the experiment led to general conclusions:

1. Carrying out tests of multi-engine propulsion system, the measurement set should be extended with additional sensors and an exhaust gases analyzer allowing for the parallel measurement of all engines.
2. Approximation with the use of neural networks gives exact results of the fit, that was confirmed by the analysis of regression residuum using the Durbin–Watson test.
3. The learned neural networks will be used to estimate the emissions of exhaust gases in a model of ship's propulsion. This solution will enable to build enhanced model that will estimate emission in port and coastal areas traffic.
4. The ship's route emissions for a vessel maneuvering at place or at low speed are higher than for vessels realizing passage at cruising speeds.
5. Active slowing down (not described in the paper) involving the operation of the propulsion system "backwards" while the unit is moving forward, also increases the value of the ship's route emissions than in the case of passive slowing down of the ship.
6. In the experiment, the position was recorded using the GPS system. The assumption was to obtain the sampling time every 1 s. In future works, the AIS system collecting information about the movement of vessels in the port will be used. This will allow the emissions of ships maneuvering in port areas to be calculated. However, the system registration time is longer than the direct GPS measurement.

The proposed methodology of dynamic emissions tests in port areas can be performed without interfering with the ship's propulsion system (in particular the fuel system of the marine diesel engine). The presented devices and methods of analysis will be further developed. It is planned to determine the influence of hydrometeorological conditions on the change the emissions of exhaust gases components from marine diesel engine. The research should be extended to ship boilers and generator sets, which also operate when the unit's propulsion system is stopped or is in idle gear while the ship is maneuvering or moored to the quay.

Author Contributions: Conceptualization, A.B.; methodology, A.B.; software, A.B.; validation, A.B.; formal analysis, A.B. and T.K.; resources, A.B.; data curation, A.B.; writing—original draft preparation, A.B.; writing—review and editing, A.B.; visualization, A.B.; supervision, T.K. All authors have read and agreed to the published version of the manuscript.

Funding: This research received no external funding.

Conflicts of Interest: The authors declare no conflict of interest.

References

1. Saxe, H.; Larsen, T. Air pollution from ships in three Danish ports. *Atmos. Environ.* **2004**, *38*, 4057–4067. [CrossRef]
2. Zhang, Y.; Yang, X.; Brown, R.; Yang, L.; Morawska, L.; Ristovski, Z.; Fu, Q.; Huang, C. Shipping emissions and their impacts on air quality in China. *Sci. Total Environ.* **2017**, *581*, 186–198. [CrossRef] [PubMed]
3. Sorte, S.; Rodrigues, V.; Borrego, C.; Monteiro, A. Impact of harbour activities on local air quality: A review. *Environ. Pollut.* **2020**, *257*, 113542. [CrossRef] [PubMed]
4. Chen, D.; Zhao, Y.; Nelson, P.; Li, Y.; Wang, X.; Zhou, Y.; Lang, J.; Guo, X. Estimating ship emissions based on AIS data for port of Tianjin, China. *Atmos. Environ.* **2016**, *145*, 10–18. [CrossRef]
5. Merk, O. Shipping Emissions in Ports. *Int. Transp. Forums Discuss. Pap.* **2014**. [CrossRef]
6. Dragović, B.; Tzannatos, E.; Tselentis, V.; Meštrović, R.; i Škurić, M. Ship emissions and their externalities in cruise ports. *Transp. Res. Part D Transp. Environ.* **2018**, *61*, 289–300. [CrossRef]
7. Moreno-Gutiérrez, J.; Pájaro-Velázquez, E.; Amado-Sánchez, Y.; Rodríguez-Moreno, R.; Calderay-Cayetano, F.; Durán-Grados, V. Comparative analysis between different methods for calculating on-board ship's emissions and energy consumption based on operational data. *Sci. Total Environ.* **2019**, *650*, 575–584.
8. Toscano, D.; Murena, F. Atmospheric ship emissions in ports: A review. Correlation with data of ship traffic. *Atmos. Environ.* **2019**, *4*, 100050. [CrossRef]
9. Aulinger, A.; Matthias, V.; Bieser, J.; Quante, M. *Effects of Future Ship Emissions in the North Sea on Air Quality*; Springer: Berlin/Heidelberg, Germany, 2014.
10. *Sea Shipping Emission 2014*; Final Report; Marin: Bilthoven, The Netherlands, 2016.
11. *Sea Shipping Emission 2017*; Final Report; Marin: Bilthoven, The Netherlands, 2019.
12. Markowski, J.; Pielecha, J.; Jasiński, R. Model to Assess the Exhaust Emissions from the Engine of a Small Aircraft during Flight. *Procedia Eng.* **2017**, *192*, 557–562. [CrossRef]
13. IMO. MARPOL 73/78: The International Convention for the Prevention of Pollution from Ships. Available online: https://www.imo.org/en/About/Conventions/Pages/International-Convention-for-the-Prevention-of-Pollution-from-Ships-(MARPOL).aspx (accessed on 6 November 2020).
14. IMO. MARPOL ANNEX VI and the Act to Prevent Pollution from Ships. Available online: https://www.epa.gov/enforcement/marpol-annex-vi-and-act-prevent-pollution-ships-apps (accessed on 6 November 2020).
15. Čampara, L.; Hasanspahić, N.; Vujičić, S. Overview of MARPOL ANNEX VI regulations for prevention of air pollution from marine diesel engines. *SHS Web Conf.* **2018**, *58*, 5–6. [CrossRef]
16. Sys, C.; Vanelslander, T.; Adriaenssens, M.; Van Rillaer, I. International emission regulation in sea transport: Economic feasibility and impact. *Transp. Res. Part D Transp. Environ.* **2016**, *45*, 139–151. [CrossRef]
17. IMO Working Group Agrees Further Measures to Cut Ship Emissions. Available online: https://www.imo.org/en/MediaCentre/PressBriefings/pages/36-ISWG-GHG-7.aspx (accessed on 6 November 2020).
18. Domínguez-Sáez, A.; Rattá, G.A.; Barrios, C.C. Prediction of exhaust emission in transient conditions of a diesel engine fueled with animal fat using Artificial Neural Network and Symbolic Regression. *Energy* **2018**, *149*, 675–683. [CrossRef]
19. *Documentation of Project 874 Hydrographic Ship*; MW: Gdynia, Poland, 2001.
20. *Testo 350 Maritime, Instruction Manual*; Testo: Lenzkirch, Germany, 2010.
21. *GPS Mouse User's Guide V1.0*; GlobalSat Technlogy Corp: New Taipei City, Taiwan, 2011.
22. *Piezotron Quartz Pressure Sensor Type 7613B for Engine Diagnostics*; The Kistler Group: Winterthur, Switzerland, 2013.
23. *Multifunction DAQ USB Module USB-47111A*; Advant. Tech. Ldt: Cincinnati, OH, USA, 2015.
24. *OCOE 02581 Datasheet*; Optom: Warsaw, Poland, 2015.
25. *Linear Potentiometer LPF Datasheet*; Ixthus Instrum.: Towcester, UK, 2019.

26. *Kubler 8.IS40.23321 Inclinometer Datasheet*; Kubler Gr.: Villingen-Schwenningen, Germany, 2015.
27. Kodosky, J. LabVIEW. *Proc. ACM Program. Lang.* **2020**. [CrossRef]
28. Kwon, S.J. Artificial neural networks. *Artif. Intel. Rob.* **2011**. [CrossRef]
29. Chatfield, C. Durbin-Watson Test. *Wiley StatsRef Stat. Ref. Online* **2014**. [CrossRef]
30. StatSoft, Inc. *STATISTICA (Data Analysis Software System), Version 6*; StatSoft, Inc.: Tulsa, OK, USA, 2001.

Article

An Ensemble-Based Approach to Anomaly Detection in Marine Engine Sensor Streams for Efficient Condition Monitoring and Analysis

Donghyun Kim [1], Sangbong Lee [2] and Jihwan Lee [3,*]

1 Korea Marine Equipment Research Institute, Busan 49111, Korea; kimdonghyun9942@gmail.com
2 Lab021 Shipping Analytics, Busan 48508, Korea; sblee@lab021.co.kr
3 Department of Industrial Data Engineering, Industrial Data Science and Engineering, Pukyong National University, Busan 48513, Korea
* Correspondence: jihwan@pknu.ac.kr; Tel.: +82-51-629-6492

Received: 22 November 2020; Accepted: 16 December 2020; Published: 18 December 2020

Abstract: This study proposes an unsupervised anomaly detection method using sensor streams from the marine engine to detect the anomalous system behavior, which may be a possible sign of system failure. Previous works on marine engine anomaly detection proposed a clustering-based or statistical control chart-based approach that is unstable according to the choice of hyperparameters, or cannot fit well to the high-dimensional dataset. As a remedy to this limitation, this study adopts an ensemble-based approach to anomaly detection. The idea is to train several anomaly detectors with varying hyperparameters in parallel and then combine its result in the anomaly detection phase. Because the anomaly is detected by the combination of different detectors, it is robust to the choice of hyperparameters without loss of accuracy. To demonstrate our methodology, an actual dataset obtained from a 200,000-ton cargo vessel from a Korean shipping company that uses two-stroke diesel engine is analyzed. As a result, anomalies were successfully detected from the high-dimensional and large-scale dataset. After detecting the anomaly, clustering analysis was conducted to the anomalous observation to examine anomaly patterns. By investigating each cluster's feature distribution, several common patterns of abnormal behavior were successfully visualized. Although we analyzed the data from two-stroke diesel engine, our method can be applied to various types of marine engine.

Keywords: marine engine; two-stroke diesel engine; onboard sensor; condition monitoring; unsupervised anomaly detection; ensemble learning; clustering analysis; anomaly analysis

---

1. Introduction

The main engine is the most important subsystem that provides the propulsion power of the vessel. Because the failure of the engine during the operation may cause a tremendous economic loss [1], the maintenance of the engine is considered as a critical activity, not for the maintenance routine, but for the vessel classification, which is a process that verifies equipment against a set of technical standards [2]. In maritime industry, the standard practice of engine maintenance follows Planned Maintenance System (PMS), where the machinery is replaced at predetermined time intervals or operating hours, regardless of its actual status [3]. From the economic point of view, however, PMS may not be an optimal strategy because it may have an unnecessary substitution of the machinery.

An alternative strategy to PMS is Conditioned-based Maintenance (CM), wherein the maintenance is carried out based on the condition of the machinery, which is detected by measuring several parameters during the vessel operation [4]. Fortunately, the recent development of IT technology has enabled real-time access to the machinery's condition and energy efficiency using data collected from onboard sensors [5–13]. Such sensor-based monitoring can be used to detect abnormal behavior of the system that may indicate the degradation or fault of the system. This study proposes a data-driven approach that enables conditioned-based monitoring of the vessel's main engine utilizing a machine learning algorithm.

We adopt an unsupervised approach in the detection of anomalous system behavior. Although most industries, including the maritime industry, continuously collect data from sensors, most of the data usually comes from the normal operating condition, and a comprehensive fault dataset is usually hard to obtain. Thus, anomaly detected from our methodology may not be directly related to the system fault because any behavior that shows a large deviation from the normal dataset can be detected as an anomaly. However, they could be used for the initial screening of engine status monitoring and can be combined with further analysis, including fault isolation and diagnosis.

Previously, several unsupervised methods have been proposed to analyze onboard sensor data in anomaly detection for the marine engine. The clustering-based approach adopts a clustering algorithm to identify clusters in the sensor data first and then check whether new data belongs to existing clusters. They assume that normal instances have stronger adherence to clusters than the anomaly. Perera and Mo [14,15] classified the most frequent operating regions of the marine engine. In their work, the Gaussian Mixture Model is adopted to represent the cluster of operating regions as a mixture of probability distributions. Brandsæter and Venem [16] proposes an efficient online method to calculate the degree of abnormality from clusters. Vanem and Brandsæter [17] compare several clustering algorithms by analyzing anomaly detection results. Although the Clustering-based approaches are intuitive and easy-to-implement, it suffers from unstable because the clustering result may be largely affected by the number of clusters, which should be specified by the user. Although there are several guidelines for determining the good cluster numbers, they cannot be applicable to the unsupervised dataset. As an alternative approach, Bae et al. [18] proposed a Statistical Process Control (SPC)-based approach to the anomaly detection of the vessel engine. To remove the distributional assumption of the dataset, a Bootstrap-based T2 multivariate chart proposed by Phaladiganon et al. [19], is adopted to determine the threshold for each sensor data. If one of the sensor values is out of the threshold, the data point is detected as an outlier. However, SPC-based approach suffers from low performance when the dataset involves high dimensional spaces [20].

To overcome the limitation of the above approaches, we propose an ensemble-based anomaly detection method to operate on a large-scale high-dimensional dataset. The idea is to apply several algorithms with varying hyperparameters to the same dataset and then combine each classifier's anomaly detection result. Because the anomaly is determined by the combination of multiple classifiers, the result is robust to the choice of hyperparameters without loss of its prediction power. To demonstrate our methodology, a data stream obtained from a 200,000-ton bulk cargo ship operated by a Korean shipping company that is collected during ten months are analyzed. The data set consists of comprehensive parameters representing engine performance, including engine rotation per minute (RPM), temperature, and pressures of lubricant oil and cooling waters. Several preprocessing steps were conducted to reduce the data size and select the informative sensor parameters. The ensemble-based algorithm then trains the preprocessed data to detect the anomalies from the input data. After detecting the anomaly, clustering analysis was conducted to the anomalous observation to examine anomaly patterns. By investigating each cluster's feature distribution, several common patterns of abnormal behavior were successfully visualized. The result shows that the proposed method can be successfully applied to the large and high-dimensional sensor streams. Although we analyzed the data from two-stroke diesel engine, our method can be applied to various types of marine engine.

The remainder of this paper is organized as follows. Section 2 explains about the target vessel and data set collected from the vessel. Section 3 addresses the procedures used for data preprocessing. Section 4 illustrates the ensemble-based machine learning model used for anomaly detection of the engine status. In Section 5, an in-depth discussion about anomaly detection results was performed. Finally, Section 6 addresses the limitation and future work of this study.

2. Data Description

This section explains the description of our target vessel and the dataset used in anomaly detection. The target vessel is a 200,000-ton bulk cargo ship, and its detailed specification is in Table 1. The data collection period spans about ten months, starting from 2019 July to 2020 April. As shown in Figure 1, its routes include main ports in Asian countries including Korea, Russia, Singapore, and Taiwan. The sensor measured the data at a one-second interval, resulting in a total of 22,513,800 observations.

Table 1. Specification of the target vessel.

Specification	
Length Overall	269.36 m
Length between perpendiculars	259.00 m
Breadth	43.00 m
Depth	23.80 m
Draught	17.3 m
Deadweight	152.517 metric t

Figure 1. Vessel operation routes over the data collection period.

The engine model used in the vessel is MAN B&W MC50, which is a slow-speed two-stroke engine [21]. The engine adopted Variable Injection Timing (VIT) systems that control the timing of the start of the fuel injection. The coolant system uses lubricant cooling for the rotating part (Crankshaft, Piston), and the fixed part (Cylinder Head, Jacket) is cooled with fresh water. The coolant is cooled by seawater in a separate heat exchanger.

In the raw dataset, more than 150 data streams were collected by onboard sensors. Some parameters were related with navigational information such as Global Positioning System (GPS) location, ground speed, wind speed, water level, etc. In contrast, others were related with subsystems' status, including engine, generator, thruster, and cargo management system, etc. In this study, we only included parameters that are attached on main engine subsystems. Other parameters that come from other subsystems were excluded because they are not in our interest. As a result, the chosen parameters are shown in Table 2.

Table 2. Description of parameters.

Sensor Name	Description
ME1 FO FLOW HOUR INLET	The consumption rate of fuel oil
ME1 FO TOTALIZER INLET	Cumulative consumption of fuel oil
ME1 FO TEMP INLET	The temperature of fuel oil
ME1 FO DENSITY INLET	The density of fuel oil
ME1 RPM ECC	Engine rotation per minute (RPM)
ME1 RPM	Same as above
ME1 SCAV AIR PRESS ECC	The Pressure of scavenging air
ME1 SCAV AIR PREE	Same as above
ME1 FO INLET TEMP	The inlet temperature of fuel oil
ME1 FO INLET PRESS	Inlet pressure of fuel oil
ME1 CYL1 PCO OUTLET TEMP ME1 CYL2 PCO OUTLET TEMP ME1 CYL3 PCO OUTLET TEMP ME1 CYL4 PCO OUTLET TEMP ME1 CYL5 PCO OUTLET TEMP	The outlet temperature of cylinder piston cooling oil
ME1 JCW INLET TEMP	The inlet temperature of jacket cooling water
ME1 JCW INLET OUTLET	The outlet temperature of jacket cooling water
ME1 CYL1 CFW OUT TEMP ME1 CYL2 CFW OUT TEMP ME1 CYL3 CFW OUT TEMP ME1 CYL4 CFW OUT TEMP ME1 CYL5 CFW OUT TEMP	The outlet temperature of cylinder block cooling water
ME1 CYL1 EXH GAS OUTLET TEMP ME1 CYL2 EXH GAS OUTLET TEMP ME1 CYL3 EXH GAS OUTLET TEMP ME1 CYL4 EXH GAS OUTLET TEMP ME1 CYL5 EXH GAS OUTLET TEMP	The outlet temperature of exhaust gas
ME1 TC1 EXH INLET TEMP	The inlet temperature of exhaust gas of turbocharger
ME1 TC1 EXH OUTLET TEMP	The outlet temperature of exhaust gas of turbocharger
ME1 TC LO OUTLET TEMP	The outlet temperature of lubricant oil
ME1 LO INLET PRESS	Inlet pressure of lubricant oil
ME1 LO INLET TEMP	The inlet temperature of lubricant oil

3. Data Preprocessing

All the data-driven approach requires representative training dataset. However, the raw data stream is not complete. It may contain out-of-range values, missing values, redundant variables, and irreverent information. If the raw data is not carefully screened, then the resulting model will not perform well on the new data. Thus, in this study, several preprocessing methods were applied to improve the quality of the dataset. The overall analysis framework is shown in Figure 2.

First, out-of-range values that exceed the acceptable sensor value ranges were removed. In some cases, sensor value shows zero or extremely large values that are out of acceptable range of sensors. Those values are usually the consequence of signal loss either from the sensors or from the communication. Because both cases were not related to the failure of the engine, it is natural to remove such outliers in the training data set.

Then, we reduce the dataset by averaging its value with a 10-min interval. One reason for this transformation is that the current data set (measured with a second interval) is too huge to train the model. Besides, the vessel engine usually undergoes slow changes during the operation compared to other vehicles such as the car or the airplane. Thus, averaging the dataset with a 10-min interval may be enough for training the model. As a result, the size of the dataset was reduced to 37,523.

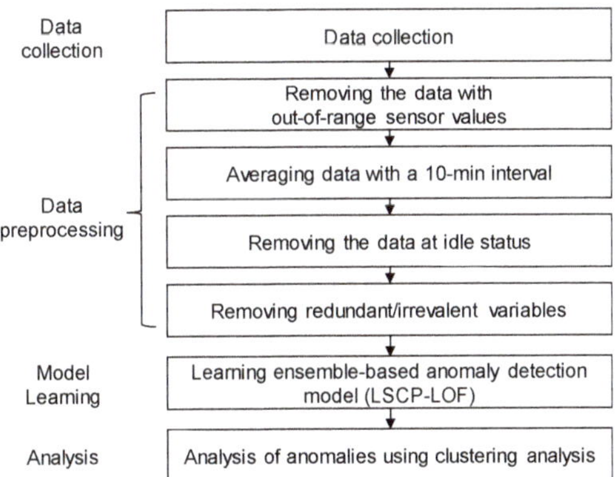

Figure 2. The overall framework for vessel main engine anomaly detection.

Next, we exclude the data collected when the vessel was idle because the vessel engine does not operate during that period. As shown in Figure 3, the vessel shows the alternating operational status (idle and normal operation) during the data collection period. This study adopts a window-based change point detection algorithm to the ground speed time-series data to distinguish the vessel's operational status. The algorithm tries to detect the rapid change points using two windows, which slide along the data stream. The statistical properties of each window are compared with a discrepancy measure. For a given cost function $c(\cdot)$, a discrepancy measure $d(\cdot,\cdot)$ as follow:

$$d(y_{u,v}, y_{v,w}) = c(y_{u,w}) - c(y_{u,v}) - c(y_{v,w}) \tag{1}$$

where y_t is the input time series at time point t and $u < v < w$ are indexes. If the discrepancy measure between two sliding windows is smaller, this indicates that there is no change point at v. On the other hand, if the sliding windows fall into two dissimilar segments, the discrepancy is significantly higher, suggesting that v is a change point. In this study, such a change point indicates the boundary between the operational status of the vessel. Because the time window is considered for change point detection, this method is less sensitive to the noise data. For more details about the methods, please refer to [22]. Figure 3 also shows change points detected by the time-window based method. In this study, we consider the area whose average ground speed is over 6 knots. Further, according to an expert opinion, we determined to consider the dataset whose RPM value is over 70.

Then, feature selection and transformation was conducted. The feature selection result and was summarized in Table 3. As shown in the table, from the original dataset, some parameters (ME1 RPM ECC, ME1 SCAV AIR PRESS ECC) were obtained from duplicated sensors of other sensors (ME1 RPM, ME1 SCAV AIR PRESS) in case of sensor failures. Because parameter values of original and duplicated sensors were exactly same throughout the data collection period, we excluded duplicated sensor parameters from the dataset. In addition, parameters related with the fuel status (ME1 FO FLOW INLET, ME1 FO DESNITY INLET, ME1 FO TEMP INLET, ME1 FO TOTALIZER INLET) were removed. Of course, the change in fuel density or temperature may affect the performance of the engine. Especially, if a vessel sailed through an emission control area, such as the western part of the United States that regulates the use of low sulfur oil, the reduced lubrication effect due to low-sulfur oil might increase the probability of accidents, such as piston sticking. However, because our vessel has not sailed through an emission control area, there was no significant change in fuel oil (such as viscosity) during the data collection period. Moreover, the vessel used fuel additives to prevent the problem that may arise from fuel status. Finally, instead of using individual sensor value of individual cylinder, we use averaged value because sensor values from five cylinders show high correlation with each other as shown in Figure 4. As a result, the variable size was reduced from 32 to 14.

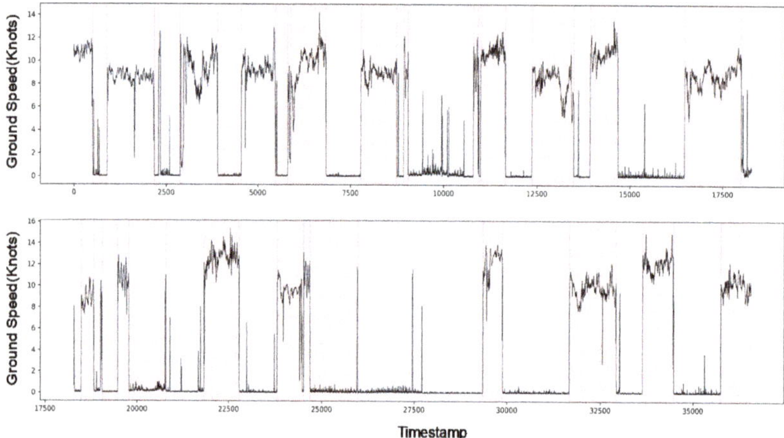

Figure 3. Time series of the vessel ground speed. Moreover, the windows-based change detection method is also applied to find operational regions among the dataset. The alternating color region indicates whether the vessel operates or not.

Table 3. Removed Parameters and Reason.

Reason	Parameters
Fuel Oil Status Indicator (not affect engine condition)	ME1 FO FLOW HOUR INLET, ME1 FO DENSITY INLET, ME1 FO TEMP INLET, ME1 FO TOTALIZER INLE
Duplicated sensors	ME1 RPM ECC, ME1 SCAV AIR PREES ECC
Aggregate value by averaging	ME1 [CYL1~CYL5] PCO OUTLET TEMP ME1 [CYL1~CYL5] PCO OUTLET TEMP ME1 [CYL1~CYL5] CFW OUTLET TEMP

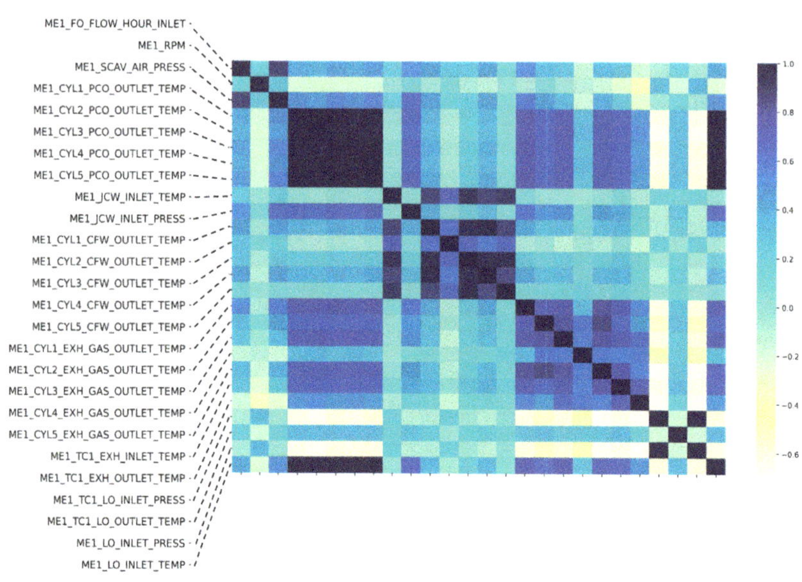

Figure 4. Correlation heatmap among entire feature set.

We may use more sophisticated variable reduction techniques such as Principal Component Analysis (PCA) or Independent Component Analysis (ICA) to reduce the variables further. However, we determine to preserve the current variables to utilize them in several analyses after detecting anomalies for examining common patterns and causes for anomalies. As a result, the comparison between dataset and preprocessed dataset is shown in Table 4.

Table 4. Comparison between original dataset and preprocessed dataset.

	Original Dataset	Preprocessed Dataset
Parameters	ME1 FO FLOW HOUR INLET ME1 FO TOTALIZER INLET ME1 FO TEMP INLET ME1 FO DENSITY INLET ME1 RPM ECC ME1 RPM ME1 SCAV AIR PRESS ECC ME1 SCAV AIR PRESS ME1 FO INLET TEMP ME1 FO INLET PRESS ME1 [CYL1~CYL5] PCO OUTLET TEMP ME1 JCW INLET TEMP ME1 JCW INLET OUTLET ME1 [CYL1~CYL5] CFW OUT TEMP ME1 [CYL1~CYL5] EXH GAS OUTLET TEMP ME1 TC1 EXH INLET TEMP ME1 TC1 EXH OUTLET TEMP ME1 TC LO OUTLET TEMP ME1 LO INLET PRESS ME1 LO INLET TEMP	ME1 RPM ME1 SCAV AIR PRESS ME1 FO INLET TEMP ME1 FO INLET PRESS ME1 CYL PCO OUTLET TEMP (Average value of 5 cylinders) ME1 JCW INLET TEMP ME1 JCW INLET OUTLET ME1 CYL CFW OUT TEMP (Average value of 5 cylinders) ME1 CYL EXH GAS OUTLET TEMP (Average value of 5 cylinders) ME1 TC1 EXH INLET TEMP ME1 TC1 EXH OUTLET TEMP ME1 TC LO OUTLET TEMP ME1 TC LO INLET TEMP ME1 TC LO INLET PRESS
Number of Observations	22,513,800 (one second interval)	37,523 (ten minutes averaging)

4. Ensemble-Based Method for Anomaly Detection

This section outlines the anomaly detection algorithm applied to the preprocessed dataset. This study adopts an unsupervised approach because we have no labeled dataset about the failure of the vessel's main engine during the data collection period. The unsupervised approach assumes that all the training data shows the normal condition [23]. Thus, if a new observation shows a large deviation from the training set, it is considered an anomaly. Unsupervised anomaly detection may show poor performance if the distribution of the normal dataset is heavy-tailed, or the normal data point is too centered to mimic the anomalous data point [24]. Moreover, the low-accuracy problem may be more severe when the feature space of the dataset is high-dimensional [23].

To remedy this problem, ensemble learning is also applied to model learning. The ensemble approach, which combines multiple base estimators in anomaly detection, is considered as a strategy to improve the model accuracy and stability because one can reduce the effect of variance on modeling accuracy by running the model multiple times [25]. Ensemble approach can be categorized as model-centric when multiple based estimators of the different hyperparameter are combined to predict anomaly score, while it is categorized as the data-centric when the different derivatives of the dataset are applied to the same model. Several ensemble-based approaches, such as feature bagging, parametric ensemble, and sub samplings, are available.

4.1. Base Anomaly Detection Algorithm: Local Outlier Factor

In this study, the Local Outlier Factor (LOF) [26] is applied as a basic anomaly detector. LOF is considered as an instance-based method because it firstly finds a relevant instance of the training data and makes a prediction using the information of these instances. Because this approach does not require the design of generic models, it is often referred to as memory-based methods. LOF [17] and the k-nearest neighborhood-based method [27] was its successful implementation. One of the problems in instance-based methods is that the performance of the anomaly detection may be severely affected by the local distributions of the data. LOF addresses this problem by using density information of its neighborhood point. For a given data point x_i, let $D^k(x_i)$ be the distance between x_i and its k-nearest neighbor, and $L_k(x_i)$ be the set of points within the k-nearest neighbor distance. Then, we calculate the reachability distance between two data points x_i and x_j $R_k(x_i, x_j)$ is calculated as follows:

$$R_k(x_i, x_j) = \max\left\{dist(x_i, x_j), D^k(x_j)\right\} \tag{2}$$

when j is in a dense region and the x_i is far from x_j, the reachability index will be equal to the true distance. If j is in sparse region, on the other hand, the reachability index will be smoothed out by its k-nearest neighbor distance. In this way, we can calculate the average reachability distance $AR^k(x_i)$ of x_i by averaging reachability distance of its k-nearest neighborhood points:

$$AR^k(x_i) = MEAN_{j \in L_k(x_i)} R_k(x_i, x_j) \tag{3}$$

the local outlier factor is the average ratio of $AR^k(x_i)$ with respect to its k-nearest neighborhood of x_i:

$$LOF_k(x_i) = MEAN_{y_i \in L_k(x_i)} \frac{AR_k(x_i)}{AR_k(x_j)} \tag{4}$$

As the LOF algorithm can detect "local" outliers regardless of the data distribution of normal behavior, it has been applied to various applications, including network intrusion detection and process monitoring [28]. Due to the computational complexity of the LOF algorithm, however, its application to large data with high dimension has been limited. This issue can be more critical for real-time application systems.

4.2. Ensemble-Based Approach to Anomaly Detection: LSCP

As an ensemble approach to anomaly detection, Locally Selective Combination in Parallel Outlier Ensembles (LSCP), which is proposed by Zhao et al. [29], is adopted. LSCP is proposed to solve the local data problem when the data consists of heterogeneous distribution, thus cannot be represented by the one generic model. The presence of the local data structure, thus, is considered as one of the main causes that lower the performance of the unsupervised anomaly detection algorithm. LSCP tries to solve this problem by identifying local regions obtained from its nearest neighbor and building competitive ensemble detectors for each local region, thus, providing more robust predictions. Moreover, LSCP utilizes a feature bagging strategy to cope with the problem arise in high-dimensional feature space.

LSCP consists of four major steps, as shown in Figure 5. In the first stage, generate pseudo-ground truth labels are generated from the ensemble. Let X_{train} be training data, and $C = \{C_1, C_2, \ldots, C_R\}$ be a collection of base detectors with different hyperparameter settings. Moreover, let $O(X_{train})$ be the matrix of the anomaly score $O(X_{train}) = [C_1(X_{train}), \ldots, C_r(X_{train})]$. Then, the pseudo ground truth denoted by the *target* is obtained by score aggregation of base estimators C as follows:

$$target = f(O(X_{train})) \tag{5}$$

In the second stage, local region is constructed. Given a data instance to test x_j, the local region ψ_j is defined as its k-nearest neighborhood defined as the follows:

$$\psi_j = \left\{x_i \middle| x_i \in X_{train}, x_i \in L_{k,ENS}(x_j)\right\} \tag{6}$$

To define the local region $L_{k,ENS}(x_j)$, t groups of [d/2, d] features are randomly selected and k nearest data object is identified. x_i is included in $L_{k,ENS}(x_j)$ when x_i are included in the neighborhood more than t/2 times.

In the Third stage, model selection and combination are conducted. For testing instance x_j, let $target^{\psi_j}$ be the pseudo ground target value from its k nearest neighborhood:

$$target^{\psi_j} = \left\{target_{x_i} \middle| x_i \in \psi_j\right\} \tag{7}$$

$$O(\psi_j) = [C_1(\psi_j), \ldots, C_r(\psi_j)] \tag{8}$$

Moreover, let $O(\psi_j)$ be the training score matrix retrieved from its anomaly score matrix:

Then, the correlation between each base detector and pseudo ground truth over the local region is calculated. Pearson correlation is applied between $target^{\psi_j}$ and $O(\psi_j)$.

In the final step, a histogram of the Pearson correlation score of each detector is constructed, and then binned with b equal intervals. Then, the collection of detectors belonging to the most frequent intervals are kept for the ensemble for the later stage. Finally, selected detectors scores are combined with the average of maximum strategy. In Zhao et al. [29], LSCP shows better performance on many real datasets. LSCP is also considered in this study because our vessel dataset is collected from several heterogeneous routes, indicating the presence of local structures.

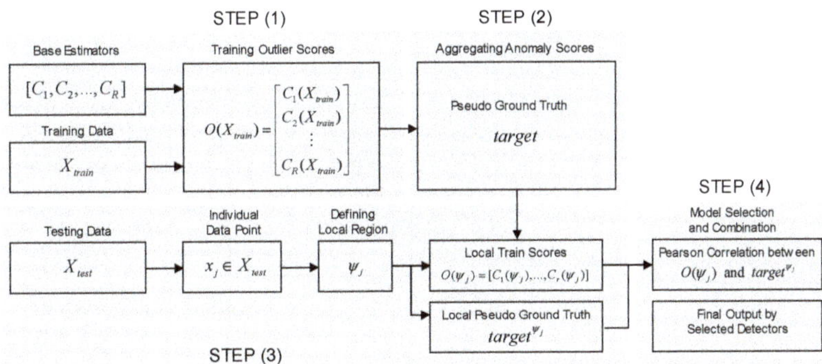

Figure 5. Locally Selective Combination in Parallel Outlier Ensembles (LSCP) Procedure (adopted from Zhao et al. [22]).

5. Experimental Result

5.1. Anomalies Detection Result

This section illustrates the result of the anomaly detection analysis. To make an ensemble anomaly detector, we combined 30 different LOF detectors. To enhance the robustness of an ensemble detector, it is required to ensure the diversity of base detectors by setting different hyperparameters. In the case of LOF, the dominant hyperparameter is k, which is the number of nearest neighbors to consider. Thus, 30 different hyperparameter set is randomly drawn from integer intervals ranging from 5 to 150. The numerical experiment was performed on Python 3.6. We used the PYOD (python toolkit for detection of outlying objects) in the implementation of LSCP [30]. The computing environment was CPU 2.2 Ghz, RAM 13 Gb.

Figure 6 shows the histogram of the anomaly scores obtained from the LSCP algorithm. The vertical line indicates the anomaly thresholds with different percentile values. Because the histogram has a very thin tail part, it seems that the anomalous data object is well separated from the normal dataset. As shown in Figure 6, we further highlighted the tail part by limiting the y-axis value to determine the proper threshold. In this study, percentile 0.998 is considered as the threshold because there is a small inflection of histogram, which may suggest the separation between different groups.

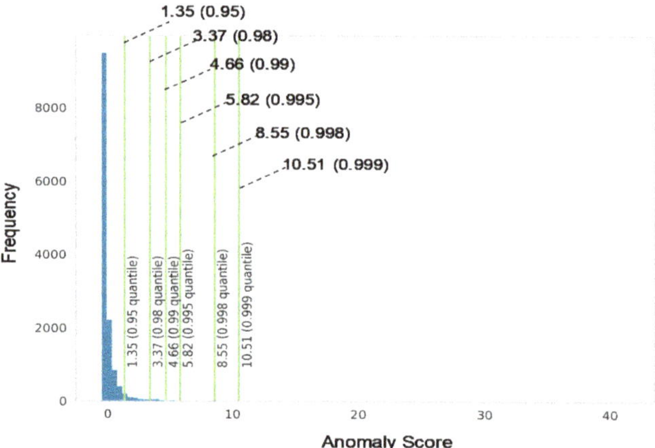

Figure 6. Histogram of anomalies score.

In Figure 7, we compared the anomalies obtained from our ensemble-based model (LSCP) with several individual detectors (LOFs) of varying hyperparameters. As shown in the figure, the anomaly detection result of an individual detector varies according to the hyperparameter. This result suggests that depending only on a single anomaly detector may be biased with the local data structure. On the other hand, anomalies detected from the ensemble-based method seems to be more robust because it includes data points that commonly appear across individual detectors.

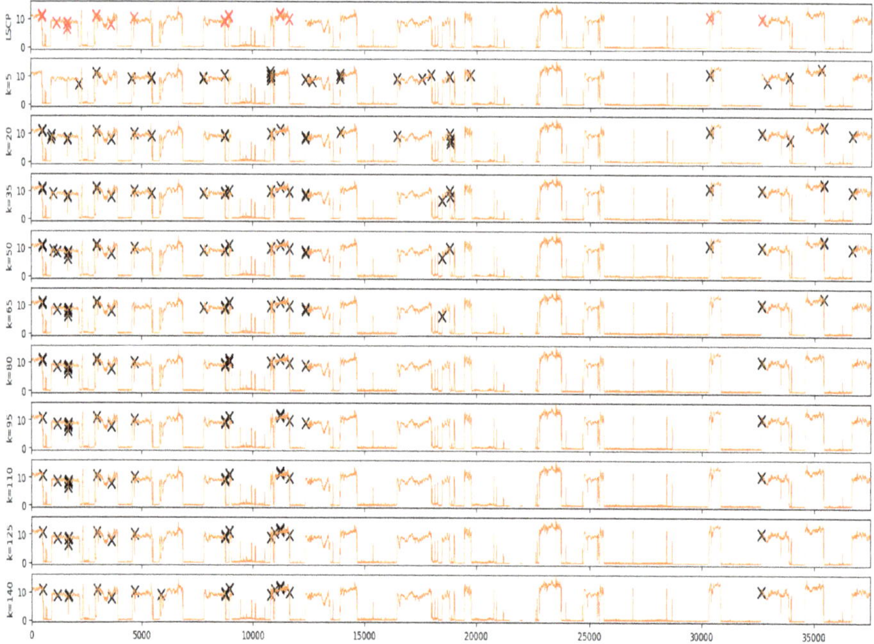

Figure 7. Comparison of anomaly detection results. The red mark indicates anomalies detected by the ensemble-based approach (LSCP), while the black mark indicates anomalies from individual detector Local Outlier Factor (LOF).

5.2. Anomalous Pattern Identification Using Clustering Analysis

We identified clusters of an anomalous dataset to examine the typical patterns of anomalous engine behavior. To this, we applied the K-means algorithm to the anomalous data points detected from our ensemble-based algorithm. As a result, we found four clusters of anomalous engine behavior. Figure 8 compares the distribution of entire variables of each anomalous data cluster. For each sensor variable, we = highlighted the cluster that shows a large deviation from the distribution of normal data points. As shown in the figure, most of the variables were highlighted by cluster 0 (highlighted by the blue line) or cluster 1 (highlighted by the green line). However, cluster 2 and 3 show little distinction from the normal data distribution. Table 5 summarizes the anomalous features associated with each cluster.

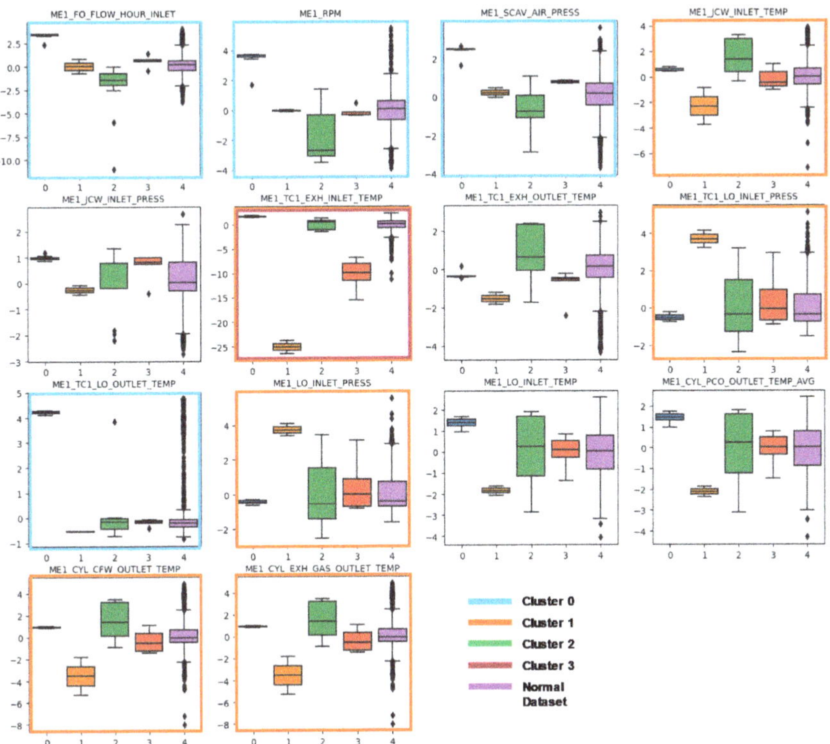

Figure 8. Comparison of boxplots among anomalous data cluster. Clusters that show large deviation from other groups are highlighted by the cluster colors.

Anomalies detected from Cluster 0 show a high value in fuel oil flow rate, RPM, scavenging air pressure, and turbocharger lubricant oil temperature. A possible cause of this anomaly may be the engine's acceleration because all of the relevant parameters seem to be the result caused by the engine acceleration [21]. On the other hand, the possible explanation about the anomalous parameters in cluster 1 seems to engine overcooling of the engine, wherein the normal temperature at which the engine operates cannot be reached. Because the engine overcooling also can damage an engine just as overheating, this region requires further investigation [31]. Cluster 3 is almost the same as cluster 0 except for lower turbocharger exhaust gas temperature. One of the possible causes of this anomaly may be the abnormal intake airflow in a marine engine [32].

Table 5. Anomalous Parameters of Each Cluster.

Clusters	Anomalous Features
Cluster 0	High fuel oil flow rate High engine RPM High scavenging air pressure High turbocharger lubricant oil outlet temperature
Cluster 1	Low jacket cooling water inlet temperature Low turbocharger exhaust gas inlet temperature High turbocharger lubricant oil inlet pressure High lubricant oil inlet temperature Low cylinder block cooling water temperature Low cylinder exhaust gas outlet temperature
Cluster 3	High scavenging air pressure Low turbocharger exhaust gas inlet temperature

5.3. Anomalous Engine Status Analysis with Vessel Operational Information

We conducted several analyses to find some potential causes of the anomalous data point. First, we analyzed the anomalous data point by examining its location on the vessel speed vs. power curve, as shown in Figure 9. Usually, there is a positive relationship between the rpm of the engine and the vessel's ground speed. If most of the anomalous data points have high speed and RPM value, then the engine's load required to operate the vessel at high speed may cause the anomaly. However, as the figure suggests, except for cluster 0, the anomalous data point in other clusters seems to spread over the speed vs. power curve. This result suggests that our anomalous data points may not have a single cause and calls for further investigation.

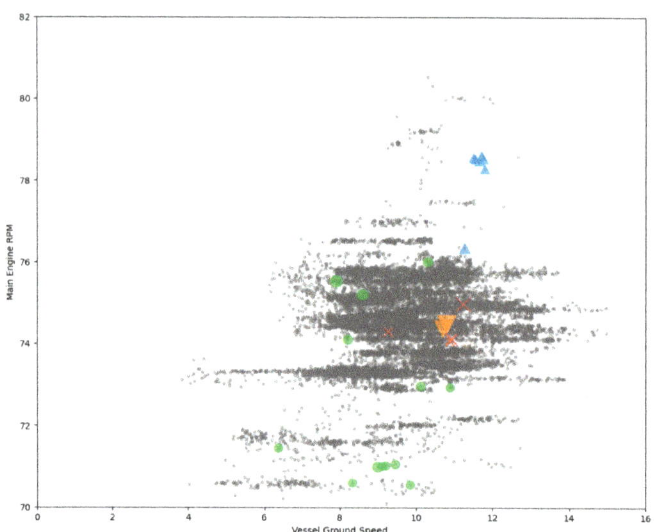

Figure 9. Anomalies plotted over speed vs. RPM scatter plot.

We also examined the anomalous data point and the time series of ground speed, as shown in Figure 10. As the figure indicated, most of the data points in cluster 1, cluster 2, and cluster 3 involved the rapid vessel speed change. This result suggests that the anomaly may be related to the acceleration or deacceleration, which may cause damage to the engine.

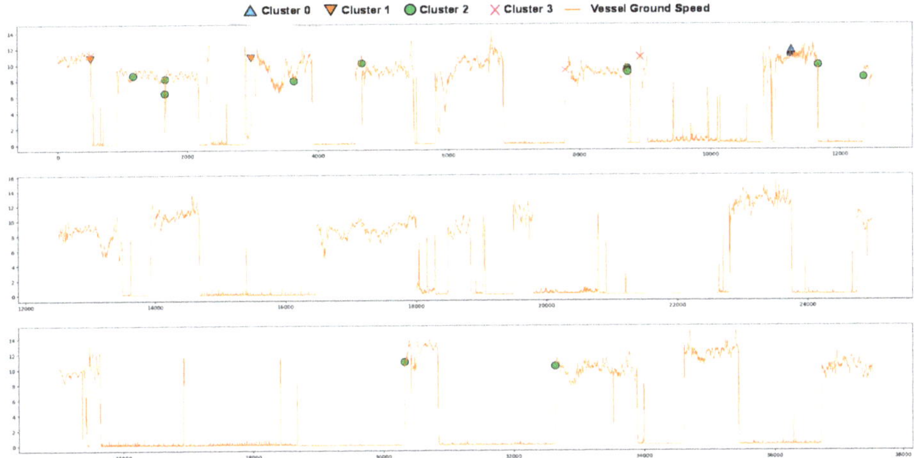

Figure 10. Anomalies plotted over the ground speed of the vessel.

Finally, we plotted the anomaly over the vessel route, as shown in Figure 11. The thin black line illustrates the navigation route of the ship for over ten months. As shown in this figure, most of the anomalies occurred in near lands, except few cases. The possible explanation for this might be that the vessel is usually driven at a low speed in the coastal waters to prevent an accident, and the engine is operated in a different pattern than usual due to frequent changes in speed. For this reason, data in the coastal waters can be classified as anomalous. We think that such information will help locate the cause of engine anomaly during ship operation in the future.

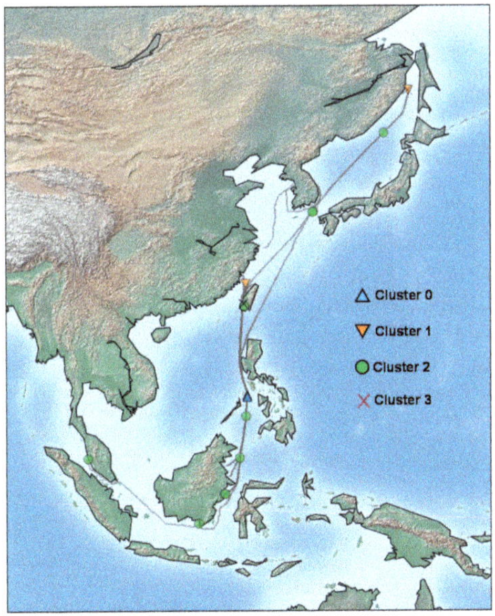

Figure 11. Anomalies plotted over the vessel routes.

6. Conclusions

In this work, a machine learning approach is adopted to detect the anomalous vessel main engine. We collected an actual dataset of a large-scale bulk carrier over ten months. This study adopted an ensemble-based algorithm to learn the large-scaled and high-dimensional engine sensor streams. As a result, each data point was successfully measured with a unified measure. With this framework, one can detect the anomalous engine behavior that shows a large deviation from the normal condition. In this study, we also conducted a clustering analysis to examine the common patterns of anomalies and which can provide information for the engine diagnosis.

The limitations of our research are as follows. First, the current dataset does not include external factors, such as the seawater temperature or air temperature. Although the cooling system in our target vessel operates to control the effect of such external factors, the relationship between external factors and engine performance should be investigated in future studies. Moreover, more rigorous preprocessing may improve the analysis result. Currently, only the outlier whose sensor value is outside of measurement range were removed from the dataset. However, some outliers may be within the measurement range, but does not satisfy the physical constraint. For example, the temperature of the exhaust gases at the inlet of the turbocharger must be higher than at the outlet. Accommodating such physical condition across sensor variables could be considered in the future works. Moreover, in future studies, we need to obtain a complete dataset. We need to increase the dataset size by extending the data collection period or combining another vessel's dataset with the same engine type.

Despite the above limitation, the unsupervised approach proposed in this paper could be used for the initial screening of the engine status monitoring and can be combined with other fault diagnosis methods. One of the natural extensions of this work is to apply an existing failure mode analysis framework to the anomalies detected by our data-driven approach. There are several works for identifying possible cause and symptoms of marine engines. Those failure modes were usually obtained from the expert knowledge [33] or the simulation experiment [32]. The development of the framework and visualization scheme for relating such failure modes and the anomalies may be helpful for fault isolation and diagnosis. Our methodology can also be extended to another subsystem. The current work only analyzed the engine-related parameters. However, the modern vessel collects sensor stream from various subsystems including cargo management, or power generation system. Considering those subsystems, thus, would be a fruitful area for future works. Another possible extension of this work is to develop more efficient method for analyzing anomalies patterns. Even though clustering analysis was conducted to explain the common cause of anomalous data, it still depends on the visual inspection, making it difficult to explain the cause of anomaly quantitatively. We can improve the analysis by adopting an Explainable Artificial intelligence (EAI) framework, such as Shapley Additive Explanation (SHAP) [34] or Local Interpretable Model-agnostic Explanations (LIME) [35], which quantifies feature contribution to an individual anomalous data point. With feature importance information, more rigorous analysis for categorizing anomalous patterns by focusing on problematic sensor values may be possible.

Author Contributions: Conceptualization, D.K.; data curation, S.L.; formal analysis, J.L.; investigation, J.L.; methodology, J.L.; project administration, D.K.; supervision, J.L.; validation, J.L.; visualization, J.L.; writing—original draft, J.L. All authors have read and agreed to the published version of the manuscript.

Funding: This research was financially supported by the Ministry of Trade, Industry, and Energy (MOTIE) and Korea Institute for Advancement of Technology (KIAT) through the National Innovation Cluster R&D program (Advancement of an open cloud platform for smart maritime convergence service_P0015306).

Conflicts of Interest: The authors declare no conflict of interest.

References

1. Lazakis, I.; Turan, O.; Aksu, S. Increasing ship operational reliability through the implementation of a holistic maintenance management strategy. *Ships Offshore Struct.* **2010**, *5*, 337–357. [CrossRef]
2. Jiang, R.; Yan, X. Condition Monitoring of Diesel Engines. In *Complex System Maintenance Handbook*; Springer: London, UK, 2008.
3. Kandemir, C.; Celik, M. A human reliability assessment of marine auxiliary machinery maintenance operations under ship PMS and maintenance 4.0 concepts. *Cogn. Technol. Work* **2019**, *22*, 1–15. [CrossRef]
4. Jimenez, V.J.; Bouhmala, N.; Gausdal, A.H. Developing a predictive maintenance model for vessel machinery. *J. Ocean. Eng. Sci.* **2019**, *5*, 358–386. [CrossRef]
5. Perera, L.P.; Mo, B. Marine engine-centered data analytics for ship performance monitoring. *J. Offshore Mech. Arct. Eng.* **2017**, *139*, 021301. [CrossRef]
6. Mak, L.; Sullivan, M.; Kuczora, A.; Millan, J. Ship performance monitoring and analysis to improve fuel efficiency. In Proceedings of the 2014 Oceans—St. John's, St. John's, NL, Canada, 14–19 September 2014.
7. Aldous, L.; Smith, T.; Bucknall, R.; Thompson, P. Uncertainty analysis in ship performance monitoring. *Ocean. Eng.* **2015**, *110*, 29–38. [CrossRef]
8. Soner, O.; Akyuz, E.; Celik, M. Use of tree-based methods in ship performance monitoring under operating conditions. *Ocean. Eng.* **2018**, *166*, 302–310. [CrossRef]
9. Kim, D.; Lee, S.; Lee, J. Data-Driven Prediction of Vessel Propulsion Power Using Support Vector Regression with Onboard Measurement and Ocean Data. *Sensors* **2020**, *20*, 1588. [CrossRef] [PubMed]
10. Chi, H.; Pedrielli, G.; Ng, S.H.; Kister, T.; Bressan, S. A framework for real-time monitoring of energy efficiency of marine vessels. *Energy* **2018**, *145*, 246–260. [CrossRef]
11. Gkerekos, C.; Lazakis, I.; Theotokatos, G. Machine learning models for predicting ship main engine Fuel Oil Consumption: A comparative study. *Ocean. Eng.* **2019**, *188*, 106282. [CrossRef]
12. Strušnik, D.; Avsec, J. Artificial neural networking and fuzzy logic exergy controlling model of combined heat and power system in thermal power plant. *Energy* **2015**, *80*, 318–330. [CrossRef]
13. Strušnik, D.; Marčič, M.; Golob, M.; Hribernik, A.; Živić, M.; Avsec, J. Energy efficiency analysis of steam ejector and electric vacuum pump for a turbine condenser air extraction system based on supervised machine learning modelling. *Appl. Energy* **2016**, *173*, 386–405. [CrossRef]
14. Perera, L.P.; Mo, B. Data analysis on marine engine operating regions in relation to ship navigation. *Ocean. Eng.* **2016**, *128*, 163–172. [CrossRef]
15. Perera, L.P.; Mo, B. Marine engine operating regions under principal component analysis to evaluate ship performance and navigation behavior. *IFAC-Pap.* **2016**, *49*, 512–517. [CrossRef]
16. Brandsæter, A.; Vanem, E.; Glad, I.K. Efficient on-line anomaly detection for ship systems in operation. *Expert Syst. Appl.* **2019**, *121*, 418–437. [CrossRef]
17. Vanem, E.; Brandsæter, A. Unsupervised anomaly detection based on clustering methods and sensor data on a marine diesel engine. *J. Mar. Eng. Technol.* **2019**, 1–18. [CrossRef]
18. Bae, Y.M.; Kim, M.J.; Kim, K.J.; Jun, C.H.; Byeon, S.S.; Park, K.M. A Case Study on the Establishment of Upper Control Limit to Detect Vessel's Main Engine Failures using Multivariate Control Chart. *J. Soc. Nav. Archit. Korea* **2018**, *55*, 505–513. [CrossRef]
19. Phaladiganon, P.; Kim, S.B.; Chen, V.C.; Baek, J.G.; Park, S.K. Bootstrap-based T 2 multivariate control charts. *Commun. Stat. Simul. Comput.* **2011**, *40*, 645–662. [CrossRef]
20. Mason, R.L.; Young, J.C. *Multivariate Statistical Process Control with Industrial Applications*; Society for Industrial and Applied Mathematics: Philadelphia, PA, USA, 2002.
21. MAN Two Stroke—Project Guide. Available online: https://marine.man-es.com/two-stroke/project-guides (accessed on 20 November 2020).
22. Chu, C.S.J. Time series segmentation: A sliding window approach. *Inf. Sci.* **1995**, *85*, 147–173. [CrossRef]
23. Chandola, V.; Banerjee, A.; Kumar, V. Anomaly detection: A survey. *ACM Comput. Surv.* **1995**, *41*, 1–58. [CrossRef]
24. Das, S.; Wong, W.K.; Dietterich, T.; Fern, A.; Emmott, A. Incorporating expert feedback into active anomaly discovery. In Proceedings of the 2016 IEEE 16th International Conference on Data Mining, Barcelona, Spain, 12–15 December 2016.
25. Aggarwal, C.C. Outlier ensembles: Position paper. *ACM Sigkdd Explor. Newsl.* **2013**, *14*, 49–58. [CrossRef]

26. Breunig, M.M.; Kriegel, H.P.; Ng, R.T.; Sander, J. LOF: Identifying density-based local outliers. In Proceedings of the 2000 ACM SIGMOD International Conference on Management of Data, Dallas, TX, USA, 16–18 May 2000.
27. Ramaswamy, S.; Rastogi, R.; Shim, K. Efficient algorithms for mining outliers from large data sets. *ACM Sigmod Rec.* **2000**, *29*, 427–438. [CrossRef]
28. Aggarwal, C.C. *Outlier Analysis*; Springer International Publishing: Cham, Switzerland, 2017.
29. Zhao, Y.; Nasrullah, Z.; Hryniewicki, M.K.; Li, Z. LSCP: Locally selective combination in parallel outlier ensembles. In Proceedings of the 2019 SIAM International Conference on Data Mining, Calgary, AB, Canada, 2–4 May 2019.
30. Zhao, Y.; Nasrullah, Z.; Li, Z. PyOD: A Python Toolbox for Scalable Outlier Detection. *J. Mach. Learn. Res.* **2019**, *20*, 1–7.
31. Huang, Y.; Hong, G. Investigation of the effect of heated ethanol fuel on combustion and emissions of an ethanol direct injection plus gasoline port injection (EDI + GPI) engine. *Energy Convers. Manag.* **2019**, *123*, 338–347. [CrossRef]
32. Rubio, J.A.P.; Vera-García, F.; Grau, J.H.; Cámara, J.M.; Hernandez, D.A. Marine diesel engine failure simulator based on thermodynamic model. *Appl. Therm. Eng.* **2018**, *144*, 982–995. [CrossRef]
33. Cicek, K.; Turan, H.H.; Topcu, Y.I.; Searslan, M.N. Risk-based preventive maintenance planning using Failure Mode and Effect Analysis (FMEA) for marine engine systems. In Proceedings of the 2010 Second International Conference on Engineering System Management and Applications, Sharjah, UAE, 30 March–1 April 2010.
34. Lundberg, S.M.; Lee, S.I. A unified approach to interpreting model predictions. In Proceedings of the 31st International Conference on Neural Information Processing Systems, Long Beach, CA, USA, 4–9 December 2017.
35. Ribeiro, M.T.; Singh, S.; Guestrin, C. "Why should I trust you?" Explaining the predictions of any classifier. In Proceedings of the 22nd ACM SIGKDD International Conference on Knowledge Discovery and Data Mining, San Francisco, CA, USA, 13–17 August 2016.

Article

Identification of Gate Turn-Off Thyristor Switching Patterns Using Acoustic Emission Sensors

Maciej Kozak [1,*], Artur Bejger [2] and Arkadiusz Tomczak [3]

1 Faculty of Mechatronics and Electrical Engineering, Maritime University of Szczecin, Wały Chrobrego 1-2, 70-500 Szczecin, Poland
2 Faculty of Mechanical Engineering, Maritime University of Szczecin, Wały Chrobrego 1-2, 70-500 Szczecin, Poland; a.bejger@am.szczecin.pl
3 Faculty of Navigation, Maritime University of Szczecin, Wały Chrobrego 1-2, 70-500 Szczecin, Poland; a.tomczak@am.szczecin.pl
* Correspondence: m.kozak@am.szczecin.pl; Tel.: +48-91501-766-199

Abstract: Modern seagoing ships are often equipped with converters which utilize semiconductor power electronics devices like thyristors or power transistors. Most of them are used in driving applications such as powerful main propulsion plants, auxiliary podded drives and thrusters. When it comes to main propulsion drives the power gets seriously high, thus the need for use of medium voltage power electronics devices arises. As it turns out, power electronic parts are the most susceptible to faults or failures in the whole electric drive system. These devices require efficient cooling, so manufacturers design housings in a way that best dissipates heat from the inside of the chips to the metal housing. This results in susceptibility to damage due to the heterogeneity of combined materials and the difference in temperature expansion of elements inside the power device. Currently used methods of prediction of damage and wear of semiconductor elements are limited to measurements of electrical quantities generated by devices during operation and not quite effective in case of early-stage damage to semiconductor layers. The article presents an introduction and preliminary tests of a method utilizing an acoustic emission sensor which can be used in detecting early stage damages of the gate turn-off thyristor. Theoretical considerations and chosen experimental results of initial measurements of acoustic emission signals of the medium voltage gate turn-off thyristor are presented.

Keywords: acoustic emission; sensor; transducer; gate turn-off thyristor; power electronics

Citation: Kozak, M.; Bejger, A.; Tomczak, A. Identification of Gate Turn-Off Thyristor Switching Patterns Using Acoustic Emission Sensors . *Sensors* **2021**, *21*, 70. https://dx.doi.org/10.3390/s21010070

Received: 25 November 2020
Accepted: 18 December 2020
Published: 24 December 2020

1. Introduction to the Subject Matter

The analysis of acoustic emission (AE) signals is widely used to detect damage in solid materials. The most popular areas of application are the supervision of fatigue phenomena and cracks in steel structures [1–4], rolling elements in bearings [5,6] or the occurrence of partial discharges in power transformers [7,8] and medium voltage switchgears [9–11]. There have been attempts at recognizing AE signals in low voltage insulated gate bipolar transistors (IGBTs) [12–16] even with changes of junction and case temperature [17]. It must be noted that AE tests were applied to small packaging low-voltage semiconductors (without electrical insulation inserted between AE sensor and device case) but it can be assumed that the propagation of elastic waves in much bigger structures will behave differently because of the extended internal volume of different types of packaging and the use of insulating inserts for medium-voltage operation.

That is the main reason gate turn-off (GTO) thyristors and newer gate-controlled thyristors are still widely used in high-power medium voltage applications such as ships main propulsion plants. The article presents possibility of using an acoustic emission sensor for diagnostics of early-stage dislocations and structural cracks inside gate turn-off thyristors.

1.1. Types of Power Electronics Devices Cases and Packaging

The silicon-controlled rectifier thyristor (SCR) can be considered as the first semi-controlled device that started medium voltage power electronics. Currently, this device in its basic not fully controlled form has largely disappeared from medium voltage applications, having gradually been replaced by exclusive GTOs and for several years now, by their upgraded version called gate controlled thyristors (GCTs) and integrated gate controlled thyristors (IGCTs). It can be assumed that the integrated gated controlled thyristor consists of the GTO structure integrated with an electronic driver circuit providing operation within a safe operation range, reduced switching losses and a short storage time [18].

Thyristors designed for medium voltage operation are usually made as a ceramic disc with metal anode and cathode plates. For proper operation they have to be efficiently cooled with means of a radiator stuck to the medium voltage plates. This is of course a main source of noises coming out of strong electric field (because of vicinity of high voltage electrodes) and electromagnetic field (coming from the current flow). Due to the switching cycle, the internal structure and inner silicon layers will be subjected to continuous mechanical stresses resulting from the repeated heating and cooling cycles.

Repeating temperature changes are the main cause of thermal stress, wear and in the end semiconductor structure failures. For designers these stresses are a quite bit of concern in the design and operation of power electronics devices, and a large number of publications have been devoted to thermal phenomena occurring inside of silicone-based electronics.

It can be assumed that the power electronics devices thermomechanical design problems can be defined in terms of the following categories [19]:

- slight temperature changes causing thermoelastic deformation,
- stress fields resulting from major changes of junction and internal temperature and displacements, along with erratic temperature distribution effects.
- elastic or elastoplastic deformation due to the effects such as thermal shock, creep, stress relaxation, stress rupture, and thermal fatigue.

To master some of aforementioned issues two main technologies are utilized in the manufacture of power thyristors: alloying and free-floating silicon technology [20].

Vacuum brazing using aluminum and silicon alloys is a commonly used method in alloy brazing to join silicon chips with molybdenum thermal compensators. Use of such technology provides a firm silicon chip-molybdenum disc junction with good cycling capacity and quite low thermal impedance.

In this case some external force is needed for installation purposes from the cathode side of thyristor is required to prove firm thermal contact. Even so, since alloying is a high-temperature process, thermomechanical stresses show up in Si-Mo structure because of the different thermal expansion coefficients of silicon and molybdenum.

However, when combining silicone chips with larger diameter outer chassis plates, this issue becomes even more important. Free-floating silicon technology introduces a semiconductor layer with the cathode and anode, metallization between them and thermal compensators. Because of the lack of soldered joints, only pressure thermal and electrical contacts between the silicon plate and thermal compensator can be distinguished. The advantage of the pressure contact design is the absence of deformation and residual stresses that occur when soldering a silicon plate with a thermal expansion joint due to the difference between the expansion coefficients. This feature is extremely important in the fabrication of semiconductor components, especially those of bigger diameter. Another important advantage of free-flowing silicon technology is that the surface layers of silicon do not dissolve during the manufacturing process. On the other hand, there is higher thermal resistance from the anode side in comparison to the soldering technology.

Another factor related to temperature expansion phenomena and changes in geometric dimensions is the type of thyristor housing [21].

There are several types of thyristor packages presented in Figure 1, ranging from low power devices enclosed in small plastic housings, then bolted medium power devices to flat pack (or press-pack) systems for high power and high voltage systems. Due to technological restrictions the thyristors enclosed in the flat-pack cases must be mounted under certain and precisely controlled pressure in order to get proper electrical and thermal contact between the semiconductor layer and the external metal electrodes. Huge diameter thyristors should not be directly soldered or glued to the large copper pole piece of the flat-pack because of the significant difference in the coefficient of thermal expansion (CTE). To avoid this issue the contact for both anode and cathode is obtained by means of pressure assembly.

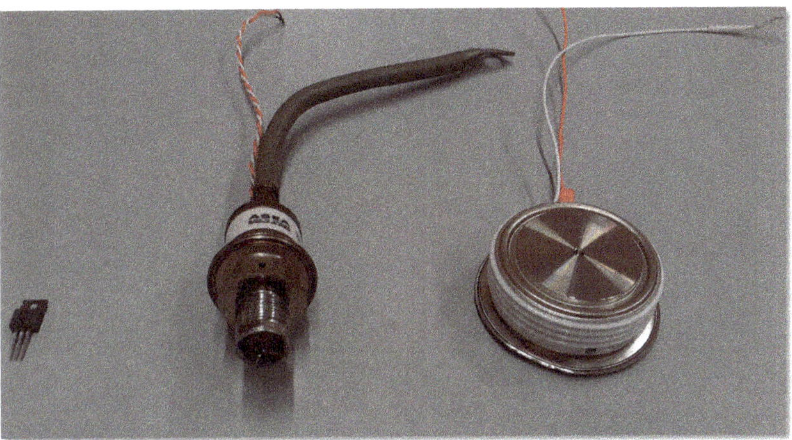

Figure 1. Example of thyristors in plastic, stud-mount and press-pack packaging.

Due to the differences in the way the inner layers are installed and connected, as well as the type of enclosure, it is expected that the acoustic emission signals will also vary during switching.

1.2. Phenomena inside of Cycling GTO Thyristor

Gate turn-off thyristors, as opposed to SCRs are fully controllable switches that allow turning on and off by applying a voltage to the gate lead. Classic silicon rectified thyristors can only be switched off by decreasing the anode current below the value of the holding current. Therefore, semi- controlled SCR thyristors are not the best choice for direct current applications. The GTO thyristor can be turned on by a certain current injected into the gate and anode gate and it conducts until this sustaining current has a proper value. It is possible to turn off this kind of thyristor simply by applying a gate current signal of negative polarity. The turn-on phenomenon in GTOs is more reliable than in a SCR thyristor and a continuous gate-anode current should be maintained in order to improve reliability.

An important issue in constructing power electronic devices is to minimize stray inductance inside the structure and to reduce the inductance value between terminals and anode and cathode plates. The mounting to the terminals should be designed in the way it minimizes inductances in order to eliminate overvoltage spikes during the switching process. Due to parasitic inductances inside the semiconductor structure, there may occur high frequency ringing when changing states fast. As occurs in every power electronics design, the capacitance of semiconductor layers and cast also creates the problem of a mutual electromagnetic interference noise between two or more devices when placed close to each other. In some situations, these capacitances may cause firing of gate circuitry charging and thus unwanted switching and serious equipment failures [21].

In addition to the aforementioned, designers of power electronic devices must primarily take into account the thermal effects occurring during conduction and switching of thyristors. In contrast to SCR thyristors, during the turn-on time the GTO thyristor needs a gate current ranging from 20 up to 30 percent of the conductive current at all times. This additionally increases the temperature of the junctions and greatly affects the deformations and dislocations within the structure. The turn-off process is initiated by applying a negative voltage between the gate and cathode of thyristor. The forward current is used to induce a cathode-gate voltage what results in a decrease of the anode forward current, and thyristor will switch off. As for the inner semiconductor structure there are identical in width and length tiny emitter mesa spots distributed inside the structure (Figure 2), which allow similar flow of the turn-off current in every path.

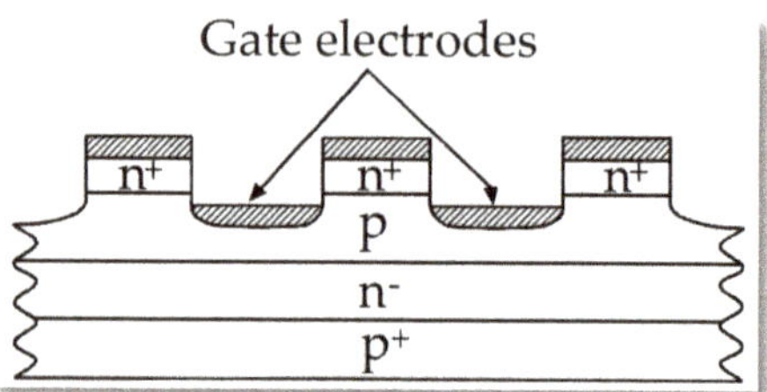

Figure 2. Cross section of a GTO showing the cathode islands and interdigitation with the gate (p-base).

In the case of no homogeneous current flow during the turn-off period can result in filamentation and with it in dynamic avalanche thus enormous silicon element heating and burn out.

The fatigue of semiconducting material comes from coefficient of thermal expansion mismatch between sticking materials of different coefficients because the switching frequency thermal cycling can significantly increase the fatigue process what can be detected with means of elastic waves inside the structure. With the temperature changes and variations, the internal wafers expand and contract at different rates what is main cause for soldered layer cracking and debonding. Some CTEs of popular materials used in power electronics devices are given in the Table 1. The extensive temperature rise also plays an important role in the chemical degradation processes such as dendritic growth and protrusion migration, therefore keeping the size of cooling plates large enough and ensuring efficient cooling is one of the most important issues related to the design and proper operation of power electronic devices [22].

Table 1. Coefficient of thermal expansion for chosen materials.

Material	CTE (mμm/mK) at 300 K
Silicon	4.1
Copper (baseplate and pole pieces)	16–16.7
Al_2O_3 (Aluminum Oxide AL98)	6.2
Tungsten (W)	4.5
Molybdenum (Mo)	4.9
Aluminum (Al)	13.1
60/40 solder (Pb/Sn eutectic)	25

The time rates of anode current during turn-on and anode-cathode voltage while turn-off are parameters needed to properly control the system and achieve reliable operation. These values should never exceed the permissible level stated by the manufacturer. When in the conducting state the areas of a device near the gate begin to conduct current sufficient time must be provided for the whole cathode area to begin conducting before the short-circuit currents become too high. If the rate of rise of forward anode-cathode voltage is too high thyristors can trigger (or self-trigger) into a conduction mode from a forward-blocking mode because of junction capacitance. To avoid any influence of switching voltage spikes additional capacitors and resistors known as snubbers are used as protection circuits. These can be found in other power semiconductor devices. A lack of snubbers can damage the silicon structure and contribute to the creation of unwanted acoustic emission signals which distort the frequency spectrum. All of aforementioned processes are the part of the thyristor wafer wear process along with natural structure aging. According to [23] due to aging some subtle changes occur such as waveform anode voltage alterations during turn on processes indicating physical changes in the thyristor gating circuit.

1.3. Acoustic Emission Signals and Analysis

Because of the cyclic nature of GTO switching the phenomena occurring inside its structure will be most notable at the working frequencies of the thyristor. There can be distinguished the following types of thyristor operation: the turning on, conducting/blocking state and turning-off events. This article covers use of AE signals obtained in the rectifying mode of operation so the expected elastic waves spreading across press-pack GTO structure would eventually contain power grid frequencies and its multiples as some of internal wave bouncing occurs. The acoustic signal analysis can be recognized as mathematical methods of signal processing in order to obtain valuable information about the inner state of a solid object. In the case of power electronics devices, the signals coming from a sensor can include a lot of "contamination" in a form of EMI or strong electric field noise which must be removed. The useful signals obtained during nominal parameter (current, voltage or junction temperature) operation can become a pattern and any abnormal state of the inner structure of the thyristor should result in change of the AE signal. When analyzing AE signals in the time domain, namely acceleration amplitude against time to quantify the strength of an elastic wave signal, a few parameters are needed (and observable): amplitude, peak-to-peak value and RMS. As was mentioned any signal coming from a semiconductor device includes a lot of additional information which can be represented as a mix of signals of different amplitudes and frequencies. The analysis of such in the time domain is not very useful so proper methods of signal analysis were introduced and now are widely used in diagnostics. Each of these have different properties and better fit various applications. These are the fast Fourier transform (FFT), frequency spectrogram and power spectral density (PSD).

The FFT decomposes the obtained signal into a Fourier series containing individual sine wave components. In numerical-based applications the fast Fourier transform in its fastest Radix-2 decimation-in-time (DIT) form is willingly used. It can be easily applied to the digital signal processor code [24] and operate in real-time. The result of Radix-2 operation over incoming signals is data containing acceleration amplitude as a function of frequency. This data enables signals analysis in the frequency domain and in diagnostic applications, the vast majority of analyses are typically done in such a domain. The FFT is fairly good for the detection and analysis of stationary state signals but with prolonged operation of the power electronics devices, the parameters like junction temperature along with geometrical dimensions (due to CTE) will change. This in turn would have an impact on the acoustic emission signals coming from the structure. In such a case it is more convenient to use the frequency spectrogram which basically creates and combines a series of FFTs and overlaps them into one plot to illustrate how the spectrum in the frequency domain changes with time. The frequency spectrogram can be very useful to illustrate how the spectrum of the acoustic emission varies in a changing environment. Another way of describing the contribution of individual frequency components to the total signal is the so-called power spectral density. A lot of acoustic signals during transient states include some noise arising from states that are dynamically changing at many frequencies at the same time. While the FFT is good enough at analyzing elastic waves signals when there is a finite number of dominant frequency components the power spectral densities are mainly used to characterize random signals. Because an ideal spectrum consists of an infinite number of components, easy calculation of the total dissipated power is not possible, so it is more convenient to denote it by power per frequency (or bandwidth) obtaining the units of V^2/Hz. Such a spectrum is called power density spectrum (PDS) and the value of the density is called power spectral density (PSD) [25]. The values of PSD are obtained by multiplying each frequency bin in a fast Fourier transform by its complex conjugate which results in the real spectrum of amplitude described in g^2. The power spectral density analysis in the case of changing, noisy signals seems to be more useful than a FFT because amplitude value is normalized to the frequency bin width which in turn leads to units described as g^2/Hz. Use of PSD has another advantage over FFT namely the dependency on bins width disappears, so comparison signals of different lengths become straightforward.

Considering its Fourier transform over the interval $\pm T/2$ as $X(\omega)$ for a time domain signal denoted as $x(t)$ the power spectral density is given by:

$$S_X(\omega) = \lim_{T\to\infty} \frac{E\left\{|X(\omega)|^2\right\}}{T} \quad (1)$$

and the area under the spectrum curve represents the total power of the signal, which is given by following equation:

$$x^{-2} = \int_{-\infty}^{+\infty} S_X(f)df = 2\int_0^{\infty} S_X(f)df \quad (2)$$

What can be considered as an advantage is that nonstationary AE signal time series analysis presents how the energy is distributed during a measurement time span. Nowadays there is widely used software which performs the aforementioned operations in real-time thus the analysis and on-line monitoring of internal state of materials is also possible for offline work. There are of course preprogrammed, compact systems created for such operations, but they are dedicated to specialized areas and the signals coming out can be heavily filtered what makes them not very practical for power electronics emission signals analysis [26].

2. Materials and Methods Used in the Experiments

In practical applications of acoustic emission sensors, they are used to detect the high frequency energy signals which are generated in inner structure cracks when part of the material is displaced or when the contacting layers have different expansion coefficients. These signals are spread in all directions inside the structure and of course they bend and bounce at the material borders of different densities. Knowing these issues there is a good chance to measure such signals (appearing as elastic waves), convert them into electrical signals and send them to a monitoring or diagnostic system. The more complex the structure is the more bounced, bent and overlaid signals must be expected but after precise filtering a lot of diagnostic information can still be obtained. The amplitude of electrical signals is quite low (up to dozens of millivolts) so the signals are amplified, and final step detectors have built-in filters. These signals can be analyzed in different ways depending on the nature of the expected phenomena. In most of the cases such signals are used for the detection of cracks, breaks and wear inside mechanical structures and the detection of electrical partial discharges. All of the aforementioned facts are well known and the detection devices have filters which are tuned in order to amplify interesting and well-known frequencies indicating the beginning of destructive processes. The nature of phenomena occurring inside multilayered press-pack type semiconductor devices is not entirely known, so it is crucial to get AE signals unaltered in any way which means that widely used front end commercial and dedicated recorders and detecting devices cannot be used. The most convenient and obvious way to get the wide spectrum of acoustic emission signals generated inside the semiconductor structure is to use a wideband AE sensor connected directly to an oscilloscope without preamplifiers and filters on. This allows the observation of pure, raw electrical signals which of course include a lot of useless information, but further analysis can reveal interesting behavior of the switching elements and can lead to the creation of unique, proper switch pattern signals. This pattern can be used for real-time observation of outgoing signals and can report early-stage malfunctions or premature failures.

As long as there are no references to informative sources covering frequencies generated inside switching thyristors the wide band frequency sensor was typed for use and detection of every kind of acoustic emission occurring in the tested semiconductor structure.

The thin disc-shaped piezoelectric material which converts material deformation into electrical signal is an important, active element of an AE sensor. To assure good electrical conductivity the piezoelectric surfaces are metalized and in order to prevent EMI interferences the whole structure is placed inside a closed cylinder made out of metal. Titanate and zirconate crystals mixed with other materials are widely used in AE transducers. The piezoelectric properties are obtained by ceramic material poling. This process involves heating the element above the Curie point in the presence of electric field and finally it produces asymmetrical internal crystal structure [27].

As the most promising and widely used device the WSα factory-calibrated sensor was chosen. The frequencies detected by WSα according to Figure 3 ranging from 0 up to 1000 kHz, thus any potentially interesting frequencies are fully covered by the chosen sensor.

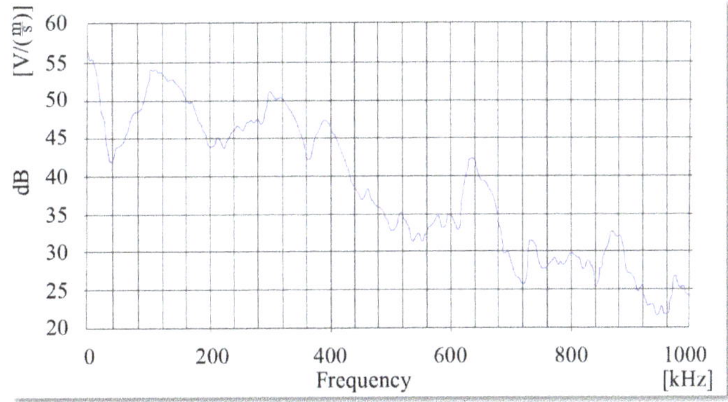

Figure 3. WSα sensor signals amplitude attenuation dependent on the frequency.

2.1. Recognition of AE Sensor Immunity to EMI Noise

The major question is what kind of signals an acoustic emission sensor detects while operating at a close distance to the conducting current object. In the article the high-power GTO thyristor is an object which produces specific AE raw signals in blocking/conducting mode and a notably high current flows through its structure. This high current produces a magnetic field which moves through the semiconducting layers, metal case and wiring harness so it will have an impact on transducer operation and final readings. Because the propagation paths and nature of AE signals in semiconductor layers is not exactly known, it seems to be reasonable to perform measurements with use of wide-band sensor and apply filtering at the very end.

The construction of the wideband AE sensor (WSα) used in our experiments consists of layers which are placed on the ceramic plate. This ceramic plate is laid on the surface of the tested material or semiconductor device (see Figure 4). In addition, it creates an insulating layer to prevent short circuits on the conducting surface.

Figure 4. The acoustic emission sensor in the metal case placed on the GTO.

Inside the metal case there is a completely enclosed crystal for RFI/EMI immunity, which converts the vibrations into electrical signals. Because of the ceramic plate width and the presence of a strong magnetic field created by the current flowing through the tested thyristor there arises questions about the magnetic field influence on the measurements. It is crucial to know how prone the sensor is to EMI noise before conducting further acoustic emission measurements. Unfortunately, acoustic emission sensor manufacturers do not provide detailed information on their resistance to magnetic and electric field interference, so it was necessary to check sensor responses in the presence of a magnetic field. To answer this question, a laboratory stand was set up, which was equipped with an autotransformer, a laboratory coil, diode, acoustic emission and magnetic field Hall-effect sensor. The latter transducer used was an AH49E type with a LM393 amplifier embedded on a PCB and powered by a battery. Both sensors were placed on top of the coil secured in the place and with means of an autotransformer the current flow across the coil was changed (Figure 5). The tests covered placing the AE sensor away from top of the solenoid just to find relationship between the near magnetic field and the acoustic emission transducer readings. The AE sensor was placed on a paper stack of different thicknesses of 0.05 mm (one sheet), 20 mm, 45 mm and 85 mm.

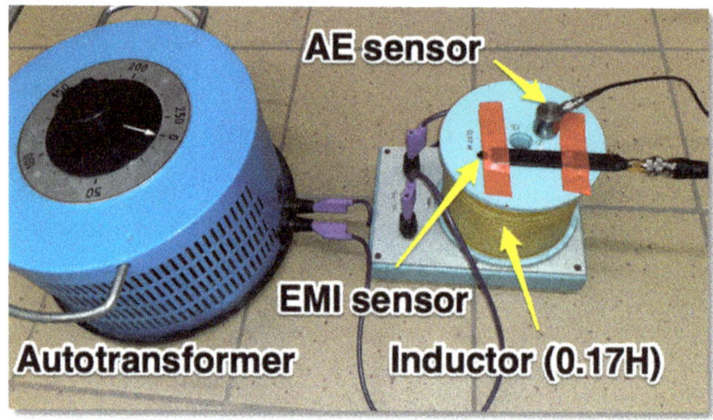

Figure 5. The sensors for AE and magnetic field signals detection.

The sinusoidal voltage coming from autotransformer was applied to the coil and acoustic emission, magnetic field and current signals were recorded on the oscilloscope.

Taking into account that the coil wire length is much larger than its diameter the magnetic field within the winding is given by $B = \mu_0 nI$ where μ_0 is the permeability constant, n—number of coil turns and I denotes current flow. With a current flowing through the coils, the magnetic field produced within the solenoid can be written as:

$$B = \mu_0 \left(\frac{N}{l} \right) I \tag{3}$$

Knowing that the magnetic flux for a given area equals to the area value multiplied by the component of magnetic field perpendicular to the penetrated area according to following equation.

This relationship can be presented as:

$$\Phi_m = BS = \frac{\mu_0 NS}{l} I \tag{4}$$

Assuming that magnetic flux depends on current flow and introducing parameter of the coil known as self-inductance $\Phi_m = L\ I$ the inductance of N turns inductor can be expressed as:

$$L_{coil} = \frac{N\Phi_m}{I} = \frac{\mu_0 N^2 S}{l} \tag{5}$$

where l is the coil wire length, S means cross-sectional area and N is a number of copper wire turns.

After applying known value of the current and coil parameters into Equations (4) and (5), the values of magnetic induction and flux were calculated. These figures have been confirmed by the results of measurement tests using a Hall sensor attached to the scope.

The solenoid chosen for the test had a self-inductance of 0.17 H with diameter equal to 11 cm, and the current flowing across coil was set to 1 A RMS (1.41 A in peak). The value of magnetic field measured directly in the center point of the coil was equal to 25 Gauss and the waveforms acquired on oscilloscope are presented in the following figures. As can be seen from Figure 6 the sinusoidal current signal creates a sinusoidal magnetic field signal (both of them are in phase) what causes oscillations. The acoustic emission signals are rapidly changing when current and magnetic field are crossing the zero value.

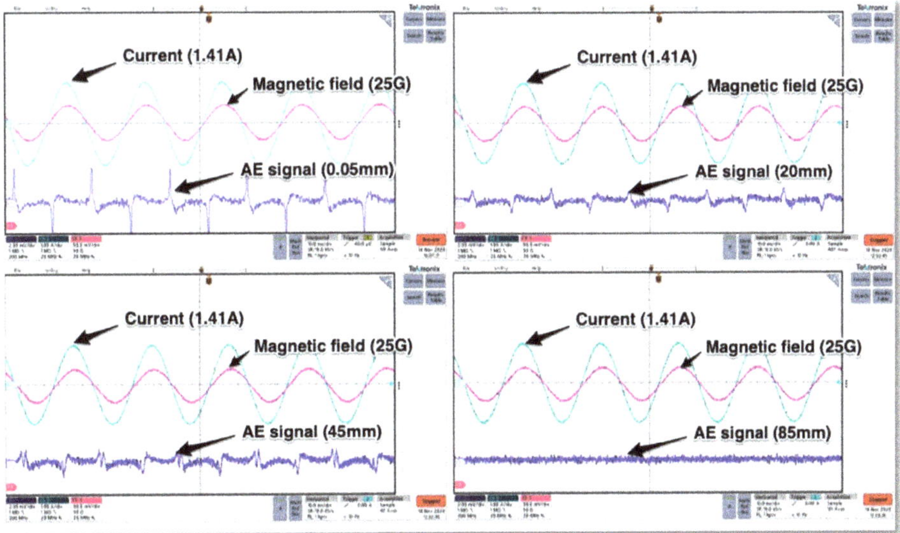

Figure 6. The magnetic field, current and AE signals obtained for different distances of the AE sensor from top of the inductor.

From the waveforms obtained it can be deduced that the signal detected by the WSα sensor depends on the magnetic field strength. In the case shown, the magnetic field decreased with the distance. When the distance between the sensor and the center of the coil was 85 mm, there was no noise interference in the signal obtained and the magnetic field was not observable.

The next tests covered AE sensor response to rectified direct current flow. Again, the same coil was used, and the power diode was placed in series connection. Similar to previous tests the magnetic field values were recorded along with the sensor—coil increasing distance. In this case, no specific parasitic signal of significant value was detected, so it can be concluded that the rectified current and the constant sign magnetic field have no impact on the AE signal readings (see Figure 7). The presented results depicting behavior of AE transducer are valid for strong magnetic fields produced intentionally with means of current flowing across massive inductor. It should be noted that in practical application, there will not be such high values of magnetic fields generated by semiconductor structures.

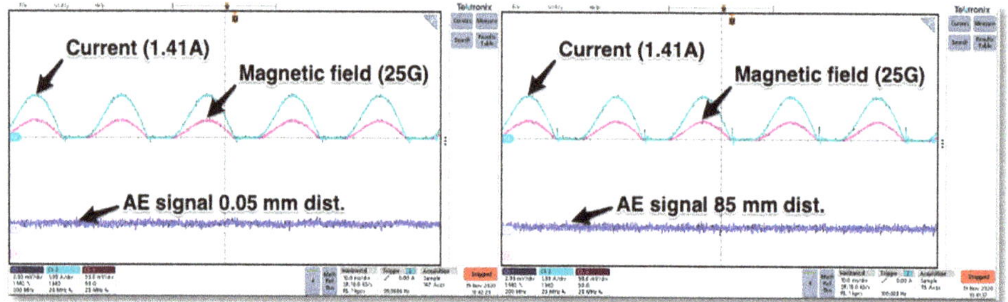

Figure 7. The test AE and magnetic field signals for rectified direct current flowing across the induction coil.

These relatively weak fields will occur mainly due to electrons vorticity transport [28] and most of it will be "intercepted" and dissipated in metal plates which are integral part of a press-pack GTO casing.

2.2. AE signals Detection of GTO in Rectifying Mode of Operation

In order to get acoustic emission signals coming out of brand-new and unused thyristor in rectifying mode the laboratory test stand was prepared. Because of the nature of acoustic emissions, the longitudinal and shear wave propagation in solid materials comes mainly from rapid movement of the material particles during cracks on a micro scale. Much bigger dislocations, cracks and fractures can be the source of low frequency signals which in turn can be detected with use of electromagnetic field detector [29,30]. The latter method was not considered in the accomplished tests but due to the necessity minimizing the influence of the magnetic fields on the readings from the transducer all elements of the investigated system were placed as far apart as possible. Similar to previous tests the Hall effect magnetic field probe along with WSα was used. Both sensors were placed on the top of metal case on the anode side just like in Figure 8.

The GTOs firing circuit was attached to 30 V, 20 A DC supply which provided enough current for the gating circuitry. Unlike most of classic SCRs used in experiments a GTO needs firing current all the time when in conducting mode and its value heavily depends on the conducted current.

The thyristor and AE transducer were pressed down firmly with a non-conductive acrylic plate. To improve wave propagation into the transducer the contact surface was coated with a silicone-based gel. The thyristor anode-cathode terminals were supplied with alternating high-current, low voltage supply.

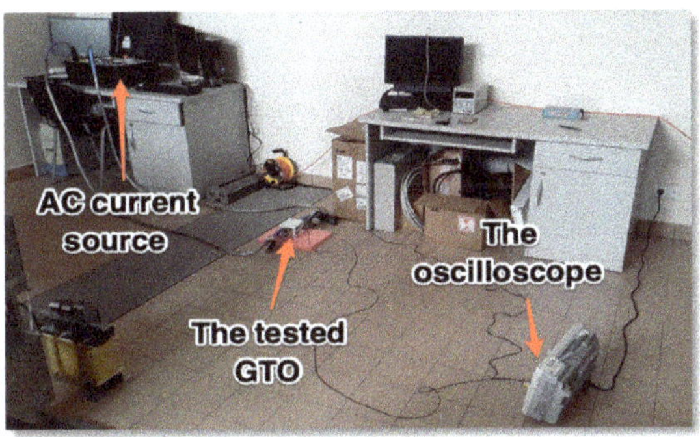

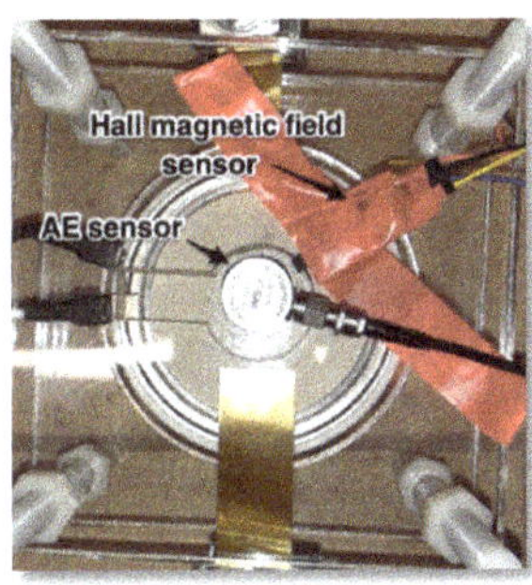

Figure 8. The laboratory test stand arrangement (**left**) and GTO 5SGS16H2500 symmetrical thyristor with AE and Hall effect sensors placed upon without a top pressing plate (**right**).

The tests carried out consisted in supplying the thyristor anode circuit with controlled by high-current autotransformer alternating voltage and firing the gate circuit by applying DC current to the gate-anode terminals. This forced the flow of rectified current of values dependent on the applied voltage, leads and the semiconductor structure anode-cathode resistance. The tests were conducted for 40, 60, 80 and 100 Amperes, respectively. Chosen results are presented in Figure 9.

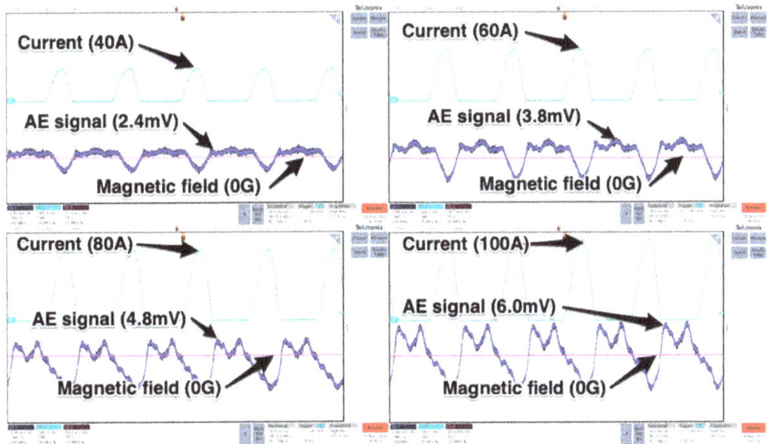

Figure 9. The chosen waveforms of raw AE signals (blue) detected by WSα transducer placed on the anode plate of conducting different currents (rectifying mode) GTO. Current and magnetic field waveforms placed for reference.

As it can be observed the magnitude of acoustic emission signals (raw data) increased with increasing current values. Because the tested thyristor is capable of long-term conducting 1200 A (with proper cooling) the magnitude of raw AE signal may reach roughly 80 mV (assuming a linear increase with current—Figure 10) so in order to get the full spectrum of acoustic emission signal the proper type of transducer should be used.

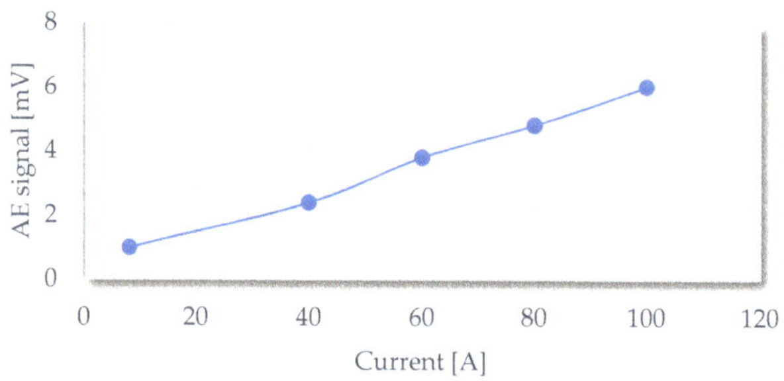

Figure 10. Graph showing the emission signal magnitude dependence as a function of the current flowing through GTO.

The oscilloscope with connected to the AE transducer, Hall effect sensor and a Dietz ammeter recorded three waveforms that were later imported into the Matlab workspace in order to perform offline signal analysis.

The last, additional test that was performed covered pencil lead break. The pencil lead has been broken on the flat, hard surface of the upper pressing plate. The objective of such test was to check if it was possible to get external emission signal in the presence of current flow and line thyristor commutation in rectifying mode. As it can be seen in Figure 11, the amplitude of the pencil break signal was so high in comparison to regular signals coming from inside of conducting thyristor the vertical scale of the scope was increased 5-fold from 2 mV/div up to 10 mV/div.

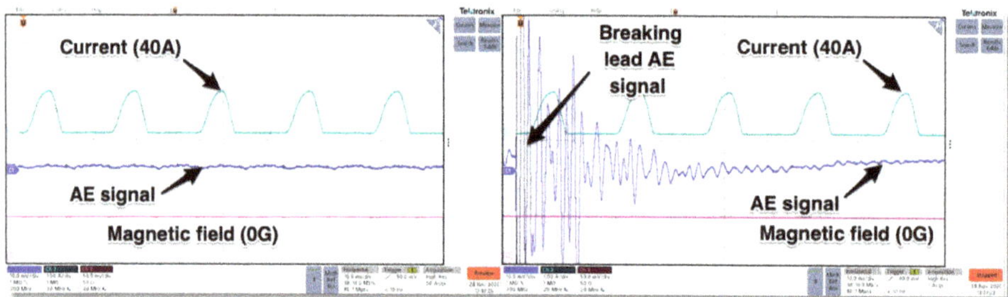

Figure 11. Waveforms of current, magnetic field and acoustic emission in regular GTO rectifying mode (**left**) and AE signal of break pencil lead test while GTO operating (**right**).

While conducting the test the broken part of the lead bounced on the surface of pressing plate thus additional emission signal occurred. The data obtained in the test were transferred into the Matlab and FFT analysis of the signals was performed.

3. Results

For the signals obtained in experimental studies in the time domain, FFT conversion was used in the frequency range of acoustic emission extended up to 5 MHz. As it turned out the only visible bins of spectrum are present in a very low frequencies range as can be seen in the Figure 12 (blue, vertical line overlapping AE signal axis).

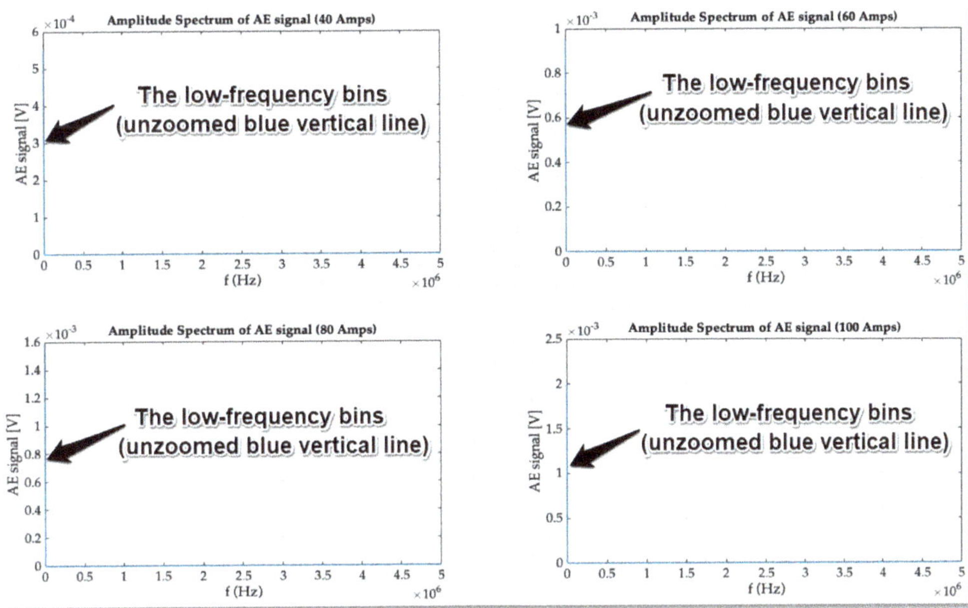

Figure 12. Wideband frequency spectrum of acoustic emission signal obtained for 40, 60, 80 and 100 amperes current.

This clearly shows that the inner structure of the GTO thyristor during regular rectifying operation is not a source of high frequency acoustic signals. The obtained results of the fast Fourier transform were magnified, and the frequency axis limited up to 2 kHz.

The magnitude of transducer signals increases with the current amplitude and the significant spectrum bins are observable up to 600 Hz. From the Figure 13 it is clear that the elastic wave frequencies inside the semiconductor structure depend only on frequency of rectified current. The displacement that takes place in the material of press-pack creates signal of base frequency (50 Hz) and wave propagates in all directions of the thyristor volume. On the basis of the spectrum analysis, it can be seen that the harmonics present in the signal are multiples of the fundamental 50 Hz harmonic. They are generated by reflecting elastic waves from the boundaries of the thyristor structure. Higher values of the signal are sensed by AE transducer located closer to the reflected waves and main source of material displacement and friction.

As it can be observed in the Figure 14 the transducer detected the pencil break elastic waves, but the energy of the signal overlapped signals coming from the internal material expansion and all additional reflected waves were present inside the GTO structure.

The FFT analysis seems to be good enough for signals of short duration analysis (up to a few cycles) but it does not take account of thermal expansion which shows up in long time operation.

In order to get the desired information about how acoustic emission signals are changing with temperature deviations the system performing the power spectral density in real-time should be chosen.

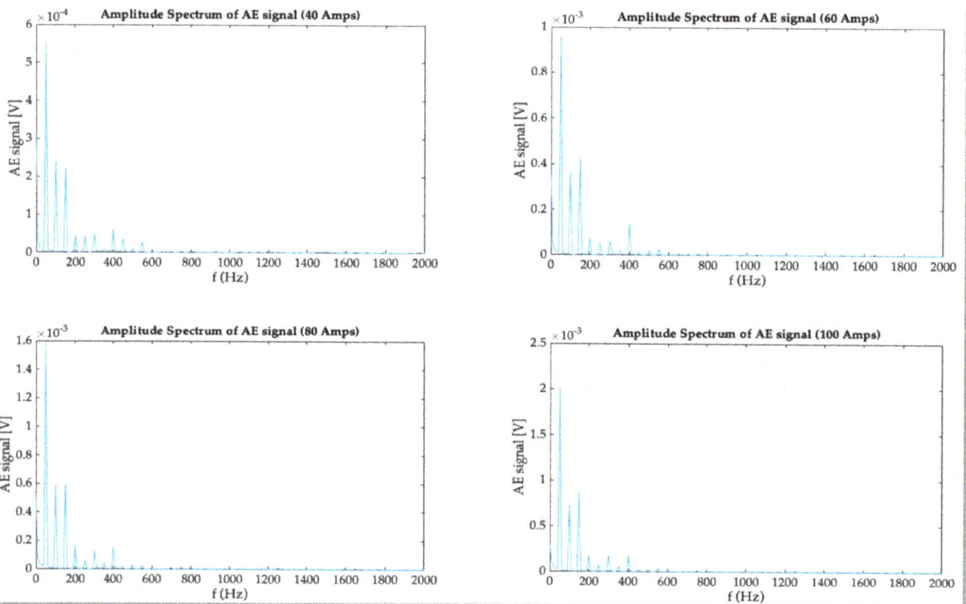

Figure 13. Narrowed frequency spectrum of signal recorded for 40, 60, 80 and 100 amperes anode-cathode rectified current.

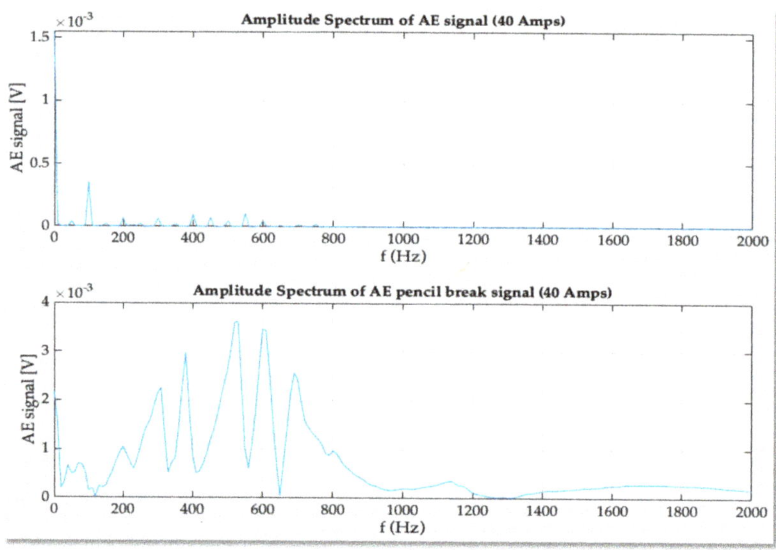

Figure 14. The frequency spectrum of signal recorded for 40 amperes anode-cathode rectified current (**upper**) and pencil lead break test effects in the presence of 40 A current (**lower**).

4. Discussion on Results

The following conclusions can be drawn from the carried out experimental tests. It is possible to use a broadband AE sensor to determine the acoustic emission signals of a brand new GTO thyristor operating in rectifier mode. The obtained waveforms are characterized by a strong dependence on the thyristor switching frequency. Regular systems equipped with preamplifier filter out low frequencies but in the case of semiconductors this information may be useful for on-line condition monitoring or diagnostics. According to the nature of flexible wave propagation, the AE sensor also detected waves reflected inside the semiconductor structure which could be used to observe changes in volume caused by abnormal conditions such as overheating, internal dislocations or a decrease of wafer plate pressure.

The conducted tests did not show the existence of signals of frequencies typical for the acoustic emission band (range from 1 kHz to 1 MHz) during regular GTO operation. The WSα sensor is sufficiently resistant to electromagnetic interference occurring on press-pack enclosure metal plates and, as the test results have shown, the magnetic fields resulting from a direct current flow up to 100 A do not interfere with the AE sensor's operation. According to our tests the use of such type of transducer in the vicinity of inductive elements generating strong magnetic fields will result in creation of parasitic signals, which may cover useful acoustic emission signals.

5. Conclusions

The presented method can be considered as a new approach to diagnose press-packed power electronics devices. As of now there is no widely available information about practical applications of acoustic emission signals and their changes with degradation, aging or any other concerning processes occurring in switching semiconducting devices. The obtained AE signals can be used as a reference in online monitoring systems which supervise the operation of the thyristor. By comparing them to the actual signals coming from a working GTO any changes noticed can be a sign of disturbances occurring in the structure. Because of possible internal semiconducting structure dislocations, cracks and fractures high frequency AE signals are expected to happen although up to now no evidences of such an effect were presented. On the other hand, due to multilayer structure aging observable, especially after prolonged time, the signal characteristics in current amplitudes coming especially from gate circuitry structures, slight but notable changes are expected to show up. The obtained results will be then some kind of a benchmark taken for good, fully operational thyristor so any signals which does not fit the pattern may be analyzed as a possible structure degradation.

In order to fully check the suitability of the presented method for testing the state of a semiconductor system, additional tests should be carried out, including prolonged exposure to higher values of currents (at least up to the nominal values) and power supply of the thyristor with the nominal medium voltage. The latter requirement makes it necessary, in order to ensure the safety of the equipment and operators, to use insulating spacers (e.g., mica plates), which will act as an electrical insulation but disperse a large amount of useful high frequency acoustic emission signals. This makes it necessary to use other types of sensors prone to strong electrical field for example made out of fiberglass.

Author Contributions: Conceptualization, M.K., A.B. and A.T.; methodology, M.K., A.B.; software, M.K.; validation, M.K., A.B.; formal analysis, A.T.; investigation, M.K. and A.B.; resources, M.K., A.B. and A.T.; writing—original draft preparation, M.K.; writing—review and editing, A.B. and A.T.; funding acquisition, A.B. and A.T. All authors have read and agreed to the published version of the manuscript.

Funding: This research was partially funded by Maritime University of Szczecin, grant number 2/S/2/KEiE/2020.

Conflicts of Interest: The authors declare no conflict of interest.

References

1. Yu, J.; Ziehl, P.; Zarate, B.; Caicedo, J.M. Prediction of fatigue crack growth in steel bridge components using acoustic emission. *J. Constr. Steel Res.* **2011**, *67*, 1254–1260. [CrossRef]
2. Roberts, T.M.; Talebzadeh, M. Acoustic emission monitoring of fatigue crack propagation. *J. Constr. Steel Res.* **2003**, *59*, 695–712. [CrossRef]

3. Mazal, P.; Vlasic, F.; Koula, V. Use of acoustic emission method for identification of fatigue micro-cracks creation. *Proc. Eng.* **2015**, *133*, 379–388. [CrossRef]
4. Krampikowska, A.; Pała, R.; Dzioba, I.; Świt, G. The Use of the Acoustic Emission Method to Identify Crack Growth in 40 CrMo Steel. *MDPI Mater.* **2019**, *12*, 2140.
5. Sharma, R.B.; Parey, A. Modelling of acoustic emission generated in rolling element bearing. *Appl. Acoust.* **2019**, *144*, 96–112. [CrossRef]
6. Miettinen, J.; Andersson, P. Acoustic emission of rolling bearings lubricated with contaminated grease. *Tribol. Int.* **2000**, *33*, 777–787. [CrossRef]
7. Shaker, Y.O. Detection of partial discharge acoustic emission in power transformer. *Int. J. Electr. Comput. Eng.* **2019**, *9*, 4573. [CrossRef]
8. Ramirez-Nino, J.; Pascacio, A. Acoustic measuring of partial discharge in power transformers. *Meas. Sci. Technol.* **2009**, *20*, 115108. [CrossRef]
9. Chu, L.; Yang, D.-G.; Ren, S.-Z. Numerical Simulation of Propagation Process of Acoustic Emission Partial Discharge Signal in Medium Voltage Switch-Gears. *DEStech Trans. Comput. Sci. Eng.* **2019**, *28798*, 522–524. [CrossRef]
10. Wang, L.; Wang, H.; Wang, L.; Lu, H.; Ning, W.; Jia, S.; Wu, J. Experimental investigation of transient earth voltage and acoustic emission measurements of partial discharge signals in medium-voltage switchgears. In Proceedings of the 2nd International Conference on Electric Power Equipment—Switching Technology (ICEPE-ST), Matsue, Japan, 20–23 October 2013; pp. 1–4.
11. Zhang, T.; Pang, F.; Liu, H.; Cheng, J.; Lv, L.; Zhang, X.; Chen, N.; Wang, T. A Fiber-Optic Sensor for Acoustic Emission Detection in a High Voltage Cable System. *Sensors* **2016**, *16*, 2026. [CrossRef] [PubMed]
12. Karkkainen, T.; Talvitie, J.P.; Kuisma, M.; Hannonen, J.; Strom, J.-P.; Silventoinen, P. Acoustic Emission in Power Semiconductor Modules—First Observations. *IEEE Trans. Power Electron.* **2013**, *29*, 6081–6086. [CrossRef]
13. Karkkainen, T.; Talvitie, J.P.; Kuisma, M.; Silventoinen, P.; Mengotti, E. Acoustic emission caused by the failure of a power transistor. In Proceedings of the 2015 IEEE Applied Power Electronics Conference and Exposition (APEC), Charlotte, NC, USA, 15–19 March 2015. [CrossRef]
14. Bejger, A.; Gordon, R.; Kozak, M. The use of acoustic emission elastic waves as diagnosis method for insulated-gate bipolar transistor. *J. Mar. Eng. Technol.* **2020**, *19*, 186–196. [CrossRef]
15. Gordon, R.; Dreas, A. Detection and Recording of Acoustic Emission in Discrete IGBT Transistors. *Multidiscip. Asp. Prod. Eng.* **2018**, *1*, 27–31. [CrossRef]
16. Li, M.; He, Y.; Meng, Z.; Wang, J.; Zou, X.; Hu, Y.; Zhao, Z. Acoustic Emission-Based Experimental Analysis of Mechanical Stress Wave in IGBT Device. *IEEE Sens. J.* **2020**, *20*, 6064–6074. [CrossRef]
17. Bejger, A.; Gordon, R.; Kozak, M. Acoustic Emission of Monolithic IGBT Transistors. *New Trends Prod. Eng.* **2018**, *1*, 755–760. [CrossRef]
18. Rashid, M.H. *Alternative Energy in Power Electronics*; Elsevier Inc.: Waltham, MA, USA, 2015; p. 223.
19. Minges, M.L. *Electronic Materials Handbook: Packaging*, 1st ed.; CRC Press: Boca Raton, FA, USA, 1989; pp. 56–57.
20. Titushkin, D.; Surma, A.; Chernikov, A.; Matyukhin, S.I. New Ways to Produce Fast Power Thyristors. *Bodo Power Syst.* **2015**, *8*, 46–47.
21. Rashid, M.H. *Power Electronics Handbook*; Academic Press: San Diego, CA, USA, 2001; pp. 32–34.
22. Pandey, R.K. *Handbook of Semiconductor Electrodeposition*; CRC Press: New York, NY, USA, 1996.
23. Cepek, M.; Krishnayya, C.P. Thyristor aging. In Proceedings of the POWERCON '98. 1998 International Conference on Power System Technology. Proceedings (Cat. No.98EX151), Beijing, China, 18–21 August 1998; pp. 649–653.
24. Kozak, M. Voltage Source Inverter Synchronization with the Use of FFT Algorithm. Mechatronics 2017—Ideas for Industrial Applications in Advances. In *Intelligent Systems and Computing*; Springer International Publishing: Cham, Switzerland, 2017.
25. Bhuyan, M. *Intelligent Instrumentation: Principles and Applications*; CRC Press: Boca Raton, FA, USA, 2011; pp. 62–63.
26. What is a Power Spectral Density (PSD)? Available online: https://community.sw.siemens.com/s/article/what-is-a-power-spectral-density-psd (accessed on 25 November 2020).
27. Ae Sensors & Preamplifiers User's Manual. Available online: https://www.vallen.de/zdownload/pdf/Sensor_Preamplifier_Description.pdf (accessed on 25 November 2020).
28. Mohseni, K.; Shakouri, A.; Ram, R.J. Electron Vortices in Semiconductor Devices. *Phys. Fluids* **2005**, *17*, 100602-1–100602-7. [CrossRef]
29. Lacidogna, G.; Carpinteri, A.; Manuello, A.; Durin, G.; Schiavi, A.; Niccolini, G.; Agosto, A. Acoustic and Electromagnetic Emissions as Precursor Phenomena in Failure Processes. *Strain* **2011**, *47*, 144–152. [CrossRef]
30. Miroshnichenko, M.; Kuksenko, V. Study of electromagnetic pulses in initiation of cracks in solid dielectrics. *Soviet Phys. Solid State* **1980**, *22*, 895–896.

 sensors

MDPI

Article

Acoustic Emission and K-S Metric Entropy as Methods for Determining Mechanical Properties of Composite Materials

Lesław Kyzioł, Katarzyna Panasiuk *, Grzegorz Hajdukiewicz and Krzysztof Dudzik

Faculty of Marine Engineering, Gdynia Maritime University, 81-225 Gdynia, Poland;
l.kyziol@wm.umg.edu.pl (L.K.); g.hajdukiewicz@wm.umg.edu.pl (G.H.); k.dudzik@wm.umg.edu.pl (K.D.)
* Correspondence: k.panasiuk@wm.umg.edu.pl; Tel.: +48-58-558-6484

Abstract: Due to the unique properties of polymer composites, these materials are used in many industries, including shipbuilding (hulls of boats, yachts, motorboats, cutters, ship and cooling doors, pontoons and floats, torpedo tubes and missiles, protective shields, antenna masts, radar shields, and antennas, etc.). Modern measurement methods and tools allow to determine the properties of the composite material, already during its design. The article presents the use of the method of acoustic emission and Kolmogorov-Sinai (K-S) metric entropy to determine the mechanical properties of composites. The tested materials were polyester-glass laminate without additives and with a 10% content of polyester-glass waste. The changes taking place in the composite material during loading were visualized using a piezoelectric sensor used in the acoustic emission method. Thanks to the analysis of the RMS parameter (root mean square of the acoustic emission signal), it is possible to determine the range of stresses at which significant changes occur in the material in terms of its use as a construction material. In the K-S entropy method, an important measuring tool is the extensometer, namely the displacement sensor built into it. The results obtained during the static tensile test with the use of an extensometer allow them to be used to calculate the K-S metric entropy. Many materials, including composite materials, do not have a yield point. In principle, there are no methods for determining the transition of a material from elastic to plastic phase. The authors showed that, with the use of a modern testing machine and very high-quality instrumentation to record measurement data using the Kolmogorov-Sinai (K-S) metric entropy method and the acoustic emission (AE) method, it is possible to determine the material transition from elastic to plastic phase. Determining the yield strength of composite materials is extremely important information when designing a structure.

Keywords: composites; recycling; acoustic emission; K-S metric entropy; mechanical properties

Citation: Kyzioł, L.; Panasiuk, K.; Hajdukiewicz, G.; Dudzik, K. Acoustic Emission and K-S Metric Entropy as Methods for Determining Mechanical Properties of Composite Materials. *Sensors* **2021**, *21*, 145. https://doi.org/10.3390/s21010145

Received: 9 December 2020
Accepted: 24 December 2020
Published: 28 December 2020

1. Introduction

In recent years, an increase in the use of glass fiber reinforced polyester laminates has been observed in many industrial branches, including shipbuilding [1–3]. The use of composite materials in shipbuilding results from their specific properties, such as resistance to rotting in the sea water environment, as well as corrosion and the effects of chemicals. These materials are characterized by high resistance to fatigue loads, especially typical for the operation of the hull or elements of the marine propulsion system on a sea wave [1–3]. Ship propulsion systems require specific technical solutions, the features of which result not only from engineering practice, but mainly from the requirements of classification societies [2–4]. Polyester laminates in conditions of salinity with air mist show high resistance to corrosion. They are also used for bearing and sealing propeller shafts. Laminates are used for sealing and bearing the rudder stocks, they are also used for such elements as deck crane bearings, washers for the sliding hatch cover, in mooring systems and in hydraulic devices for operating pressures ranging from 25 to 100 MPa [5].

Due to their anisotropic properties, composite materials constantly ask designers to define the parameters of mechanical properties with high precision. One such problem is determining the transition of a material from elastic to plastic. Such a problem does not exist in elastically plastic materials. Currently, the researcher has more and more perfect machinery and equipment facilities, which creates new opportunities for a more detailed understanding of the properties of anisotropic materials, which are undoubtedly polymer composites.

Having the appropriate equipment facilities, the authors set themselves the goal of determining the yield point of the composite material with the greatest possible accuracy. The knowledge of this material parameter allows for greater use/effort of the material, which is of great importance when designing the structure. For composite materials, the standards do not provide for the determination/knowledge of the yield point. For this reason, the calculations of composite-based structures are most often underestimated. One of the methods allowing to determine the yield point of the tested polymer composites is the method of acoustic emission. Basically, this method, classified as a non-destructive method, is used to monitor the process of damage to materials (structures). Monitoring can be carried out under both operational and laboratory conditions. The tested object is subjected to a load that is recorded. The applied amount of force cannot lead to the complete destruction of the structure [6]. Acoustic emission (AE) is a phenomenon in which elastic waves are generated, forced by stresses resulting from structural changes (fiber cracking, matrix cracking, etc.) in solids. The frequency of AE waves in the range from a few kHz to 1 MHz is detected on the structure surface by piezoelectric sensors that convert the strain energy into an electrical signal [6–17].

The AE method is especially used to test fiber-reinforced polymer structures [6,7,9,12–17]. The AE bending test described in [6] contributed to the development of the AE signal identification methodology. This resulted in the identification of three characteristic time intervals: initiating damage (DI), cumulative damage (DA), and localization damage (DL) [6].

The static tensile test of bars reinforced with polymer fibers, during which various damages were registered (delamination, matrix cracking, fiber detachment and cracking)- contributed to the determination of the size and frequency of each signal [13].

The paper [14] describes the mechanical damage of a thick laminated carbon-epoxy composite subjected to static and fatigue tests. Three types of signals were related to three failure modes (matrix cracking, delamination, fiber breakage). Parameters such as amplitude, duration and energy were analyzed [14]. The papers [15–17] describe the use of acoustic emission parameters to identify the type of damage.

Various mathematical methods are used to study the properties of materials. Modern measurement tools and methods ensure their very high accuracy and obtain large data sets. In order to organize these sets and obtain results, one can use the methods used, among others, in mathematical statistics. In search of deviations of experimental data from their mean values, an analysis is performed, and the interpretation of measurement errors and distortions is made. The end result is specific procedures which include, among others, limits, frames, and divisions of numerical values to which further experimental data should be assigned. In engineering practice, there are numerous obstacles resulting from a significant dispersion of mean values. Spreading measurement data across studies can take complex forms, and it can be difficult to identify a single parameter to control a particular process. Examples are the graphs of the static tensile test, on the basis of which the internal dynamics of the material deformation process cannot be observed. The dynamics of the tensile deformation process of the sample in the form of chaotically arranged measurement points (especially in the elastic range of the material) can only be observed by significantly enlarging a part of the graph [18–22].

The occurrence of deterministic chaos in experimental data sets, characterized by the behaviour of materials under given loads in real structures or in laboratory conditions on strength machines, has not yet been sufficiently described [22].

Earlier tests carried out on materials showing a clear yield point showed a high agreement of the yield strength values obtained in accordance with the standard and the specified K-S entropy method [18–23]. On the basis of a series of laboratory tests carried out by the authors of the study for isotropic and anisotropic materials, it was assumed that the application of the K-S method allows to determine the transition of the material from the elastic to the plastic phase. The method of visualizing the changes in the dynamics of data obtained from strength tests in laboratories or from measurements on real objects may be of great importance for the future of design and modelling of composite materials.

In the article, on the example of a polyester-glass composite with and without recyclate, research was undertaken to determine the yield point. For this purpose, the acoustic emission (AE) method and the K-S metric entropy method were used.

2. Materials and Methods

The glass polyester recyclate used for testing is a fragment of the main deck sourced from one of the ships. This scrap was pre-crushed using a mechanical hammer, and then ground in a dedicated plastic waste processing station. Crushed this way, the material was screened on sieves to obtain various fractions. A 1.2 mm sieve mesh was used. The 1.2 mm granulate obtained is referred to as the recyclate and was used as the filler of a newly produced glass polyester composite matrix.

Producing a proper grain-sized recyclate was followed by producing a composite on the basis of resin, glass mat used as reinforcement, and the recyclate-used as the filler. Manual contact lamination was used to produce the new composite. Polyester, structural, orthophthalic, rigid POLIMAL 109A resin manufactured by Organika-Sarzyna S.A. and a hardening and booster agent were used as matrix for all materials produced. Two different composite materials with different weight additives were produced: basic material without recyclate (pure), material with a 10% recyclate content [24–29]. The contents of% by weight of the reinforcement, matrix and recyclate are presented in Table 1.

Table 1. The contents of% by weight of the reinforcement, matrix and recyclate in prepared materials.

	Number of Layers of Reinforcement	Matrix, % by Weight	Resin, % by Weight	Recyclate, % by Weight
0%, without recyclate	12	40	60	0
10% recyclate, ≤1.2 mm granulation	12	29	61	10

Prepared materials are shown on Figure 1.

(a) (b)

Figure 1. Cross-sections of composite materials with recyclate content of 1.2 mm granulation and percentage content: (**a**) 0%, (**b**) 10%, magnification ×3 (confocal microscope).

The mechanical properties of composites with the addition of glass polyester recyclate were determined according to PN-EN ISO 527-4_2000P. For this purpose, each of the two composites was used to prepare samples for static tensile testing. The shape and dimensions of the samples are shown in Figure 2.

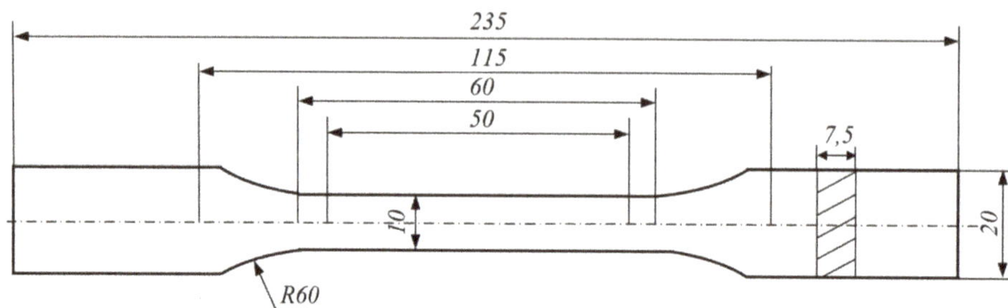

Figure 2. Dimensions of samples for static tensile test in accordance with PN-EN ISO 527-4_2000P.

The static tensile test of the samples from the tested composites was carried out on a universal testing machine with hydraulic drive, type MPMD P10B, with TestXpert II software, version 3.61. by Zwick & Roell with the Epsilon model 3542 extensometer (Epsilon Technology Corp., Jackson, WY, USA). The tests were carried out on tests at ambient temperature (22 °C) at the strain rate equal to $2 \times 10^{-5}\ s^{-1}$.

2.1. Acoustic Emission Method

The research was carried out at the universal hydraulic testing machine Zwick&Roell MPMD P10B with TestXpert II software. Additionally, Epsilon 3542 extensometer was used for measuring elongation during test. For monitoring tensile test of chosen specimens Physical Acoustics Company (Physical Acoustic Corporation, 195 Clarksville Rd, Princeton Junction, NJ, USA) acoustic emission system was used. A general view of the laboratory stand is presented in Figure 3 [30].

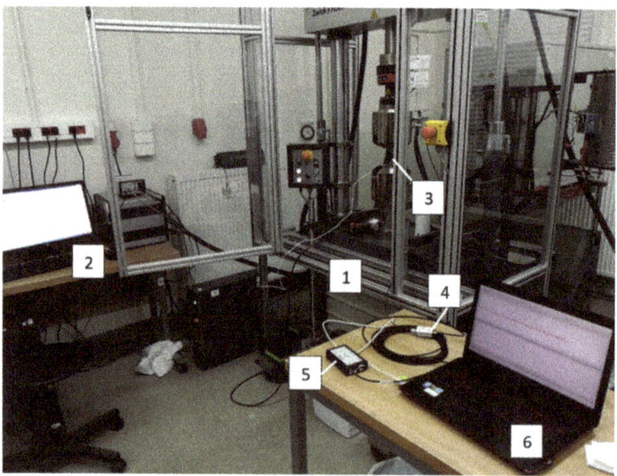

Figure 3. General view of laboratory stand: 1–tensile stress machine, 2–tensile stress machine computer, 3–AE sensor, 4–preamplifier, 5–AE recorder, 6–AE computer [30].

A view of the specimen fixed into tensile testing machine grips during research, with AE sensor and extensometer, is shown in Figure 4.

Research AE was performed using set consisted of a single channel recorder USB AE Node (Physical Acoustic Corporation, Princeton Junction, NJ, USA), type 1283 with bandpass 20 kHz–1 MHz, preamplifier with bandpass 75 kHz–1.1 MHz, AE-Sensor vs. 150 M (with a frequency range of 100–450 kHz), computer with AE Win for USB Version E5.30 software for recording and analysing AE data.

Between the sensor and a surface of the specimen a coupling fluid was used. AE Sensor was fixed to specimen by elastic tape.

Figure 4. View of specimen fixed into tensile testing machine grips: 1–tensile testing machine grips, 2–specimen, 3–extensometer, 4–AE sensor [30].

2.2. Metric Entropy Method

Figure 5 shows an example graph of a static test on stretching the composite with a 10% recyclate content. This kind of a graph doesn't present the inside dynamics of the process of distorting materials. The chaotic arrangement of the measurement points is presented on the Figure 6 by means of the consecutive enlargements of the fragments marked of the graph on the Figure 5.

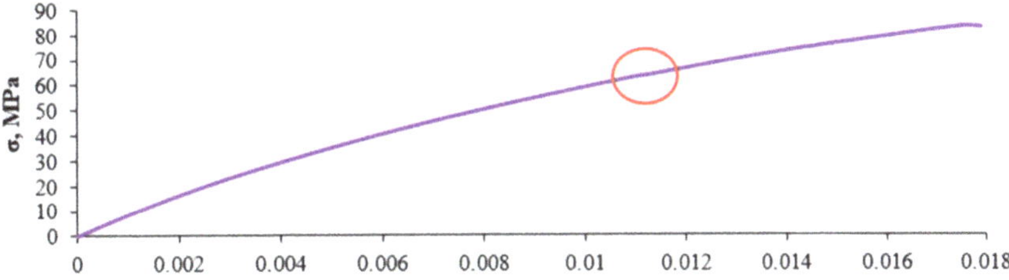

Figure 5. An example graph of a static test on stretching the composite with a 10% recyclate content.

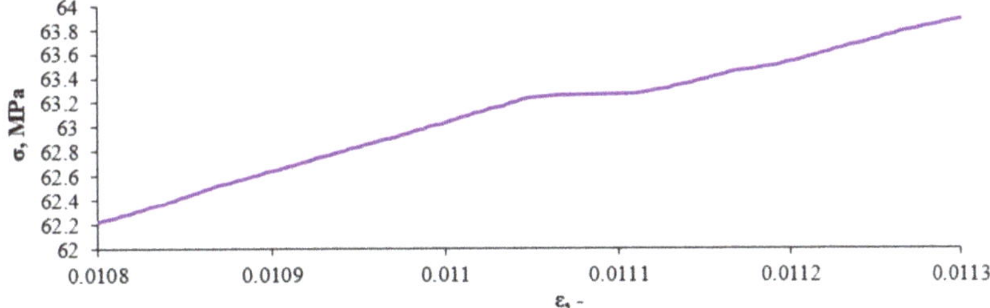

Figure 6. Chaotic arrangement of the measurement points.

Figure 6 shows a fragment of a static tensile test of the composite containing 10% of recyclate.

The most important factor that influences the precision of a test concerns the type of a specimen that generates measurement distortion, depending on the starting point state of a specimen's material and the conditions of its fixing. Anisotropy and distribution of stresses within the material, the structure and material's crystallographic defects distort the homogeneity of the stresses' distribution and strains in the specimen's measurement field. They lead to the development of local fields of plastic strains also when treated with small forces, which don't exceed the elastic limit. Resolution, in other words, the precision of the measurement, is defined on the basis of the noise measurement. Other factors that influence the test's results are related to one another in a system machine-specimen, that is controlled in a loop of feedback. The behaviour of this system depends on hydraulic system's quality and efficiency as well as the characteristics of a specimen being tested. Parameters of the feedback between the specimen and the machine must be regulated according to the specimen's predicted characteristics. Any distortions in the system machine-specimen, are visible especially when the characteristic of a certain material changes, i.e., during the transition from an elastic into a plastic range, when transition into plastic strains occurs and rigidity of the measured system undergoes a sudden decrease. Damage of the specimen initiates a local process, which has got an effective impact on the dynamic characteristic of the whole system. During this transition, we can observe an increase of force error in relation to the applied force, but it is not only due to the measurement precision, but the total effect of the measurement track's resolution, the efficiency of hydraulic system, the setting of the steering, and the degree of changes of the material's characteristics.

Classic thermodynamic entropy entails the flow energy from large to small scale [31–34]. According to the second law of thermodynamics, the total entropy of an isolated system can never decrease over time, and will thus always approach positive values. A statistical interpretation of the second law of thermodynamics was coined by L. Boltzmann. He demonstrated that macroscopic state entropy S is proportional to the thermodynamic probability of this state. This interrelation was expressed by formula:

$$S = k \ln W \tag{1}$$

This means that the dynamics of a system leads to the formation of increasingly probable states as it approaches the maximum value of thermodynamic probability. K-S metric entropy [22,31–34] joins opposite-direction energy flow, i.e., transitions from small-scale to large-scale dispersion. Energy dispersion is never a continuous process. As a dynamic notion, K-S entropy is the "entropy of a time unit" [22], and is therefore non-negative, but can both increase and decrease. The relationship between statistical mechanics and the chaos theory reflects the notion of K-S metric entropy, which was introduced by Kolmogorov [35].

The essence of metric entropy is that it is dynamic by nature, as it describes system movement typical for chaotic processes [22,31–34]. Metric entropy is a value that measures the instability of the dynamics of a system, i.e., expresses a method for the numerical description of chaos.

Works [18–23] present an overview of a measurement collection and a diagram for calculating K-S entropy for successive positions of measurement collections containing an n number of items, relative to measurement data.

For a discrete probability distribution, K-S metric entropy [35,36] is described using the following formula:

$$S = -\sum_{i=1}^{N} p_i \ln p_i \tag{2}$$

where:

N denotes number of partitions, into which the set of all possible results was divided and p_i is the probability of occurrence of results in the ith partition (according to $p \ln p = 0$, if $p = 0$).

If the partitions are equally probable, which means that $p_i = 1/N$ for all i, then entropy is defined by $S = \ln N$ and assumes its maximum value. However, if the results are known for specific partitions, then entropy will assume its minimum value of $S = 0$, since $p_i = 1$. The conditional extremes of the entropy function were determined using the Lagrange multiplier method [35,36].

The K-S metric entropy method assumes that the qualitative changes which take place at the structural threshold separating the elastic state from the plastic state correspond to a specific measurement point. Energy dissipation takes place in the system, and the deterministic chaos of data related to this phenomenon causes the variability of entropy.

In this method, it is essential to prepare the input data used to calculate metric entropy correctly. In the sample stretching process, the tester must record the stress and the strain, and a column of ε σ quotients must be created at the same time. The sequence of successive strains, stresses and ε σ quotient remainders is formed by deducting successive rows. As the first derivative of the time sequence, it reflects the local dynamic of measurement data. For input data prepared in this manner, a measurement collection occupies successive positions in a column of numbers, shifting by a single row every entropy calculation step. Determined this way, entropy is them recorded on a diagram.

The minimum value of metric entropy in the vicinity of the passage from elastic to plastic state is marked by the point that separates individual states of the process. The value of stress corresponding to the passage of the tested material from its elastic to its plastic state is determined on the basis of the "critical" point [18–23].

To determine the "critical" point which corresponds to the transition of the material from one state to another, we need to have a declared number of constituents k in the measurement collection, and a number of sub-partitions N, into which the measurement collection will be divided. Based on the data from tensile testing, the effect of k and N values on the shape of the entropy diagram and the position of the minimum value relative to the tensile curve were analyzed. Measurement data includes 1000 to 2000 points in the elastic scope. The N number was adopted according to a formula applied in mathematical statistics in the construction of histograms: $N \leq 5 \, log \, k$ [18–23].

An analysis of determination of passage from the elastic state to the plastic state, applying the "minimum entropy" method on tested materials, has indicated that the correct selection of the k number of constituents of the measurement collection and the N number of sub-partitions, is key to drawing the K-S metric entropy fluctuation diagram, marking a clear minimum which confirms the passage from one state to another. Correctly prepared data can be then entered to calculation software [18–23].

3. Results

Table 2 shows all results of a static tensile test of samples of composite materials tested are given.

Table 2. Mean values of a static tensile strength test carried out on composite materials obtained from three samples taken from each material.

Recycled Content, % Weight, mm		E, MPa	R_m, MPa	ε, -
0	**Mean Value**	8982	135	0.0216
	Standard Deviation	0.53	0.87	0.0057
10%, 1.2	**Mean Value**	7830	110	0.0201
	Standard Deviation	0	0.39	0.0057

Table 2 presents the mean values for two composite materials. Three samples were taken from each material for the static tensile tests.

The study of the structure of the composites was carried out on samples by optical microscope (Axiovert 25, Carl Zeiss AG, Oberkochen, Germany). The samples were cut across (in thickness) and the surface was analyzed. Figure 7 shows the structures of composite materials.

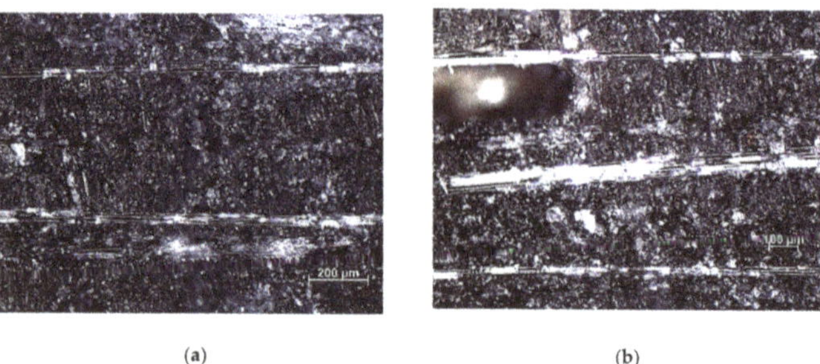

(a) (b)

Figure 7. Composite structures: (**a**) without recyclate, (**b**) 10% of recyclate, magnification ×100.

In general, the mechanical properties of the composite depend greatly on the ability of the matrix and filler to adhere well to each other. The structures of composites presented in Figure 7 slightly reflect their mechanical properties. In the case of a material without recyclate (Figure 7a), evenly distributed glass fibers are visible, which results in high adhesion of the resin to the glass fiber. This guarantees a high physical contact of the fiber with the polyester, which results in high strength properties of the composite.

The introduction of polyester-glass recyclate to composites reduced their tensile strength. In the structure of the composite with 10% recyclate content (Figure 7b), air pores and areas in the form of recyclate clusters are visible. The result is a reduction in the strength properties of the material. Figure 8 shows the results of a static tensile test for composite materials.

A static tensile test carried out on composite samples without recyclate content and with 10% recyclate content pointed to a clear impact of recyclate content on the mechanical properties of the composites tested. The results of the static test pointed to a deterioration of the mechanical properties of the composite, proportional to the increase in recyclate content.

Obtained results from the static tensile test of composite materials with the addition of recyclate served as the input material for the application of metric entropy K-S to determine the yield point of the composite materials tested.

Entropy is calculated using stress and strain data measured in the process of stretching the samples. In this study, the samples were stretched by applying fixed force (fixed movement speed is also possible). The resultant diagram of K-S entropy in the function of time is closely related to the stress diagram, which also entails the function of time.

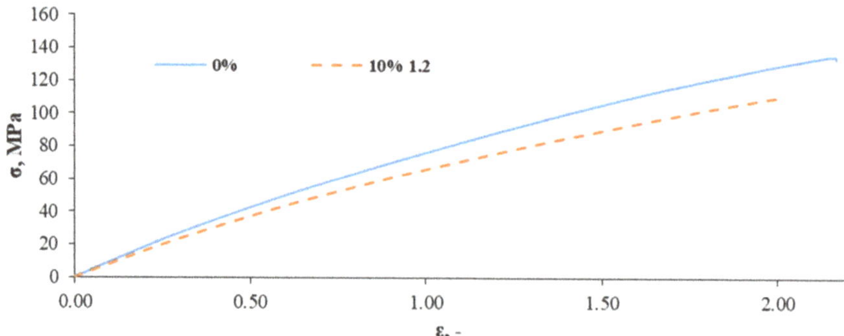

Figure 8. Results of the static tensile test for composite samples.

Before proceeding to the primary calculations, the number of n-elements of the data collection divided into an N number of partitions must be determined. The ultimate n

and N values are heuristically determined. The correct selection of n-elements and N-partitions affects the correctness of the problem to be solved, and this on the correctness and transparency of the resultant diagrams. Correctly solved, the problem can be used to point the local minimum of K-S entropy, which indicates the passage from the elastic state to the plastic state.

To calculate metric entropy, deformation data is used during a static tensile test. In the example below the first 60 deformation results were used. (Table 3). Their minimum and maximum for the selected 60 values (n-elements) have been established (min = 0.00002711, max = 0.00014899). The minimum value was subtracted from the maximum value and this value was divided into 4 (N—number of subsets). Then this value (0.00003046) was added to the minimum value, thus obtaining the first subset, subsequent subsets were created while adding the calculated value in sequence. Then the probability of finding the deformation value in the next subset is determined. The next step is to calculate the logarithm of probability and multiply the probability and logarithm of probability. Substitution of the obtained values, using the K-S entropy formula, allows calculating the values for 30 deformation values, i.e., half.

$$S = -\sum_{i=1}^{N} p_i \ln p_i = -(-0.3535 - 0.3380 - 0.3466 - 0.3466) = 1.3847 \quad (3)$$

Table 3. Procedure of calculating K-S metric entropy.

	I	II	III	IV
min	0.00002711	0.00005758	0.00008805	0.00011852
max	>0.00005758	>0.00008805	>0.00011852	≥0.00014899
	0.00002711	0.00005880	0.00008959	0.00012039
	0.00002902	0.00006131	0.00009106	0.00012227
	0.00003102	0.00006385	0.00009264	0.00012416
	0.00003303	0.00006645	0.00009452	0.00012596
	0.00003495	0.00006902	0.00009647	0.00012759
	0.00003673	0.00007160	0.00009849	0.00012943
	0.00003847	0.00007398	0.00010057	0.00013141
	0.00004022	0.00007624	0.00010277	0.00013364
	0.00004198	0.00007846	0.00010498	0.00013593
	0.00004370	0.00008050	0.00010739	0.00013817
	0.00004553	0.00008252	0.00010979	0.00013997
	0.00004745	0.00008415	0.00011209	0.00014220
	0.00004961	0.00008603	0.00011432	0.00014466
	0.00005190	0.00008779	0.00011646	0.00014675
	0.00005407		0.00011848	0.00014899
	0.00005637			
p_i	0.27	0.23	0.25	0.25
$\ln p_i$	−1.30933332	−1.46967597	−1.38629436	−1.38629436
$p_i \ln p_i$	−0.35352000	−0.33802547	−0.34657359	−0.34657359
S	1.38469265			

This is exemplified by a study [18–23], in which the application of correct procedures and an authorial program yielded correct results. Figure 9 presents a metric entropy diagram applied on a strain diagram in the function of time for a sample of composite without recyclate.

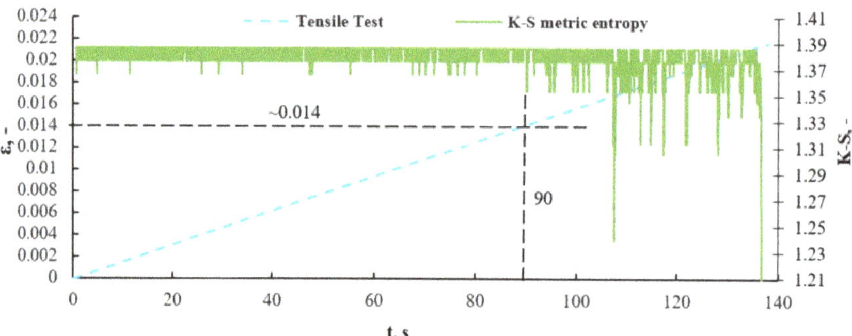

Figure 9. A metric entropy diagram applied on a strain diagram in the function of time, for a sample of composite without recyclate.

In the 90 s of time, the lowest entropy value corresponded to strain ε = 0.014. At this point, according to the K-S metric entropy theory, structural changes occur in the material. Analyzing previous test results for isotropic materials with a clear yield point [18–21], the mathematical statistics method showed exactly the same point as the yield point on the tensile graph. It can be stated that irreversible changes occur in the material at this point and can be defined as the transition from the elastic to plastic phase in the composite material. Figure 10 presents the diagram ε-σ with the strain of 0.017 marked and corresponding to a stress of 116 MPa.

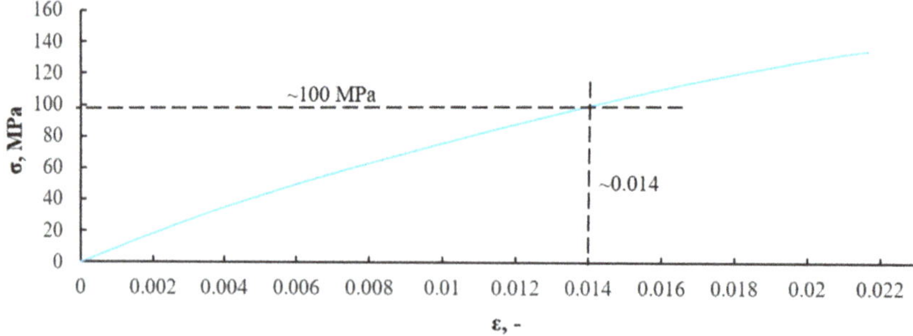

Figure 10. Determination of stress value on a composite without recyclate sample corresponding to strain ε = 0.014 (90 s of time in Figure 9).

After transferring the dependence of strain and K-S entropy from time, the strain values in the strain stress diagram, we obtain the stress value for the transition from the elastic (linear) phase to the nonlinear phase. In the case of the material without recyclate content, the structural change, defined as the transition from the elastic phase to the non-linear range phase, occurred at a stress value of about 116 MPa and a strain of about 0.014. Figure 11 presents a graph of the value of the effective acoustic emission signal as a function of time, plotted on the stress–strain graph for a composite without recyclate.

The diagram presented in Figure 11 shows that the beginning of structural changes in the material occurred in the range of 78–80 s, which corresponded to the stress of 89 ÷ 92 MPa. The calculations carried out by the K-S method showed that it corresponds to the transition of the material from the elastic to the plastic phase. Figure 12 presents a metric entropy diagram applied on a strain diagram in the function of time, for a composite sample with a 10% recyclate content and 1.2 mm in mesh size.

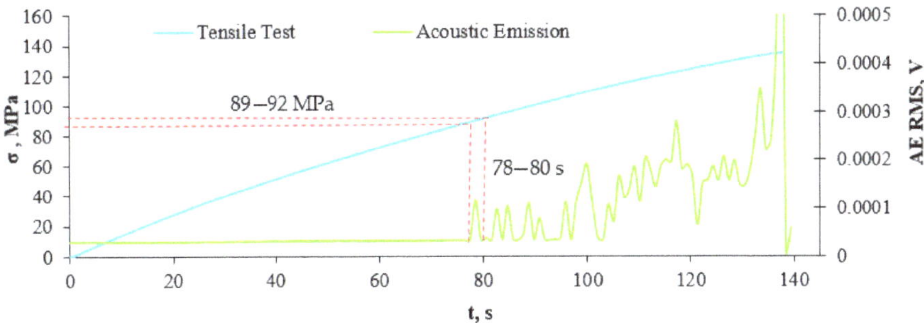

Figure 11. Graph of the value of the effective acoustic emission signal as a function of time, plotted on the stress-strain graph for a composite without recyclate.

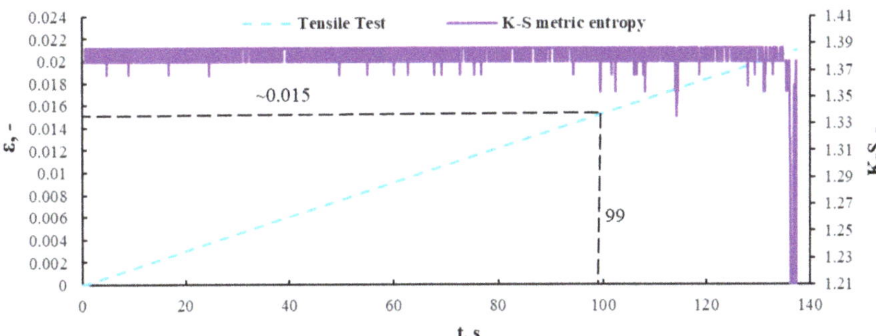

Figure 12. A metric entropy diagram applied on a strain diagram in the function of time, for a composite sample with a 10% recyclate content and ≤1.2 mm in mesh size.

In the 99 s of time, the lowest entropy value corresponded to strain $\varepsilon = 0.015$. For a material with a 10% recyclate content, the deformation, with a significant decrease in entropy, metric K-S, was 1.34. Structural changes have occurred in the material with less deformation than in the material without recyclate content, which results directly from the mechanical properties of the material. Figure 13 presents the diagram $\sigma(\varepsilon)$ with the strain of 0.015 marked and corresponding to a stress of 82 MPa.

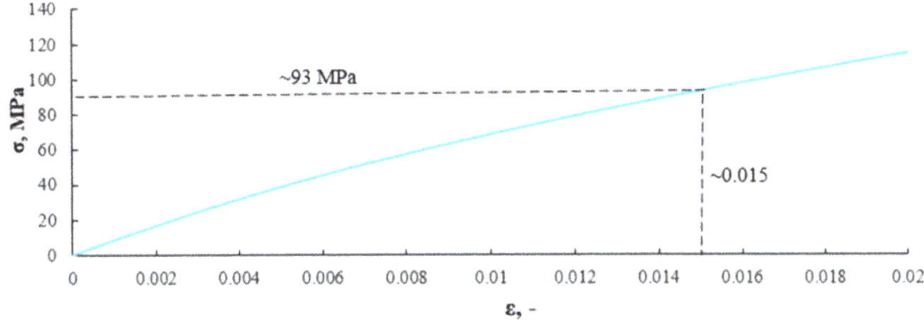

Figure 13. Determination of stress value on a composite sample with a 10% recyclate content and 1.2 mm in grain size corresponding to strain $\varepsilon = 0.015$ (99 s of time in Figure 12).

As a result of the transfer of the dependence of strain and entropy K-S on the time of the strain value on the stress strain graph, we obtain the stress value for the transition from the elastic phase to the non-linear range phase. In the case of a 10% recyclate content, a structural change occurred, defined as the transition from the elastic to the non-linear range phase at a stress of about 93 MPa and a deformation of about 0.015. Figure 14 presents a graph of the value of the effective acoustic emission signal as a function of time, plotted on the stress–strain graph for a composite with 10% content of recyclate.

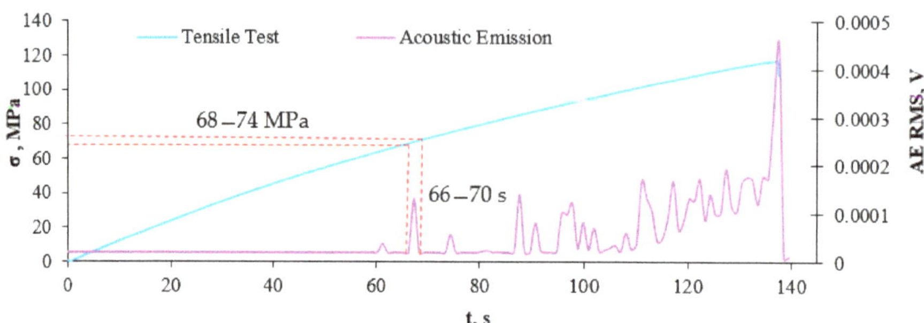

Figure 14. A metric entropy diagram applied on a strain diagram in the function of time, for a composite sample with 10% recyclate content.

Based on the results obtained by the AE method, a significant increase in the RMS value of the signal appeared in 68–74 s, which corresponds to a stress of 66–70 MPa. The values of stresses obtained by the method of K-S metric entropy and acoustic emission are similar, which indicates the occurrence of a qualitative change in the material, i.e., a transition from elastic to non-linear range phase. Table 4 presents a summary of the results obtained using the K-S and AE method.

Table 4. Summary of results for analysed samples obtained with the AE method and the K-S metric entropy.

Recycled Content, % Weight, mm	Transition from Elastic to Nonlinearity by the Method of K-S Metric Entropy, $R_{K\text{-}S}$, MPa	Transition from Elastic to Nonlinearity (Range) by AE Method, R_{AE}, MPa
0%	~100 MPa	89–92 MPa
10%, ≤1.2 mm	~93 MPa	68–74 MPa

The results obtained by the K-S entropy method and the acoustic emission (AE) method show good agreement. Interpretation of the charts presented in Figures 8–14 allows for the assessment of qualitative changes in the material. In particular, the transition of the composite without recyclate from the elastic phase to the nonlinear phase, determined by the K-S method, took place at a stress of approx. 100 MPa and the AE method in the range of 89–92 MPa. In the case of a composite with the addition of 10% recyclate, the appropriately determined stress values by the K-S method are 93 MPa, and by the AE method are 68–74 MPa. Characteristically, the emission shows results that are on average 15 MPa lower than the metric entropy. In each analyzed case, using the acoustic emission method, changes are detected earlier and at lower stress values. The K-S metric entropy method is based on elongation data (in the case of constant force growth rate) and it is a mathematical statistical method. Metric entropy is a value that measures the instability of the dynamics of a system, i.e., it expresses a method of numerically describing chaos. This means that the metric entropy method, using the machine-sample relationship, allows you to determine the point (local minimum K-S metric entropy) at which there is chaos in the obtained measurements and at the same time the transition from elastic to nonlinear phase. The acoustic emission analyzes the transient elastic waves resulting from the processes taking place in the composite during loading, such as the cracking of fibers, matrix, etc. The shift in time results from the fact that the acoustic emission 'detects' changes in the material before the elastic to nonlinear. The AE method is a very accurate method that allows you to directly record test results. However, the K-S metric entropy method can also serve as a tool to assist in the description of material changes under stress. This method should be improved and successfully used in material testing.

4. Discussion

The results of the research showed that the addition of a larger number of granules (over 10%) has an unfavorable effect on composites, as it significantly reduces their strength. The mechanical strength of the composite depends largely on the ability of the matrix and filler and their good adhesion to each other. Basically, the addition of milled polyester/glass granules to composites resulted in a reduction in tensile strength due to poor interfacial adhesion of the granule and matrix particles.

Due to their granular nature, higher contents of granules result in a rougher surface with numerous microcracks, making the composite surface rougher. Research has shown that composites containing granules have a rough surface with many microcracks (Figure 7a,b). The surface of the composite without granules shows a smooth surface with few microcracks. The smooth surface is associated with strong intermolecular bonds, which in turn leads to an increase in tensile strength. The microstructure tests showed that in the tensile test, stronger adhesion between the matrix granules was observed, which gave the composite higher tensile strength.

An analysis of the diagrams presented in Figures 9, 10, 12 and 13 points to a close relationship and regularity between metric entropy diagrams and tensile diagrams. Applying the values of this strain on the diagram $\sigma(\varepsilon)$ (Figure 10), we can define the value of stress (100 MPa). Analogously, in Figure 13, minimum entropy value was determined for a sample with a 10% recyclate content and 1.2 mm in grain size, for which the strain was ε = 0.015. The values of this strain correspond to a stress of σ = 93 MPa (Figure 13).

The tensile strength of the composite without recyclate is 135 MPa (Table 2), and the yield point is in the range of 89–100 MPa. In the case of a composite with 10% recyclate content with a tensile strength of 110 MPa, the yield point is in the range of 74–93 MPa. This requires clarifying the research methods. Research results indicate that research should be continued and research methods improved.

In addition to the decrease in tensile strength, the value of the composite elasticity modulus also decreases with the increase in the granulate content. Due to the lower content of granules, the bonds between the granulate particles and the polyester matrix have a positive effect on their interfacial adhesion and prevent damage to the composite.

Two research methods presented in the paper, aimed at determining the transition of a material from the elastic phase to the nonlinearity phase, have shown their practical usefulness. While the AE method is commonly used in material research, the K-S metric entropy method requires elaboration. The difference between the results obtained when using these methods is approx. 10–20%. It is small, and so, when calculating marine structures made of composite materials, the safety factor should be taken into account, the value of which is within a large range. The accuracy of the results obtained may not be of great importance. The use of modern instrumentation and new calculation methods creates great opportunities for detailed values of material parameters. It is of great importance considering the increasing use of composite materials for very complex and complicated ship structures.

5. Conclusions

1. The use of modern measuring devices and calculation methods enables the determination of material parameters with extremely high accuracy.
2. The method of metric entropy (K-S and acoustic emission AE) allows to determine the transition of the material from elastic to non-linear range phase.
3. K-S metric entropy and acoustic emission method are an excellent engineering tool useful for determining material constants.

4. Both methods create new possibilities for testing materials, especially in terms of fatigue strength.
5. Research should be continued in order to improve research methods of materials due to more and more modern test stands and measuring instrumentation.

Author Contributions: Formal analysis, K.D., K.P., and L.K.; funding acquisition, L.K.; investigation, K.P., L.K., and G.H.; methodology, K.P., L.K., K.D., and G.H.; project administration, L.K.; resources, K.P. and L.K.; software, G.H. and K.D.; supervision, L.K. and K.P.; validation, K.P., L.K., and G.H.; visualization, K.P.; writing—original draft, K.P., L.K., K.D., and G.H.; writing—review & editing, K.P., K.D., and L.K. All authors have read and agreed to the published version of the manuscript.

Funding: This research received no external funding.

Institutional Review Board Statement: Not applicable.

Informed Consent Statement: Not applicable.

Conflicts of Interest: The authors declare no conflict of interest.

References

1. Abrate, S. *Impact on Composite Structures*; Cambridge University Press: Cambridge, UK, 1998.
2. Lankford, B.W.; Angerer, J.F. Glass reinforced plastic developments for application to minesweeper construction. *Nav. Eng. J.* **1971**, *83*, 13–29. [CrossRef]
3. Smith, C.S.; Pattison, D. Design of structural connections in GRP ship and boat hulls. In Proceedings of the Conference on Designing with Fiber Reinforced Materials, Mechanical engineering, London, UK, 27–28 September 1977.
4. Kyzioł, L. Możliwośći Wykorzystania Tworzyw Kompozytowych Do Wałów Okrętowych. In *Zeszyty Naukowe AM Gdynia*; Wydawnictwo Uniwersytetu Morskiego w Gdyni: Gdynia, Poland, 2016.
5. Orkot Marine Bearings. *Engineering Manual Version 4*; Trelleborg Sealing Solutions Rotherham: Trelleborg, Sweden, 2014.
6. Friedrich, L.; Colpo, A.; Maggi, A.; Becker, T.; Lacidogna, G.; Iturrioz, I. Damage process in glassfiber reinforced polymer specimens using acousticemission technique with low frequency acquisition. *Compos. Struct.* **2021**, *256*, 113105. [CrossRef]
7. Shiotani, T.; Fujii, K.; Aoki, T.; Amou, K. Evaluation of progressive failure using AE sources and improved b-value on slope model tests. *Prog. Acoust. Emiss.* **1994**, *529*, 534–536.
8. Kurz, J.H.; Finck, F.; Grosse, C.U.; Reinhardt, H.-W. Stress Drop and Stress Redistribution in Concrete Quantified Over Time by the b-value Analysis. *Struct. Health Monit.* **2006**, *5*, 69–81. [CrossRef]
9. Barile, C.; Casavola, C.; Pappalettera, G.; Vimalathithan, P.K. Acousto-ultrasonic evaluation of interlaminar strength on CFRP laminates. *Compos. Struct.* **2019**, *208*, 796–805. [CrossRef]
10. Carpinteri, A.; Lacidogna, G.; Corrado, M.; Di Battista, E. Cracking and crackling in concrete-like materials: A dynamic energy balance. *Eng. Fract. Mech.* **2016**, *155*, 130–144. [CrossRef]
11. Gómez Muñoz, C.Q.; García Márquez, F.P. A New Fault Location Approach for Acoustic Emission Techniques in Wind Turbines. *Energies* **2016**, *9*, 40.
12. Rescalvo, F.J.; Valverde-Palacios, I.; Suarez, E.; Roldán, A.; Gallego, A. Monitoring of Carbon Fiber-Reinforced Old Timber Beams via Strain and Multiresonant Acoustic Emission Sensors. *Sensors* **2018**, *18*, 1224. [CrossRef]
13. Ghaib, M.; Shateri, M.; Thomson, D.; Svecova, D. Study of FRP bars under tension using acoustic emission detection technique. *J. Civ. Struct. Health Monit.* **2018**, *8*, 285–300. [CrossRef]
14. Djabali, A.; Toubal, L.; Zitoune, R.; Rechak, S. An experimental investigation of the mechanical behavior and damage of thick laminated carbon/epoxy composite. *Compos. Struct.* **2018**, *184*, 178–190. [CrossRef]
15. Beheshtizadeh, N.; Mostafapour, A.; Davoodi, S. Three point bending test of glass/epoxy composite health monitoring by acoustic emission. *Alex. Eng. J.* **2019**, *58*, 567–578. [CrossRef]
16. Huguet, S.; Godin, N.; Gaertner, R.; Salmon, L.; Villard, D. Use of acoustic emission to identify damage modes in glass fibre reinforced polyester. *Compos. Sci. Technol.* **2002**, *62*, 1433–1444. [CrossRef]
17. Barré, S.; Benzeggagh, M.L. On the use of acoustic emission to investigate damage mechanisms in glass-fibre-reinforced polypropylene. *Compos. Sci. Technol.* **1994**, *52*, 369–376. [CrossRef]
18. Garbacz, G.; Kyzioł, L. Application of metric entropy to determine properties of structural materials. *Polím. Ciênc. Tecnol.* **2020**, *29*, 1–9. [CrossRef]
19. Panasiuk, K.; Kyziol, L.; Dudzik, K.; Hajdukiewicz, G. Application of the Acoustic Emission Method and Kolmogorov-Sinai Metric Entropy in Determining the Yield Point in Aluminium Alloy. *Materials* **2020**, *13*, 1386. [CrossRef] [PubMed]
20. Garbacz, G.; Kyzioł, L. Determination of yield point of structural materials with using the metric entropy. *J. KONES* **2014**, *21*, 97–104. [CrossRef]
21. Garbacz, G.; Kyzioł, L. Application of metric entropy for results interpretation of composite materials mechanical tests. *Adv. Mater. Sci.* **2017**, *17*, 70–81. [CrossRef]

22. Garbacz, G. Przetwarzanie danych doświadczalnych z uwzględnieniem ich chaotycznego charakteru na przykładzie testów mechanicznych, (Processing of experimental data taking into account their chaotic nature). Ph.D. Thesis, Instytut Podstawowych Problemów Techniki Polskiej Akademii Nauk, Warszawa, Poland, 2009.
23. Kyzioł, L.; Hajdukiewicz, G. Application of the Kolmogorov-Sinai Entropy in Determining the Yield Point, as Exemplified by the EN AW-7020 Alloy. *J. KONBiN* **2019**, *49*, 241–269. [CrossRef]
24. Kyzioł, L.; Panasiuk, K.; Barcikowski, M.; Hajdukiewicz, G. The influence of manufacturing technology on the properties of layered composites with polyester–glass recyclate additive, Progress in Rubber. *Plast. Recycl. Technol.* **2020**, *36*, 18–30.
25. Panasiuk, K. Analiza właściwości mechanicznych kompozytów warstwowych z recyklatem poliestrowo-szklanym. Ph.D. Thesis, UMG Gdynia, Gdynia, Poland, 2019.
26. Kubik, J.; Mraczny, K. *Kompozyty Warstwowe Z Tworzyw Odpadowych (Composites from Waste Plastics)*; Oficyna Wydawnicza Politechniki Opolskiej: Opole, Poland, 2001.
27. Kanemasa, N.; Nakagawa, T. FRP recycling in Europe, North America and the Pacific Rim area. *JEC Compos.* **2010**, *61*, 37–42.
28. Jacob, A. Composites can be recycled. *Reinf. Plast.* **2011**, *55*, 45–46. [CrossRef]
29. Panasiuk, K.; Hajdukiewicz, G. Production of composites with added waste polyester-glass with their initial mechanical properties. *Sci. J. Marit. Univ. Szczec.* **2017**, *52*, 30–36.
30. Panasiuk, K.; Kyzioł, L.; Dudzik, K. The use of acoustic emission signal (AE) in mechanical tests. *Prz. Elektrotech.* **2019**, *95*, 8–11. [CrossRef]
31. Amigó, J.M.; Kennel, M.B.; Kocarev, L. The permutation entropy rate equals the metric entropy rate for ergodic information sources and ergodic dynamical systems. *Phys. D* **2005**, *210*, 77–95. [CrossRef]
32. Ying-Qiana, Z.; Xing-Yuana, W. A symmetric image encryption algorithm based on mixed linear–nonlinear coupled map lattice. *Inf. Sci.* **2014**, *273*, 329–351.
33. Keller, K.; Sinn, M. Kolmogorov-Sinai entropy from the ordinal viewpoint. *Phys. D* **2010**, *239*, 997–1000. [CrossRef]
34. Latora, V.; Baranger, M. Kolmogorov-Sinai Entropy Rate versus Physical Entropy. *Phys. Rev. Lett.* **1999**, *82*, 520. [CrossRef]
35. Kolmogorov, A. Entropy per unit Time as a Metric Invariant of Automorphism. *Dokl. Russ. Acad. Sci.* **1959**, *124*, 754–755.
36. Sinai, Y. On the Notion of Entropy of a Dynamical System. *Dokl. Russ. Acad. Sci.* **1959**, *124*, 768–771.

 sensors

MDPI

Article

Spectral Analysis of Torsional Vibrations Measured by Optical Sensors, as a Method for Diagnosing Injector Nozzle Coking in Marine Diesel Engines

Sebastian Drewing * and Kazimierz Witkowski

Faculty of Marine Engineering, Gdynia Maritime University, 81-225 Gdynia, Poland; k.witkowski@wm.umg.edu.pl
* Correspondence: s.drewing@wm.umg.edu.pl

Abstract: The study aimed to verify whether it is possible to diagnose the coking of a marine diesel engine injector nozzle by performing a spectral analysis of the crankshaft's torsional vibrations. The measurements were taken using laser heads, clocked at 16 MHz. The reasons for selecting this type of optical sensors are described as well. The tests were carried out under laboratory conditions, using a test stand with a Sulzer 3AL 25/3 engine, operating under a load created by a Domel GD8 500–50/3 electric generator. A unique method is presented in the paper, which enables the measuring and calculation of torsional vibrations of engine crankshafts. The method was developed at the Chair of Marine Power Plants at the Maritime University of Gdynia. It has been proven that the distribution of differences in the values of individual harmonic components depends on the location of a defective injector nozzle in the cylinder.

Keywords: marine propulsion; marine power plants; condition monitoring; torsional vibration spectra; diagnostics; marine diesel engines; coked injector; frequency; harmonic orders

Citation: Drewing, S.; Witkowski, K. Spectral Analysis of Torsional Vibrations Measured by Optical Sensors, as a Method for Diagnosing Injector Nozzle Coking in Marine Diesel Engines. *Sensors* **2021**, *21*, 775. https://doi.org/10.3390/s21030775

Academic Editor: Leszek Chybowski
Received: 28 December 2020
Accepted: 20 January 2021
Published: 24 January 2021

1. Introduction

Internal combustion engines with reciprocating pistons are one of the most complex pieces of machinery operating in ship engine rooms. In the course of their operation, frequent defects affecting the fuel system, including injector nozzles, are experienced. The reasons for this are the long-term supply of the engine with fuels of poor quality or with high contents of bio-additives, and poor technical condition of the system piston, piston rings and cylinders, resulting in the combustion of engine oil. The frequency of incidents of this type has been increasing since 2012, mainly due to the introduction of low-sulfur fuels used that are used in the marine industry to ensure compliance with ISO 8217:2012 and ISO 8217:2017 standards mandated by the International Maritime Organization. Low sulfur fuel oils (LSFOs) are characterized by a higher catalytic mud (consisting, inter alia, of very hard aluminum and silicon compounds) content (up to 60 ppm), compared to fuels with a high sulfur content. This is caused by the fact that fine particles end up in low-sulfur by-products of the refining process which are then mixed with residual fuels to reduce sulfur content. Marine engine manufacturers, such as MAN and Wärtsilä, recommend a maximum content of these hard fractions of 15 ppm [1].

The current methods used for diagnosing the operation of fuel systems of internal combustion engines are mainly based on indirect parametric methods that rely on the analysis of variability in selected parameters, induced by damage to the injection system. On the other hand, methods based on injection system pressure measurements seem to be much more effective [2]. The main disadvantage of this solution is the fact that it cannot be usually applied in engine rooms of sea-going vessels and ships operating on inland waters. This is caused, inter alia, by the high cost of piezoelectric pressure sensors, the requirements of classification societies that prohibit any welded and soldered joints, and the need to use certified covers on the pipelines concerned [3,4]. Research performed over the years, as well as experience gathered while evaluating practical applications, continuously increases the degree of efficiency and diagnostic suitability of information concerning machine vibrations. The operational diagnostics of ship propulsion shafts, based on the measurements of mechanical vibrations, consists of measuring specific physical quantities that characterize vibrations of selected driveline components [5]. Such quantities include the displacement, velocity, and acceleration of vibrations. However, diagnosing machinery and mechanisms aboard a vessel, using vibroacoustic methods, creates a number of problems. This stems primarily from a significant concentration of machinery and equipment in a confined area. Other factors include common power sources, the connection of machine foundations by fixed elements of the hull, and, finally, the fact that individual components operate, simultaneously, inside and outside the hull. This results in the overlapping of vibrations from different sources [6–9]. Torsional vibrations of drive shafts, which are least susceptible to such phenomena, seem to be the most difficult to measure. The crankshaft of an internal combustion engine is a flexible component that is exposed to periodic forces generated by gases and mass. These forces serve as impulses that generate different forms of forced vibrations of the shaft. In a piston engine, these forces generate—in addition to bending and longitudinal oscillations—torsional vibrations as well [10]. A defect of the injector nozzle of an internal combustion engine injector reduces the gaseous forces in a given cylinder, thus changing the distribution of torque affecting the crankshaft.

The search for a reliable method allowing the recreation of the pressure values developing inside a cylinder, based on indirect measurements, has been ongoing for over 30 years. The focus is two-fold: on the one hand, measurements of instantaneous angular velocity of the crankshaft are performed [11–14], while on the other, measurements of the engine's lateral vibrations [15–18] are taken.

These studies prove that both lateral vibrations and angular velocity contain information about the pressure inside engine cylinders but pertain to different frequency ranges. For the reasons described above, i.e., due to the fact that vibrations overlap, the decision was made to rely, in our study, on instantaneous angular velocity measurements.

Rotational speed fluctuations are mainly caused by low-frequency portions of the pressure curve and, therefore, angular velocity is much less sensitive to sudden pressure changes, compared to lateral engine vibrations.

There are some commonly known ways to diagnose engine operation based on the fluctuations of angular velocity of the flywheel. Under specific engine operating conditions, changes in instantaneous angular velocity may be a source of information on the incorrect operation of specific cylinders [19–22]. Considering the fact that instantaneous velocity is a derivative of displacement in time (displacement is the function of time, thus it is a classic example of a derivative function based on its argument), a decision was made to check whether this signal may be relied upon for diagnostic purposes.

In consideration of the planned future application of the experiment's results in diagnosing interference/damage of medium- and high-speed marine diesel engines in actual engine room conditions, the selection of an appropriate measurement method was crucial. Vibrations present in a vessel's engine room are caused both by operating machinery and auxiliary devices, as well as by the ship's hull that is exposed to waves, precluded the use of most contact-based (i.e., rotating together with the shaft) and contactless torsional vibration measurement methods.

High cost and difficulties experienced while ensuring rigid installation of the measuring head (making sure that it does not move in relation to the rotating shaft), preclude the use of the contactless method based on laser interferometers. Unfortunately, the use of cheap and popular passive magneto-resistive or magneto-inductive sensors was ruled out as well—as was the use of induction sensors relying on the Hall effect, because they generate an insufficient number of pulses per revolution, need to be positioned very close to the shaft (less than ~5 mm), and are sensitive to lateral vibrations [23].

Contact-based methods relying on the use of relatively cheap and commonly applied piezoelectric vibration acceleration transducers could not be used either. Installation of accelerometers poses a risk of the instrumentation being disconnected due to excessive centrifugal loads. The centrifugal force affecting a typical AT3/500 piezoelectric accelerometer that weighs 90 g and is mounted on a shaft with the diameter of 0.2 m equals 222 N at 1500 rpm (described method for diagnosing injector nozzle coking is also planned for low power generators). Additionally, an expensive telemetric system or "sensitive" slip rings are required to transmit the acceleration signals obtained. Acceleration values are measured. The angle of torsion is obtained by integration. The absolute reference position is not available, and thus processing in the domain of angles is impossible as well [24].

For reasons similar to those applying to accelerometers, another contactless method based on tensometric bridges was excluded as well. A method using the newest sensors with fiber grating (FBG—fiber Bragg grating) could not be applied either [25].

Having excluded the above-mentioned methods, a contactless method relying on the use of optical sensors was the only option left. There are many types of optical sensors available on the market, but the majority of them are designed to detect objects. In order to measure angular velocity, encoders which use two types of tapes are most commonly used, depending on whether they are to be glued around the shaft (with zebra tape) or attached to the shaft (disc/zebra disc). Zebra disks and tapes are available in a variety of strip widths in order to adjust the number of pulses per revolution to the diameter of the shaft.

This method was rejected as well due to an insufficient number of pulses per revolution. A decision was made to choose a contactless method that relies on incremental encoders, taking into account the following:

(a) Measurement accuracy resulting from a large number of pulses per revolution;
(b) Absolute reference to the accurate identification of the phase and processing in the domain of angles;
(c) Ability to mount on the free ends of the shaft of a generator set.

Manufactured for industrial applications by the leading brand of Leine Linde, incremental encoders generate up to 10,000 pulses per revolution. Attempts were made to use the encoders of this particular manufacturer for taking the measurements. Unfortunately, due to the length of the free end of the vessel's engine shaft and, consequently, the rather considerable transverse vibration amplitude values, frequent ruptures of the encoder's nylon driving shafts were experienced. Enamor Sp. z o.o., a Polish manufacturer of electronic ship optimization and monitoring gear, offered similarly priced upgraded ETNP-10 encoders with laser heads clocked at 16 MHz. Those heads were used (based on our proprietary method) to determine displacement changes of two shaft ends (i.e., to measure torsional vibrations) [26,27].

2. Materials and Methods

The measurements were taken on a laboratory test-stand at the Chair of Ship Power Plants of the Maritime University of Gdynia, equipped with a diesel–electric unit (DEU) operating at the speed of 750 rpm. The unit consisted of a three-cylinder Sulzer 3AL25/30 diesel piston engine and a Domel GD8 500-50/3 three-phase synchronous generator. The electricity generated is released by the generator to a water blade resistor. The engine is of the supercharged variety and is equipped with a VTR 160 Brown-Boveri turbocharger with an intercooler. The characteristics of the generator are presented in Table 1. An electronic, stationary Unitest 2008 indicator was used to measure and record pressure waveforms. The measuring system included a recorder with a power supply, three Kistler 6353A24 piezoresistive combustion pressure sensors (reads the pressure with an error $< \pm 0.75$, operating range from 0 to 20 MPa), three Kistler 4067E piezoresistive injection pressure sensors (reads the pressure with an error smaller than ± 0.8, operating range from 0 to 300 MPa) (Figure 1), and an angular position decoder with an integrated sensor operating with the resolution of 720 pulses per crankshaft revolution.

Table 1. Test-stand technical and nominal parameters.

Sulzer 3AL25/30 Four-Stroke Engine		
Piston diameter	250	(mm)
Piston stroke	300	(mm)
Nominal effective power	408	(kW)
Mean effective pressure	1.47	(MPa)
Injector opening pressure	25	(MPa)
Fuel delivery advance angle	17	(deg)
Nominal rotating speed	750	(rpm)
Number of cylinders	3	(-)
Firing order	3-2-1	(-)
GD8 500-50/3 Synchronous Generator		
Power	500	(kVA)
Rotating speed	750	(rpm)
Stator voltage	400	(V)
Stator current	723	(A)
Frequency	50	(Hz)

Figure 1. Installation of the Kistler 4067E injection pressure sensor (in the background, the electric cable of the Kistler 6353A24 combustion pressure sensor protruding from the indicator cock).

The recorder communicated with a PC via a USB 2.0 interface. The indicator recorded combustion and fuel injection pressure every 0.5° of crankshaft rotation, based on sixteen full engine cycles, i.e., 32 crankshaft rotations, rendering 1440 pressure measurements per one engine cycle (720 degrees of crankshaft rotation).

A modified ETNP-10 redundant measuring system was used to measure torsional vibrations of the DEU shaft. It consisted of the following:

(a) Two laser heads;
(b) An electronic block, converting the voltage signal from the measuring heads into digital records;
(c) A Saia Burgess Controls programmable logic controller (PLC) for data recording.

The signal was processed and recorded in the measuring and control block (Figure 2). The laser heads, being the source of the shaft torsion signal, tracked the movement of two perforated discs with 180 symmetrical slots along their perimeter (Figure 3).

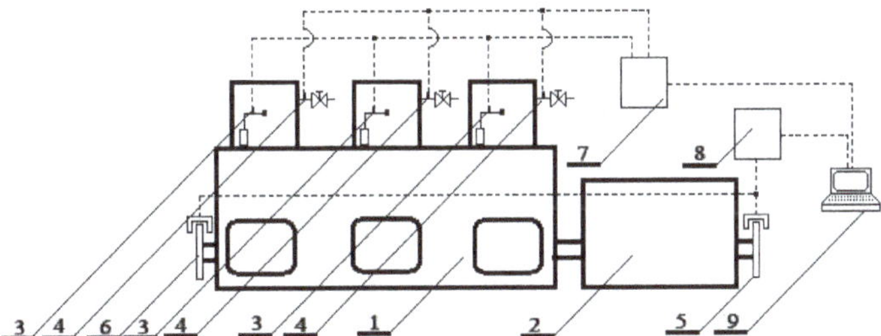

Figure 2. Block diagram of the test stand: 1—SULZER 3AL25/30 marine diesel engine; 2—Domel GD8 500-50/3 synchronous generator; 3—Kistler 4067E sensors for measuring pressure in the injection system; 4—Kistler 6353A24 sensors for measuring combustion pressure; 5,6—ETNP-10 laser heads tracking the movement of the perforated disc; 7—Unitest 2008 indicator; 8—ETNP-10 measuring and control block; 9—computer system for recording measurement data.

Figure 3. ETNP-10 laser heads are mounted on both free ends of the shaft and track the movement of perforated discs with 180 symmetrical slots along their perimeter ((**a**)—view from the engine side, (**b**)—view from the generator side)).

The diesel–electric unit (DEU), operating in the capacity of a vibration signal generator, is a very complex system. This results partly from the following:

(a) As the load of the generator changes, a phase shift occurs between its electromotive force and the voltage in the mains to which electric energy is generated. Any further increase in the load on the unit elevates the value of the phase shift, which also affects the distribution of torsional moments of the drive shaft;

(b) In order to achieve the run uniformity factor of $\leq 1/250$, the unit was equipped with a heavy flywheel. The rotor of the generator is heavy as well;

(c) The shaft is of the resilient variety.

The laser head emitted a laser beam with a frequency of 16 MHz, which was directed onto a photodiode. Assuming sensitivity of photodiode at level ∓10 impulses, the accuracy of measurement depends on the engine's revolutionary speed, and medium speed diesel engines typical for marine electro-generators reach the level of 0.015% [28]. The gaps and teeth, when passing through the light beam, create groups of signals in the form of a number of pulses with the value of "1" as the light passes through a gap, and of "0" when the light is covered by a tooth. The measured values must be related to angular positions of the shaft. For this purpose, an additional gap was created in the tooth of the first disc. The disc was positioned in such a way that the additional gap corresponded to the top dead-center of the first cylinder (Figure 3a). In addition, the gap serves as a signal that triggers the measurement. The electronic system recognizes two types of signals, so both the gap and the tooth provide information about the instantaneous angular velocity of one of the discs. Then, in order to calculate the torsion of the DEU shaft, the method developed at the Chair of Ship Power Plants was applied, consisting of:

(a) Counting the pulses (i_{1i}) generated by the first measuring head while the first perforated disc moved by two teeth and two gaps (i.e., by 4 degrees of the crankshaft's rotation). Because the validation of results is based on indicator charts, a gradual measure of the angle was adopted;

(b) Counting the pulses (i_{2i}) generated by the second measuring head while the second perforated disc moved by two teeth and two gaps (i.e., by 4 degrees of the crankshaft's rotation);

(c) Calculating the time (t_{1i}) in which two teeth and two gaps of the first disc moved by 4 degrees of the crankshaft's revolution:

$$t_{1i} = \frac{i_{1i}}{f} \tag{1}$$

where f is the frequency of the laser beam emitted by the measuring head (16,000,000 Hz);

(d) Calculating the time (t_{2i}) in which two teeth and two gaps of the second disc moved by 4 degrees of the crankshaft's revolution:

$$t_{2i} = \frac{i_{2i}}{f} \tag{2}$$

(e) Calculating the mean angular velocity (ω_{1i}) for the movement of the first disc by 4° of the crankshaft's revolution:

$$\omega_{1i} = \frac{4^\circ}{t_{1i}} \tag{3}$$

(f) Calculating the mean angular velocity (ω_{2i}) for the movement of the second disc by 4° of the crankshaft's revolution:

$$\omega_{2i} = \frac{4^0}{t_{2i}} \tag{4}$$

(g) Calculating the displacement of the second disc (φ_{2i}), assuming that the displacement of the first disc (φ_{1i}) was increasing every 4° of the crankshaft's revolution (i.e., it equaled 4; 8; 12; 16° ... of the crankshaft's revolution), meaning that the displacement of the second disc was the product of the second disc's velocity and the time during which the two teeth and two gaps of the first disc moved by 4° of the crankshaft's revolution:

$$\varphi_{2i} = \omega_{2i} \cdot t_{1i} \tag{5}$$

(h) Adding all partial displacements of the second disc in order to obtain the total displacement value of the second disc (φ_2). The system measures displacements by one section, consisting of two teeth and two gaps, which is equal to 4° of the crankshaft's revolution. This means that one full rotation is divided into 90 sections. The system measures 10 crankshaft revolutions, so the total number of sections equals 900:

$$\varphi_2 = \varphi_{2i} + \sum_{i=1}^{900} \varphi_{2i} \tag{6}$$

(i) Adding all partial movements of the first disc (φ_1):

$$\varphi_1 = \sum_{i=0}^{900} \varphi_{1i} \tag{7}$$

(j) Calculating torsion fluctuations (φ) by subtracting the sum of the first disc's displacements from the sum of the second disc's displacements:

$$\varphi^o = \sum_{i=0}^{900} \varphi_{1i} - \left(\varphi_{2i} + \sum_{i=1}^{900} S_{2i} \right) = \varphi_1 - \varphi_2 \tag{8}$$

The data obtained in the course of the experiment were collected at equal time intervals (determined by the frequency of the measuring head). They are periodic and continuous, and can therefore be subjected to spectral analysis [29]. As proven in [30], in the case of lateral vibrations of intermediate and helical shafts, amplitudes of the frequency components, and their changes recorded during engine operation may provide a detailed information about local resonance phenomena. This allows the identification and localization of a defect of a specific element, e.g., a bearing, or to detect excessive misalignment between shaft axes.

In consideration of the above, a decision was made to verify whether it was possible to detect fuel system defects based on the spectral analysis of torsional vibrations in a diesel–electric unit. Due to the limited computing power of the hardware available and due to the fact that digital signal analysis is based on the discrete Fourier transform (DFT), the decision was made to use the fastest version based on the Cooley and Tukey algorithm, known as FFT (fast Fourier transform) [31]. The Hamming window was used as the smoothing window (being a modified version of the Hanning window), because it allows the obtaining of a good amplitude accuracy and frequency resolution [32].

3. Results and Discussion

The engine was adjusted statically before the laboratory tests commenced. Standby and prime-rated diesel generator sets are designed to operate between 50 and 85% of the full nameplate, while continuous-rated diesel generator sets are optimized between 70 and 100% maximum continuous rating (MCR). Due to the fact that:

(a) The use of the method is planned for use as an on-line diagnostic system on autonomous and unmanned ships;
(b) Failures introduced on one cylinder increase the load on other defect-free cylinders, which often results in exceeding the alarm thresholds of permissible exhaust gas temperatures, the most universal level 70% of loading of the diesel–electric unit was adopted.

Three main research tasks were performed for the generator power rating of 250 kW, which corresponded to 70% of its MCR.

Task 1. Measurement of pressure in cylinders, pressure in the fuel injection system, and torsional vibrations in the diesel–electric unit's shaft—performed in a defect-free ship engine.

Task 2. Measurement of pressure in cylinders, pressure in the fuel injection system, and torsional vibrations in the diesel–electric unit's shaft—performed in a marine engine with one coked injector nozzle (Figure 4) moved from one cylinder to the next. The factory-assigned cylinder number system was relied upon, meaning that cylinder one was nearest to the timing system drive. The task was divided into the following stages:

stage 1—coked injector nozzle in cylinder one,
stage 2—coked injector nozzle in cylinder two,
stage 3—coked injector nozzle in cylinder three.

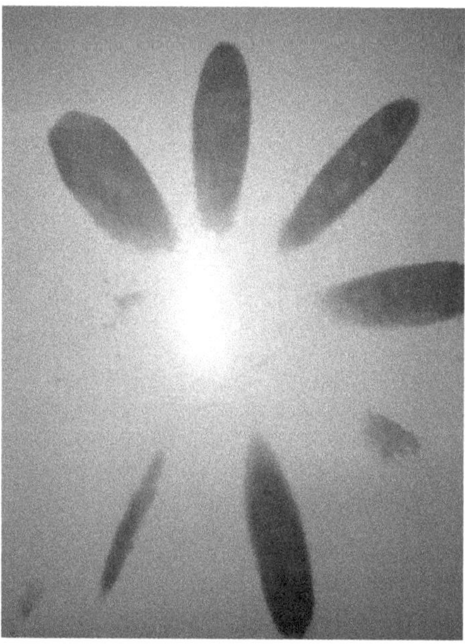

Figure 4. Fuel stream sprayed by a coked injector nozzle (naturally coked obtained from the engine department of the repair shipyard). Three out of nine ports are clogged.

Task 3. Developing the spectra of the shaft's torsional vibrations and analyzing these.

The ship's diesel–electric unit is a generator of vibration signals leading to time-varying shaft torsions. It was assumed that the shaft torsion signal would be the highest at the time the fuel ignites, i.e., once per two crankshaft revolutions. In the case of a three-cylinder engine, this harmonic component should be tripled to obtain the combustion harmonic component of $1\frac{1}{2}$ (Table 2).

Table 2. Orders of selected harmonic components and corresponding frequencies.

Order of a Harmonic (k)	Frequency (Hz)
$\frac{1}{2}$	6.25 (one cylinder combustion)
1	12.5 (basic harmonic component)
$1\frac{1}{2}$	18.75 (combustion harmonic component)
2	25
$2\frac{1}{2}$	31.25
3	37.5
$3\frac{1}{2}$	43.5
4	50 (polar pulsation for four pairs of poles of a single voltage phase)

3.1. Results of Tests of a Diesel–Electric Power Unit with a Defect-Free Ship Engine

As indicated by the diagrams (Figure 5), fuel injection characteristic curves $p_{inj} = f(\alpha)$, as well as variations in instantaneous cylinder pressure $p_{cyl} = f(\alpha)$, reveal similar waveforms and do not differ significantly in terms of their values.

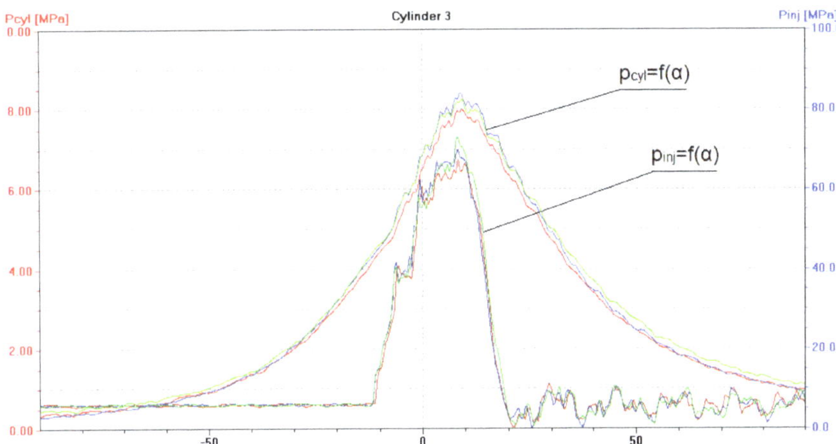

Figure 5. The diagram (screenshot of Unitest 2008) illustrates the $p_{cyl} = f(\alpha)$ function with fuel injection characteristics of $p_{inj} = f(\alpha)$ for a diesel–electric unit (DEU) with a defect-free/healthy ship engine loaded at 70% of MCR. Waveforms for cylinder: — No. 1, — No. 2, — No. 3.

The maximum fuel injection pressure values p_{inj} are slightly different, while the angles at which the injection of fuel commences have the same value. This proves that the static adjustment of the fuel injection system was correct. Torsional vibration waveforms in the case of a unit operating under such a load are shown in Figure 6.

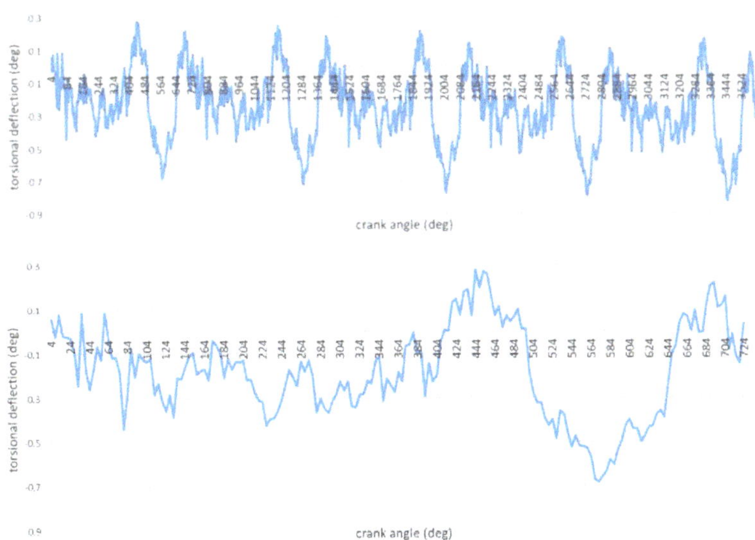

Figure 6. Torsional vibrations of the shaft as a function of the shaft rotation angle. DEU with a defect-free/healthy ship engine (ten turns and one cycle).

In order to generate the spectra, torsional vibrations recorded were subjected to the discrete Fourier transform (DFT). The spectra obtained are shown in Figures 7 and 8.

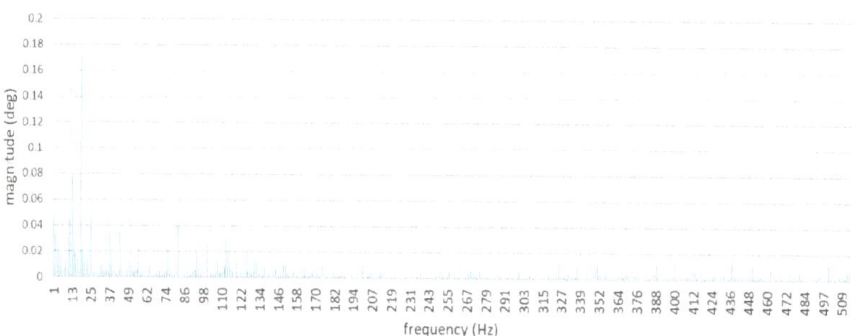

Figure 7. Full spectrum of torsional shaft vibrations. Diesel–electric unit with a defect-free ship engine.

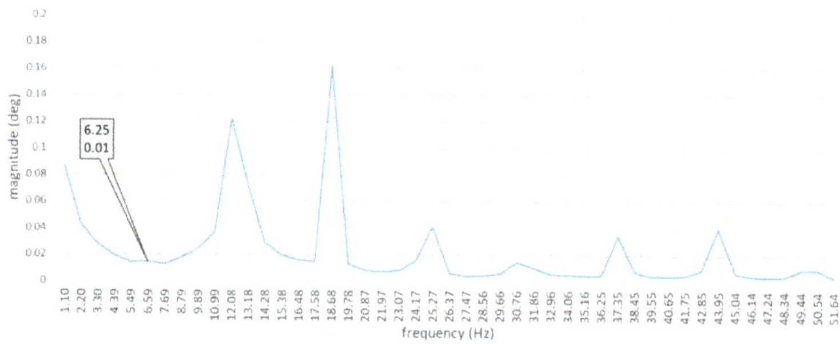

Figure 8. Mean value of torsional shaft vibration spectra. The diagram shows the 8th harmonic component only. Diesel–electric power unit with a defect-free ship engine.

For the total of 124 measurements of the shaft torsional vibrations and the prepared amplitude-frequency spectra, the matrices were developed, and the average Pearson's correlation coefficients were calculated. Comparisons were made for full spectra, and also truncated to the first sixteen harmonics. The obtained results are characterized by a strong correlation (Table 3).

Table 3. Sample Pearson's correlation coefficient.

Engine Condition	Sample Pearson's Correlation Coefficient Full Spectra/First Sixteen Harmonics
Defect-free/healthy engine	0.84/0.83
Coked injector nozzle in the first cylinder	0.80/0.78
Coked injector nozzle in the second cylinder	0.92/0.93
Coked injector nozzle in the third cylinder	0.82/0.81

For the speed of the tested engine equaling 750 rpm, the value of the combustion harmonic component k = $1\frac{1}{2}$ was 18.75 Hz, while the frequency of the basic harmonic component was 12.5 Hz. (Table 2). After decomposition of the signal, these two main harmonic components are clearly visible in the spectrum (Figure 8). The diagram also shows other harmonic components of orders 2 (25 Hz), $2\frac{1}{2}$ (31.25 Hz), 3 (37.5 Hz), and $3\frac{1}{2}$ (43.5 Hz). In accordance with the previous assumption that was supported by the literature [20], it was the harmonic component of combustion which achieved the highest amplitude value.

3.2. Results of Tests with a Diesel–Electric Power Unit and a Coked Injector Nozzle

After installing a coked injector nozzle (in one cylinder at a time), the following have been recorded:

(a) An increase in the pressure of fuel injected by the coked nozzle, to approximately 100 MPa (Figures 9–11). Injectors in a defect-free engine spray fuel at the pressure of approximately 70 MPa (Figure 5). The difference was 30 MPa and maximum permissible error in this case was 0.8 MPa;

(b) An increase in the pressure of fuel sprayed by the two remaining, non-defective injectors, to approximately 75 MPa (Figures 9–11);

(c) A drop of the maximum combustion pressure in the cylinder with the coked injector, to approximately 7 MPa (Figures 9–11). The maximum combustion pressures in cylinders of a defect-free engine were almost equal and amounted to approximately 8 MPa (Figure 5). The difference was 1 MPa and the maximum permissible error in this case was 0.075 MPa;

(d) An increase in the maximum combustion pressure in the two remaining, defect-free cylinders, up to approximately 9 MPa (Figures 9–11).

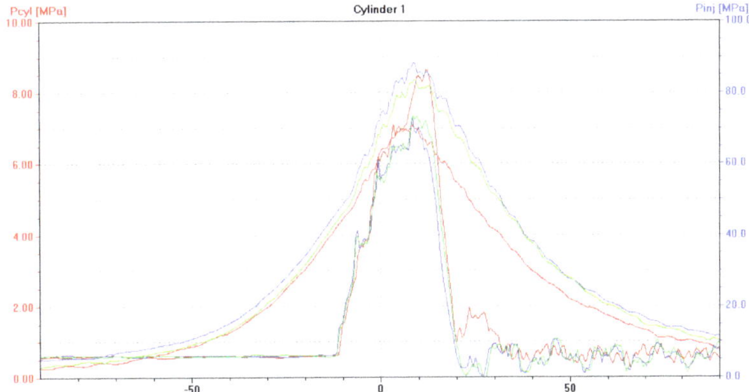

Figure 9. The diagram (screenshot of Unitest 2008) illustrates p_{cyl} = f(α) with the fuel injection characteristics of p_{inj} = f(α) for a DEU with a defective ship engine-coked injector nozzle in the first cylinder, loaded at 70% MCR. Waveforms for cylinder: — No. 1, — No. 2, — No. 3.

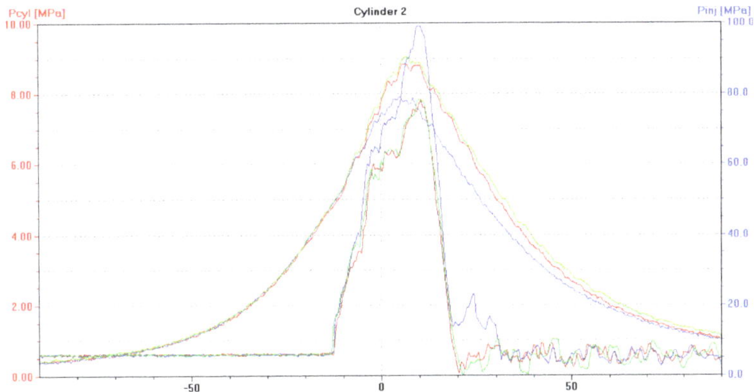

Figure 10. The diagram (screenshot of Unitest 2008) illustrates p_{cyl} = f(α) with the fuel injection characteristics of p_{inj} = f(α) for DEU with a defective ship engine-coked injector nozzle in the second cylinder, loaded at 70% MCR. Waveforms for cylinder: — No. 1, — No. 2, — No. 3.

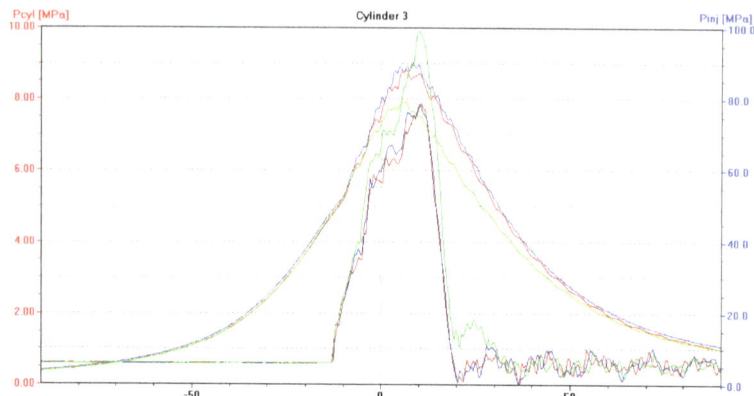

Figure 11. The diagram (screenshot of Unitest 2008) illustrates p_{cyl} = f(α) with the fuel injection characteristics of p_{inj} = f(α) for DEU with a defective ship engine-coked injector nozzle in the third cylinder, loaded at 70% MCR. Waveforms for cylinder: — No. 1, — No. 2, — No. 3.

This means that the differences in torque values generated by each cylinder will be greater and, consequently, the pulsation/distribution of torsional forces affecting the drive shaft will be changed. In this case, the recorded waveforms of torsional vibrations were also subjected to the discrete Fourier transform (DFT) in order to obtain the spectra. The spectra obtained are shown in Figure 12. The spectra of torsional vibrations of a unit with a coked injector nozzle, installed in one cylinder at a time, show a clear domination of the combustion harmonic component's amplitude. After a preliminary analysis, it was found that, unlike in the case of spectra obtained for a defect-free ship engine, a harmonic component of the order of $\frac{1}{2}$ (6.25 Hz) was clearly visible. This component corresponds to the combustion of a fuel–air mixture in a single cylinder (Table 2).

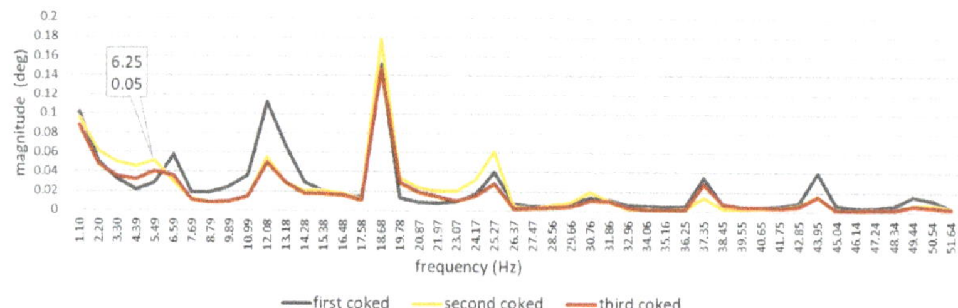

Figure 12. Mean value of torsional shaft vibration spectra. The diagram is limited to the 8th harmonic component only. Ship engine with a coked injector nozzle.

Torsional vibration spectra of the DEU with a defective injector nozzle were different from those obtained for a defect-free engine (Figure 13):

— With the first injector nozzle defective, the largest differences in amplitude values were observed for seven harmonic components of the following orders: $\frac{1}{2}$ (6.25 Hz), 1 (12.5 Hz), $1\frac{1}{2}$ (18.75 Hz), 4 (50 Hz), 6 (75 Hz), $6\frac{1}{2}$ (81,25 Hz), and 8 (100 Hz);

— With the second injector nozzle defective, the largest differences in amplitude values were observed for twelve harmonic components of the following orders: $\frac{1}{2}$ (6.25 Hz), 1 (12.5 Hz), $1\frac{1}{2}$ (18.75 Hz), 2 (25 Hz), 3 (37.5 Hz), $3\frac{1}{2}$ (43.5 Hz), $4\frac{1}{2}$ (56.25 Hz), 6 (75 Hz), $6\frac{1}{2}$ (81,25 Hz), 7 (87.5 Hz), $7\frac{1}{2}$ (93.75), and 8 (100 Hz);

— With the third injector nozzle defective, the largest differences in amplitude values were observed for nine harmonic components of the following orders: $\frac{1}{2}$ (6.25 Hz),

1 (12.5 Hz), $1\frac{1}{2}$ (18.75 Hz), 2 (25 Hz), $3\frac{1}{2}$ (43.5 Hz), $4\frac{1}{2}$ (56.25 Hz), 6 (75 Hz), $6\frac{1}{2}$ (81,25 Hz), 8 (100 Hz).

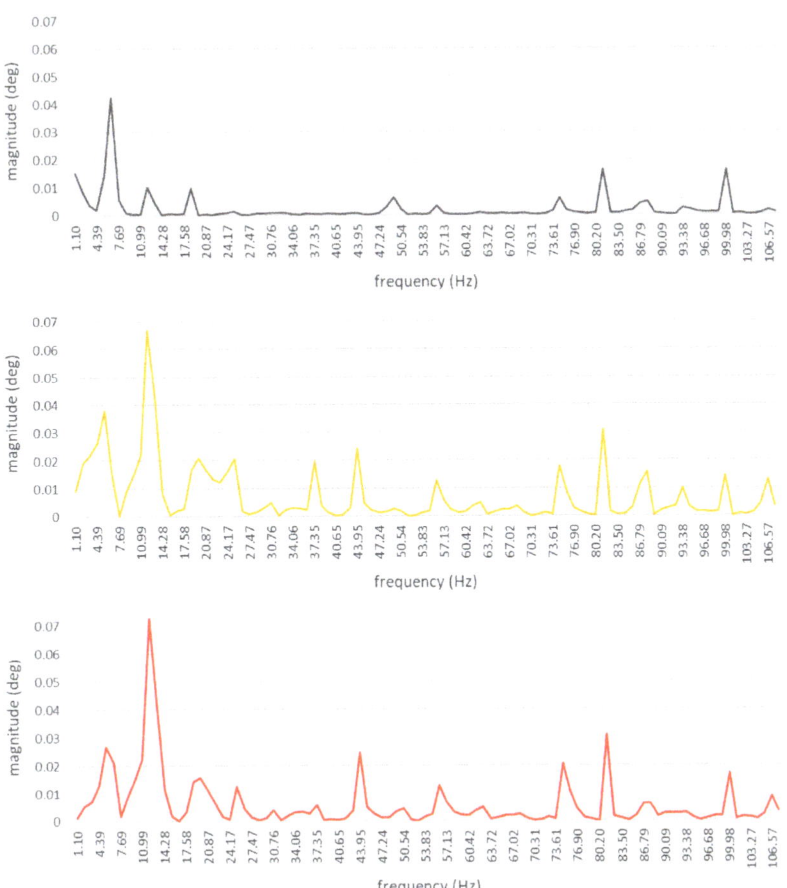

Figure 13. Absolute values of differences in the spectra related to a diesel–electric unit with a defect-free engine and with a defective injector nozzle. The diagram is limited to the 16th harmonic component only. Waveforms for: differences between defect-free and first coked; differences between defect-free and second coked; differences between defect-free and third coked.

4. Conclusions

The results obtained while testing the system for measuring torsional vibrations, designed and built at the Maritime University of Gdynia, allow us to conclude that:

(a) The assumptions adopted for the proprietary algorithm used for calculating torsional vibration values were correct;
(b) The data, recorded by 16 MHz laser heads, are sufficient to determine torsional vibrations of the diesel–electric unit's shaft;
(c) The spectra obtained for the defect-free/healthy ship engine are strongly correlated (Table 3). It proves the high repeatability of the results for a given sample;
(d) The spectra obtained for the engine with a particular coked injector are strongly correlated (Table 3), which also proves the high repeatability of the results for a given sample;

(e) The values of harmonic component orders obtained are clearly visible in the spectra (Figure 8);
(f) Torsional vibration spectra of the DEU with a defective injector nozzle were different from those obtained for a defect-free engine (Figure 13);
(g) The distribution of differences in the values of the first sixteen harmonic components depend on the cylinder in which the defective injector nozzle was installed.

The observations made lead to a conclusion that it is possible to diagnose coking of a ship diesel engine injection nozzle by relying on spectral analysis of the shaft's torsional vibrations which are measured by optical sensors.

Author Contributions: Conceptualization, S.D. and K.W.; methodology, S.D.; software, S.D; validation, S.D. and K.W.; formal analysis, S.D. and K.W.; resources, S.D.; data curation, S.D.; writing—original draft preparation, S.D. and K.W.; writing—review and editing, S.D. and K.W.; visualization, S.D. and K.W.; supervision, K.W. All authors have read and agreed to the published version of the manuscript.

Funding: This research received no external funding.

Institutional Review Board Statement: Not applicable.

Informed Consent Statement: Not applicable.

Data Availability Statement: Not applicable.

Conflicts of Interest: The authors declare no conflict of interest.

References

1. Report Joint Hull Committee. *Marine Engine Damage Due to Catalytic Fines in Fuel*; Report Joint Hull Committee: London, UK, 2013.
2. Pawletko, R. Możliwości diagnozowania wybranych uszkodzeń aparatury wtryskowej silnika z zapłonem samoczynnym w oparciu o przebieg wykresu indykatorowego. In *Diagnostyka Maszyn Roboczych i Pojazdów*; Polskie Towarzystwo Diagnostyki Technicznej: Warszawa, Poland, 2005.
3. Polski Rejestr Statków. *Rules for the Classification and Construction of Sea-Going Ships: Part VII, Machinery, Boilers and Pressure Vessels*; Polski Rejestr Statków: Gdańsk, Poland, 2019.
4. Polski Rejestr Statków. *Rules for the Classification and Construction of Inland Waterways Vessels: Part VI, Machinery and Piping Systems*; Polski Rejestr Statków: Gdańsk, Poland, 2016.
5. Huang, Q.; Zhang, C.; Jin, Y.; Yuan, C.; Yan, X. Vibration analysis of marine propulsion shafting by the coupled finite element method. *J. Vibroeng.* **2015**, *17*, 3392–3403.
6. Bejger, A.; Burnos, T. Time-Frequency Analyze of Some Acoustic Emmision Signals. In Proceedings of the III International Scientifically-Technical Conference, Explo-Diesel & Gas Turbine '03, Lund, Sweden, 5–9 May 2003.
7. Bielawski, P.; Burnos, T. Diagnostyka zużyciowa z zastosowaniem systemu ENTEK. In Proceedings of the V Krajowa Konferencjia, Diagnostyka Techniczna Urządzeń i Systemów Diagnostycznych, Ustroń, Poland, 13–17 October 2003; pp. 307–308.
8. Bielawski, P. Maintenance diagnosis of turbo-machines on board m/f Polonia. In Proceedings of the III International Scientifically-Technical Conference, Explo-Diesel & Gas Turbine '03, Lund, Sweden, 5–9 May 2003.
9. Lin, T.R.; Pan, J.; O'Shea, P.J.; Mechefske, C.K. A study of vibration and vibration control of ship structures. *Mar. Struct.* **2009**, *4*, 730–743. [CrossRef]
10. Popenda, A.; Nowak, M. Analiza drgań giętnych wału z wykorzystaniem modelu polowo-obwodowego. *Przegląd Elektrotechniczny* **2017**. [CrossRef]
11. Brown, T.S.; Neill, W.S. Determination of engine cylinder pressure from crankshaft speed fluctuations. *SAE Trans.* **1992**, *101*, 771–779.
12. Citron, S.J.; O'Higgens, J.E.; Chen, L.Y. Cylinder by cylinder engine pressure and pressure torque waveform determination utilizing speed fluctuations. *SAE Trans.* **1989**, *101*, 933–947.
13. Moro, D.; Cavina, N.; Ponti, F. In-cylinder pressure reconstruction based on instantaneous engine speed signal. *J. Eng. Gas Turbines Power* **2002**, *124*, 220–225. [CrossRef]
14. Rizzoni, G.; Connolly, F.T. Estimate of IC engine torque from measurement of crankshaft angular position. *SAE Trans.* **1993**, *102*, 1937–1947.
15. Antoni, J.; Daniere, J.; Guillet, F. Effective vibration analysis of IC engines using cyclostationarity, Part II—New results on thereconstruction of the cylinder pressures. *J. Sound Vib.* **2002**, *257*, 839–856. [CrossRef]
16. Ball, J.; Bowe, M.; Stone, C.R.; McFadden, P. Torque estimation and misfire detection using block angular acceleration. *SAE Tech. Paper Ser.* **2000**, *109*, 622–643.

17. Koşnig, D.; Toşrk, C.; Boşhme, J. DetroitDesign of optimum periodic time varying filters for applications in combustion diagnosis of car engines. In Proceedings of the International Conference on Acoustics Speech and Signal Processing, Detroit, MI, USA, 9–12 May 1995; pp. 1924–1927.
18. Lyon, R.H. Vibration based diagnostics of machine transients. *Sound Vib.* **1988**, *22*, 18–22.
19. Gruca, M. Relationship between angular vibrations of the engine block, rotational speed and work of individual cylinders piston engine. *Logistyka* **2014**, *6*, 4216–4225.
20. Gawande, S.H.; Navale, L.G.; Nandgaonkar, M.R.; Butala, D.S.; Kunamalla, S. Detecting power imbalance in multi-cylinder inline diesel engine genset. *J. Electron. Sci. Technol.* **2010**, *8*, 273–279.
21. Chybowski, L.; Kazienko, D. Quantitative indicators of the instantaneous speed of a ship's main engine and its usability in assessing the quality of the combustion process. *Multidiscip. Asp. Prod. Eng.* **2020**, *3*, 93–106. [CrossRef]
22. Kazienko, D.; Chybowski, L. Instantaneous rotational speed algorithm for locating malfunctions in marine diesel engines. *Energies* **2020**, *13*, 1396. [CrossRef]
23. Siemens. *Identifying Best Practices for Measuring and Analyzing Torsional Vibration*; Siemens: Munich, Germany, 2014.
24. Palermo, A.; Britte, L.; Janssens, K.; Mundo, D.; Desmet, W. Various torsional vibration measurement methods for optimal trade-off between high accuracy and ease of instrumentation. In Proceedings of the Torsional Vibration Symposium, Salzburg, Austria, 21–23 May 2014.
25. Wang, J.; Wei, L.; Ruiya, L.; Liu, Q.; Yu, L. A fiber bragg grating based torsional vibration sensor for rotating machinery. *Sensors* **2018**, *18*, 2669. [CrossRef] [PubMed]
26. Dereszewski, M.; Drewing, S. Decomposition of harmonic wavelets of torsional vibrations as basis for evaluation of combustion in compression-ignition engines. *J. KONES Powertrain Transp.* **2018**, *25*, 129–135.
27. Drewing, S.; Dereszewski, M. Amplitude changes of component harmonic of torsional vibration of the shaft of the petrol fuel unit on the influence of changes of fuel doses. *Sci. J. Gdyn. Marit. Univ.* **2018**, *108*, 34–42.
28. Dereszewski, M. Monitoring of torsional vibration of a crankshaft by instantaneous angular speed observations. *J. KONES Powertrain Transp.* **2016**, *23*. [CrossRef]
29. Bonnier, J.S.; Tromp, C.A.J.; Klein, W.J. Decoding torsional vibration recordings for cylinder process monitoring. In Proceedings of the 22nd CIMAC International Congress on Combustion Engines, Copenhagen, Denmark, 18–21 May 1998.
30. Rudnicki, J.; Korczewski, Z. Diagnostic testing of marine propulsion systems with internal combustion engines by means of vibration measurement and results analysis. *Combust. Engines* **2013**, *154*, 308–313.
31. Kim, H.J.; Lekcharoen, S.A. Cooley-Tukey modified algorithm in fast fourier transform. *Korean J. Math.* **2011**. [CrossRef]
32. Harris, F.J. Time domain signal processing with the DFT. In *Handbook of Digital Signal Processing*; Elliot, D.F., Ed.; Academic Press: Cambridge, MA, USA, 1987; Chapter 8; pp. 633–699.

Article

Application of Measurement Sensors and Navigation Devices in Experimental Research of the Computer System for the Control of an Unmanned Ship Model

Tadeusz Szelangiewicz [1,*], Katarzyna Żelazny [1], Andrzej Antosik [2] and Maciej Szelangiewicz [3]

[1] Faculty of Navigation, Maritime University of Szczecin, Wały Chrobrego 1-2, 70-500 Szczecin, Poland; k.zelazny@am.szczecin.pl
[2] Company Smart-Electronics, 72-002 Dołuje, Poland; andrzej.antosik@smart-electronics.eu
[3] Company Conmar, 71-075 Szczecin, Poland; maciej.szelangiewicz@icloud.com
* Correspondence: t.szelangiewicz@am.szczecin.pl

Abstract: Unmanned autonomous transport vessels (MASS) are the future of maritime transport. The most important task in the design and construction of unmanned ships is to develop algorithms and a computer program for autonomous control. In order for such a computer program to properly control the ship (realizing various functions), the ship must be equipped with a computer system as well as measurement sensors and navigation devices, from which the recorded parameters are processed and used for autonomous control of the ship. Within the framework of conducted research on autonomous ships, an experimental model of an unmanned ship was built. This model was equipped with a propulsion system not commonly used on transport vessels (two azimuth stern thrusters and two bow tunnel thrusters), but providing excellent propulsion and steering characteristics. A complete computer system with the necessary measuring sensors and navigation devices has also been installed in the model of the ship, which enables it to perform all functions during autonomous control. The objective of the current research was to design and build a prototype computer system with the necessary measurement sensors and navigation devices with which to autonomously control the unmanned ship model. The designed computer system is expected to be optimal for planned tasks during control software tests. Tests carried out on open waters confirmed the correctness of the operation of the computer system and the entire measurement and navigation equipment of the built model of the unmanned transport vessel.

Keywords: autonomous unmanned vessel (MASS); computer control system; measuring sensors and navigation devices; model tests of unmanned vessel in open water; recording of movement parameters and images from lidars and cameras

Citation: Szelangiewicz, T.; Żelazny, K.; Antosik, A.; Szelangiewicz, M. Application of Measurement Sensors and Navigation Devices in Experimental Research of the Computer System for the Control of an Unmanned Ship Model. *Sensors* **2021**, *21*, 1312. https://doi.org/10.3390/s21041312

Academic Editor: Enrico Meli

Received: 29 December 2020
Accepted: 6 February 2021
Published: 12 February 2021

1. Introduction

In maritime transport, since the beginning of the 21st century, new technological solutions have been designed and researched to lead to:

- Improvements and further increases in the safety of maritime transport: the analyses of maritime accidents and catastrophes carried out indicate that their cause is not a failure of the ship's equipment, but only human error (depending on the type—collision, grounding, fire—man is responsible for from 80–96% of all accidents, [1–3];
- Reduction in ship operation costs: about 20–30% of these costs is the maintenance of the crew and shipowners' services dealing with the crew;
- Environmental protection: although maritime transport accounts for about 3% of the world's CO_2 emissions to the atmosphere, new ship designs are expected to have completely green propulsion systems;
- Improvements in other indicators affecting the economic side of maritime transport: new hull design with less weight will result in better use of ship displacement;

- More accurate running of ships on the shipping lines, which can lead to a reduction in energy expenditure, reduced voyage time, or improved punctuality of entry into the port of unloading.

Such a new solution in maritime transport, which will meet higher expectations, is unmanned autonomous vessels. The most important task during the design and construction of autonomous ships is to develop computer control systems. Such a system has to be equipped with special software that allows the determination of the route of the voyage and the realization of various tasks (maneuvers) during the autonomous control, as well as starting of the remote control during e.g., change in the autonomy level because of various emerging threats or difficult tasks during the operation of the ship, e.g., voyage through areas with a high intensity of ship traffic or entry into ports.

In order for the software installed in the computer system to be able to perform various tasks during autonomous control, the ship must be equipped with various measuring and navigation devices. This paper will present an experimental model of an autonomous transport ship, equipped with a computer control system and the necessary measurement and navigation equipment. A prototype of the measurement and navigation system will be presented, as well as research and tests of the use of measurement sensors and navigation devices for the autonomous control of an unmanned ship model.

2. Status of Research on Autonomous Control Systems for Unmanned Ships

The first research work on the possibility of building unmanned autonomous transport vessels began in the early 21st century.

In 2013, Rolls-Royce Marine began working on the design of an autonomous ship. The Advanced Autonomous Waterborne Applications Initiative (AAWA) project was launched to develop the design and technical solutions (navigation, safety, monitoring, collection, and processing systems) for building autonomous vessels. The AAWA project has also developed a virtual center for controlling the fleet of unmanned vessels [4].

Currently, research and experimental work on MASS's are being carried out worldwide in two directions:

- Construction and testing of remote and then autonomous control systems on existing crew ships (usually small car-passenger ferries operating at short distances);
- Design, construction and experimental research of unmanned transport vessels in the first stage, these are physical models of ships with a length of several meters and then the construction of target unmanned vessels.

Examples of conducted tests of autonomous control systems installed on unmanned ships are listed below:

- Tests of autonomous control on Falco's Finnish car ferry—December 2018 [5]: the ferry performs autonomous maneuvers, is supervised by an operator from the land-based center in Turku. No information about the applied measurement and navigation devices, algorithms, and control programs is available.
- Tests of autonomous control of the ferry Suomenlinna II—Finland, December 2018 [6]: tests monitored from the land-based center in Helsinki. No information about the applied measuring and navigation devices, algorithms, and control programs.
- Autonomous control tests on the supply vessel SeaZip 3, the Netherlands, in the North Sea—March 2019 [7]: no information about the applied measuring and navigation devices, algorithms and control systems.

Examples of projects and studies for new unmanned autonomous vessels are listed:

- The Re-Volt—remote container ship with electric drive (Norway) [8]: designed from 2017 by DNV GL, it is to carry 100 TEU; there is no information about the ship's construction and propulsion and no information about the applied, for autonomous control, measuring and navigation devices, and about the effects of experience from the realized project.

- Ship Yara Birkeland—designed since 2017 unmanned container ship (Norway) [9]. Figure 1 shows model tests in 2018.

(a) (b)

Figure 1. Model of the ship Yara Birkeland for research and testing (**a**) [9], and the autonomous container ship Yara Birkeland launched in February 2020 (**b**) [10] (120 TEU, L = 79.5 m, B = 14.8 m).

Apart from the pictures of the model and the launched ship (Figure 1), there is no information about the results of the model tests, about the applied measuring and navigation devices, about the construction of the built ship, and about the algorithms and systems of remote and autonomous control of the ship.

There are also other projects, such as the ShippingLab Project (Denmark), SeaShuttle Project (Norway and the Netherlands), Autoship Project (United Kingdom R-R and Norway Kongsberg), but also in these cases, there is no practical information available.

The information about the implemented unmanned ship transport projects is placed on the Internet in the form of notes and not in scientific publications. Apart from modest information about conducted trials, they do not contain any important information about specialist equipment, construction of a control system, and about the algorithms and computer control programs.

In addition to online information on research or tests of autonomous control systems for unmanned transport vessels, a few articles have been published on other unmanned waterborne objects, e.g., small, unmanned hydrodrones for measurement or research. The publication [11] presents a concept and electronic system for planning bathymetric measurements in shallow waters using an autonomous unmanned MASS. The article lacks data concerning the vehicle structure and control algorithms. The article [12], consisting of two parts, presents the classification of floating autonomous objects, a proposal of autonomy levels, and an extensive review of prototypes of autonomous units without information about the design, construction, and results of autonomous control tests.

The most information on the structure and equipment of hydrodrones for hydrographic research is presented in [13]. The article presents a functional diagram of the hydrodrone with a description of the possibilities of realizing different levels of autonomy, hardware architecture with a list of measuring and navigation devices used during operation at selected autonomy levels, and software architecture. The construction of the hydrodrones with the measurement and navigation sensors installed was not included, nor were the results of tests and studies conducted in the open water.

Publications [14–17] present the results of tests in the basin model carried out on remote-controlled models equipped with navigation devices and a computer system for autonomous execution of some maneuvers. Publications [14] and [15] examine the possibility of cruising the model ship "Essp Osaka" on a given trajectory and performing circulation tests. The publication [16] presents a control and measurement system for real time maneuvering of the transport ship model. The publication [17] presents a computer system for research and evaluation of autonomous maneuvers to avoid collision at sea. In these publications, a model of a transport vessel with classic propulsion was examined:

one propeller and one finned rudder placed behind the propeller. In addition to these publications, there are also articles on dynamic systems of vessel positioning and movement along a given trajectory. These publications include control algorithms, a description of the control and measurement system installed, and the results of tests carried out. The information from these publications can also be useful for the design, construction, and testing of unmanned transport vessels with autonomous control.

Subsequent publications [18–21] presented the results of computer simulation studies and experiments of autonomous ship models. The studies presented there concerned the detection of obstacles at sea by means of radar, lidar, sonar, or cameras and the determination of a safe navigation route. The configuration of the computer systems used, the technical parameters of these systems, and the measurement and navigation equipment were not included; neither are the methods of measurement data processing nor their registration and transmission to land-based centers monitoring the movement of the autonomous vessels presented.

3. Purpose, Scope, and Test Method

The analysis of the available literature shows that although research and tests of autonomous control systems for floating objects are being carried out, there is very little relevant information given in these publications useful for practical use in the design and construction of autonomous control systems for unmanned transport vessels.

The requirements and functions to be performed by an unmanned, autonomously operated transport ship are defined mainly in IMO (International Maritime Organization) regulations and in regulations of Classification Societies. It follows from these regulations that the ship must be equipped with a computerized control system enabling the change of autonomy level—switching to remote control by the operator (navigator) from the land-based center. In addition, the equipment of the ship with measuring sensors and navigation devices must provide full navigational information and diagnostic information about the operation of all devices on the ship (mainly propulsion). An example of the complexity of navigation equipment is the proposal of an integrated navigation bridge on an autonomous ship presented in the publication [22]. The anticipated navigation and measurement equipment on an actual ship are very sophisticated; there are various devices and sensors measuring the same parameters (there must be an appropriate level of redundancy required by the relevant regulations).

Since it is very expensive to build an unmanned transport vessel and to carry out tests on the autonomous controlling system, such tests are carried out on physical models of unmanned vessels.

The analyzed scientific publications lack information on the configuration of the computer system with measurement and navigation equipment. There is also no information on the technical parameters of the measurement and navigation equipment and the on-board computer, in particular its computing power necessary for autonomous control of the unmanned ship model.

The aim of the conducted research on unmanned transport vessels is the design and construction of a prototype model of the ship together with a computer system and the development of software for autonomous control. The autonomous control system must perform the following functions:

- Automatic mooring/unmooring to the quay;
- Selection and optimization of the shipping route from one port to another;
- Maritime detection of fixed and mobile objects (other vessels);
- Possibility to switch to remote control (change of autonomy level);
- Performing anti-collision maneuvers;
- Electronic communication between ships;
- Other functions (after the formulation of rules for autonomous vessels).

Checking if the developed computer program will correctly perform the listed functions will be carried out by the experimental method using a model of an unmanned vessel and by the simulation method (computer simulation). The whole research task has been divided into two stages. The first stage of the research is the design and construction of an experimental model of an unmanned vessel and its equipment with: a propulsion system, computer control system, and all those necessary for autonomous control sensors and navigation devices. The second stage is the development of algorithms and a computer program for autonomous control.

The objective of stage I is to develop and build a computer control system for a model of an unmanned vessel under the following assumptions:

- The propulsion system of the model of the ship must be ecological and emission-free with smooth regulation of control parameters;
- Computer system parameters (computing power, memory capacity, speed of processing measurement data, their recording in the memory, and transmission to the Ground Control Station (GCS) must be adjusted to the volume (number) and frequency of recorded parameters;
- The set of measurement sensors and navigation devices must be able to perform all of the above functions;
- Technical and operational parameters of the computer system components (e.g., dimensions, range, measurement accuracy, etc.) must be appropriate for the ship model (dimensions, speed of movement of the ship model—Table 1);
- The cost of building the system should be as low as possible.

Table 1. Technical and operational parameters of an unmanned vessel.

Parameter	Unit	Ship	Model (Scale 1:25)
Length over all L_{oa}	m	78.75	3.15
Length between perpendiculars L_{PP}	m	75.00	3.00
Breadth B	m	11.10	0.47
Draft T	m	4.33	0.17
Displaced value ∇	m^3	2500.00	0.16
Block coefficient c_B	-	0.6589	0.6589
Velocity V	m/s	8.22	1.64

The scope of the study for the implementation of stage I performed by the experimental method in open water is to check:

- Correct operation of the computer system, measurement and navigation sensors, and registration of all parameters;
- Correctness of processing and sending received signals from these sensors to the computer control system;
- Computing capacity of a computer made in industrial technology receiving and processing the recorded and measured parameters from measurement sensors and navigation devices;
- The correctness of sending the recorded parameters from the measurement and navigation sensors in real time to GCS;
- Correct detection of obstacles set on the water and performing anti-collision maneuvers;
- The correctness of switching to manual or remote control with GCS or when changing the autonomy level.

4. Experimental Model of an Unmanned Transport Vessel Controlled Autonomously

Source data analyses suggest that the first unmanned transport autonomous vessels will be small container ships operating in internal waters or dedicated waters [10], equipped with environmentally friendly electric propulsion systems.

Based on these projections, in 2018, the project of an unmanned container ship (bulk carrier, Table 1), autonomously controlled, was made (Figure 2). The model of the unmanned ship in scale 1:25 (Figure 3) was built for testing of the autonomous control system (Table 1).

Figure 2. Project visualization of an unmanned container ship (bulk carrier), controlled autonomously.

(**a**) (**b**)

Figure 3. Unmanned vessel model propulsion system: (**a**) azimuthal stern thrusters, (**b**) bow tunnel thrusters.

The experimental model of an unmanned ship has been equipped with an ecological, electric propulsion system of the same design as the actual ship. The propulsion system was designed so that the actual ship (and its model) would have excellent propulsion and maneuvering characteristics and could autonomously (or remotely) perform all possible maneuvers. The propulsion system consists of (Figure 3):

- Two stern azimuthal thrusters;
- Two bow tunnel thrusters.

This type of propulsion of the ship (or model) allows for very accurate and safe execution of all necessary maneuvers during autonomous control without human intervention.

The hydrodynamic characteristics of the model propellers were tested experimentally in the model basin (description of the research and measured characteristics are included in [23]).

Planned research and tests of the autonomous control system (automatic mooring, selection and optimization of the navigation route, obstacle detection at sea and anti-collision, and other maneuvers) require a land-based system for controlling and monitoring the movement of the ship model and a computerized control system installed in the ship model.

The computer control system installed in the model consists of:

- PC-class computer;
- I/O cards (analog–digital and digital–digital);
- Individual controllers for each drive motor;
- RC receiver (remote control in emergency situations or change of autonomy level);
- Internal Ethernet network;
- External antennas for communication on different frequencies;
- A set of measuring sensors and navigation devices adequate for the model of an unmanned transport vessel and planned tests of autonomous control.

The control computer is an industrial-grade PC mini equipped with a passive cooling system, SSD drive, and Intel i7 processor (Advantech, Warszawa, Poland). The parameters of the control computer was sufficient for:

- Processing data from all sensors;
- Performing calculations resulting from ship model control algorithms;
- Machine learning using artificial intelligence.

All these computing tasks must be performed in real time.

The computer works with an I/O module whose purpose is:

- Conversion of data sent by the computer into motor control signals—PWM;
- Sensor readings for measuring the power consumed by motors;
- Reading the system supply voltage;
- Reading data from the ship's emergency manual control system;
- Reading data from the direct control receiver.

The I/O module uses an ARM® 32-bit Cortex®-M3 single-chip microcontroller running at 72 MHz.

When controlling the drive motors, the value and direction of rotation is determined by applying a PWM signal from the I/O module at a frequency of 400 Hz and a pulse width of 1000–2000 µs (neutral, motor stop 1500 µs). The PWM signal is generated in the I/O module by an 8-channel PWM generator with a resolution of 12 bits controlled by a bidirectional I2C digital bus controlled by the microcontroller.

The use of a separate I/O module with its own software instead of directly controlling the motors and reading their parameters via a PC was dictated by safety and functionality considerations:

- The I/O module has a Fail Safe mode—in case of suspension of the PC software during testing new algorithms, it allows one to take over manual control of the ship model;
- Independent power system monitoring and response to battery charge status;
- Prevents accidental startup of drive systems during software testing;
- Provides a long-range emergency communication channel;
- It is possible to modify the I/O module software to implement a simple software controlling system, which, in the event of a Fail Safe condition, will allow the ship model to return to a preset position.

The I/O module uses several Fail Safe systems for the following events:

- Suspension or lack of communication between the PC and I/O module: the software sends cyclically, in addition to control and read commands, a data frame called heart beat—a frame of a few bytes sent at a software-controlled frequency of 1–0.1 Hz as a request–response. If the frame does not arrive in the expected time, the I/O module stops the thrusters and switches the control mode to manual.
- I/O Module Software Suspend: Software Watchdog 500 ms (can be reduced if necessary).
- Loss of the manual control signal (priority—the system provides for manual control of the ship model in every situation): then, the thrusters are stopped (worst case scenario) and the ship model has to be recovered by itself. It is possible to extend the I/O module by a function for automatic ship model return.
- Supply level too low: A two-stage action is provided. When the first level is exceeded, the I/O module will not allow a high-power task. When the second level is exceeded, the thrusters are stopped, but the system can be switched to manual mode and the thrusters can be forced to work manually until the batteries are completely discharged—this is an extreme emergency situation and destructive for the batteries, but in some situations, it must be possible to move the model ship away from a dangerous place, e.g., to avoid a collision. All values are configured in the I/O module. The operator can view these data in the GCS and is informed in advance when the critical values are approached.

The system uses a 115,200 bit/s serial link for communication between the PC and the I/O module. Communication based on commands issued in ASCII format has been developed for the system. The time delay for the I/O module to respond to a query from a PC is between 50 and 100 ms. To speed up communication, it is possible to change communication from ASCII to binary format.

Due to the variety of data exchange systems between subsystems, the computer has a driver software, processing data from sensors and I/O module to a common JSON protocol, with which the software controlling the ship model communicates. In addition, the driver has a module to log data for further analysis.

The communication system with the Ground Control System (GCS) uses a broadband network operating in the 2.4 and 5 GHz bands, and the 433 or 868 MHz Long Range System (LRS) low-bandwidth band for emergency communications. Two omnidirectional antennas with 3.5 dBi gain for the 2.4 GHz band and 5 dBi gain for the 5 GHz band and a dual-band directional panel antenna with 9 dBi gain for the 2.4 GHz band and 11 dBi gain for the 5 GHz band in GCS were used for MASS communications. In addition to broadband communications, basic exchange of telemetry and control data with the MASS is conducted at all times via a narrowband 433 or 868 MHz LRS-type link, which allows for control of the MASS when other communications systems fail.

By using a PC-class computer and internal LAN MASS, it is possible to install any other communication systems.

The GCS system is based on a built-in industrial PC-class computer with an LCD monitor. Due to the extensive communication systems and safety systems (Fail Safe), the MASS manual control system is based on a separate proportional control controller, a separate LRS narrowband communication module, and a dual 2.4 and 5 GHz broadband communication system, as well as having a built-in LTE modem.

5. Measuring Sensors and Navigation Devices Used in the Autonomous Control System

The developed computer system of autonomous control, during the ship's model task (e.g., cruise on a given route and performing an anti-collision maneuver), must receive the necessary information concerning not only the vessel's movement parameters (position, speed, course), but also information or warnings about an emerging fixed or moving obstacle (another floating vessel).

The model of the unmanned vessel has been equipped with the following measurement and navigation devices, connected to the I/O cards of the computer system (Figures 4 and 5):

- Lidars working in the 360° range (bow and stern);
- Global Navigation Satellite Systems (GNSS) receivers (bow and stern);
- Electronic compass;
- HD cameras (bow and stern).

The block diagram of the computerized control system of the unmanned vessel transport model is shown in Figure 5.

The parameters of the measuring and navigation sensors installed in the ship's model have been selected so that the measured values are sufficiently accurate for the model with specified dimensions and speed of sailing (Table 1) and for the planned tests on the water during calm weather (no undulations and high wind speed).

Laser lidars for measuring the distance and angular position of the ship from an obstacle in the 2D plane in the range of 360°, measurement frequency from 5 to 15 Hz, time of a single measurement 0.25 ms, range of the rangefinder from 0.15 to 12 m, and data transmission—115.2 kbit/s via USB port.

Cameras: for observation of the environment in the bow of the model and behind the stern, recording and transmitting HD video using IP protocol, and data transmission is via Ethernet, using an encoding system (H264 or H265 codec).

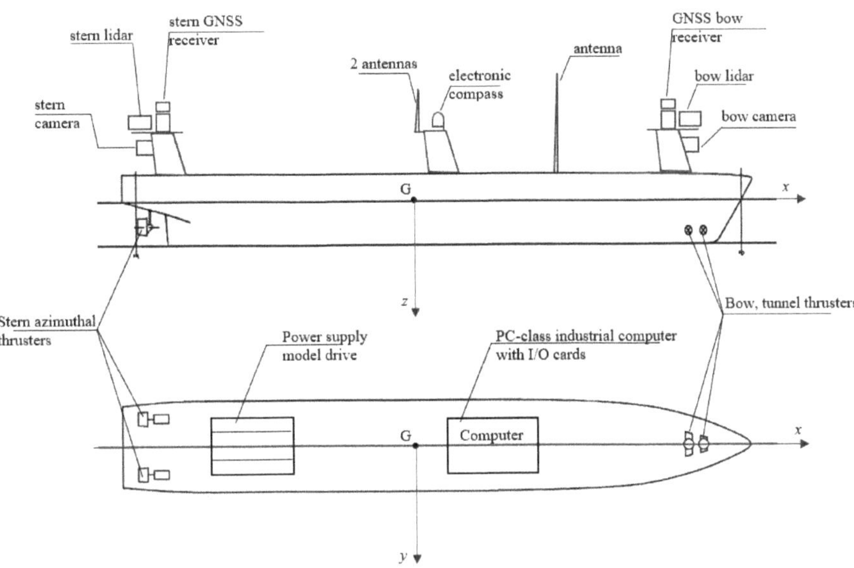

Figure 4. Computer system and navigation and measurement devices in the transport model of an autonomous ship.

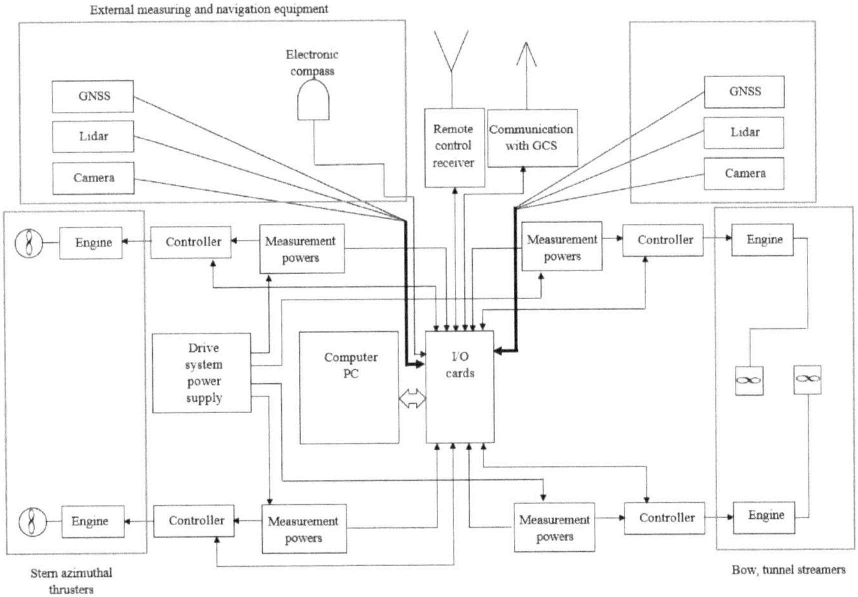

Figure 5. Block diagram of the unmanned ship control system, with navigational measurement devices.

GNSS receivers: for measuring the position and speed of the model ship, data transmission 9.6 kbit/s, via NMEA protocol, USB port.

Electronic compass: for ship course measurement, a three-axis compass with electronic inclination compensation, data transmission 4.8 kbit/s, NMEA protocol, USB port.

Parameters (images) from measurement and navigation devices are recorded and transmitted in real time to the I/O module and to GCS. To integrate all elements of the computer system, an Ethernet switch was used.

6. Results of Tests of the Computer System and Measurement and Navigation Devices

Experimental research on the model of an unmanned ship equipped with a computer system for autonomous steering was carried out in December 2020 on "Głębokie" Lake in Szczecin (Figure 6).

Figure 6. "Głębokie" Lake, which has been tested on for the system of autonomous control of an unmanned ship model [24].

During testing of the computer system and measurement and navigation equipment of the unmanned transport vessel, the model was controlled remotely and wirelessly from GCS. Figure 7 shows a commissioning GCS, which displays an image from two cameras and lidars and a model of the ship during the tests.

In order to achieve the assumed objectives of the experimental research, many tests and performance tests of the computer system and the measurement and navigation sensors installed in the model of unmanned ship were performed. One of the tests with the registration of all parameters is shown on the drawings and pictures below.

An example of the trajectory of the unmanned vessel model's movement, drawn on a map of the basin on the basis of measuring the model's position from the GNSS antenna, is shown in Figure 8.

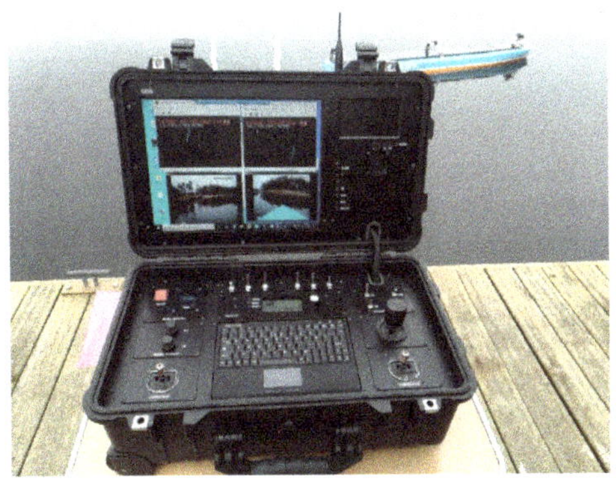

(**a**) (**b**)

Figure 7. Experimental model of an unmanned, autonomous transport vessel with navigation equipment—tests on open water ("Głębokie" Lake): (**a**) land-based control station GCS; (**b**) unmanned ship model with computer system and navigation equipment.

Text Console


```
11:17:32   $GNRMC,111732.00,A,5328.35014,N,01429.05897,E,0.108,,151220,,,D*62
11:17:32   $GNVTG,,T,,M,0.108,N,0.199,K,D*30
11:17:32   $GNGGA,111732.00,5328.35014,N,01429.05897,E,2,06,1.91,17.7,M,39.6,M,,0000*76
11:17:32   $GNGSA,A,3,02,07,09,30,13,05,,,,,,,2.72,1.91,1.93*11
11:17:32   $GNGSA,A,3,,,,,,,,,,,,,2.72,1.91,1.93*19
11:17:32   $GPGSV,4,1,15,02,18,242,37,04,01,095,,05,51,291,40,06,00,204,*75
11:17:32   $GPGSV,4,2,15,07,71,089,38,09,36,097,36,13,19,272,37,14,10,163,18*77
11:17:32   $GPGSV,4,3,15,16,10,026,09,18,00,349,,27,04,053,,28,03,167,*78
11:17:32   $GPGSV,4,4,15,30,65,195,41,36,27,159,34,49,28,192,35*44
11:17:32   $GLGSV,1,1,00*65
11:17:32   $GNGLL,5328.35014,N,01429.05897,E,111732.00,A,D*77
11:17:33   $GNRMC,111733.00,A,5328.35021,N,01429.05894,E,0.060,,151220,,,D*69
11:17:33   $GNVTG,,T,,M,0.060,N,0.111,K,D*3F
11:17:33   $GNGGA,111733.00,5328.35021,N,01429.05894,E,2,06,1.91,17.7,M,39.6,M,,0000*72
11:17:33   $GNGSA,A,3,02,07,09,30,13,05,,,,,,,2.72,1.91,1.93*11
11:17:33   $GNGSA,A,3,,,,,,,,,,,,,2.72,1.91,1.93*19
11:17:33   $GPGSV,4,1,15,02,18,242,37,04,01,095,,05,51,291,41,06,00,204,*74
11:17:33   $GPGSV,4,2,15,07,71,089,38,09,36,097,36,13,19,272,37,14,10,163,20*7C
11:17:33   $GPGSV,4,3,15,16,10,026,10,18,00,349,,27,04,053,,28,03,167,*70
11:17:33   $GPGSV,4,4,15,30,65,195,41,36,27,159,32,49,28,192,35*42
11:17:33   $GLGSV,1,1,00*65
11:17:33   $GNGLL,5328.35021,N,01429.05894,E,111733.00,A,D*73
11:17:34   $GNRMC,111734.00,A,5328.35023,N,01429.05897,E,0.035,,151220,,,D*6F
```


mooring from the quay (platform);

the model is approaching an obstacle;

the model, after the maneuver, floating along the obstacle.

(**a**) (**b**)

Figure 8. Registered (GNSS antennas and electronic compass) example of the movement trajectory of an unmanned ship model during one of the tests (the ship model is not drawn on the appropriate scale in relation to the quay (platform): (**a**) the movement trajectory of the ship model from one of the tests; (**b**) parameters recorded and saved from the GNSS antenna.

In the ship model, during the tests, the parameters of the propulsion system were also measured in real time (model speed, power (voltage and current) of individual propulsion engines, battery capacity). An example record of the measured parameters is shown in Figure 9.

During the tests, the correctness of the operation of the cameras and laser lidars was checked, as well as the processing of the received image in the computer system for model control of the ship, image recording, and sending it to the land station. Example images from the cameras and lidars recorded during the test for the trajectory points from Figure 8 and the corresponding ship model images are shown in Figures 10 and 11 (point 1), Figures 12 and 13 (point 2), and Figures 14 and 15 (point 3).

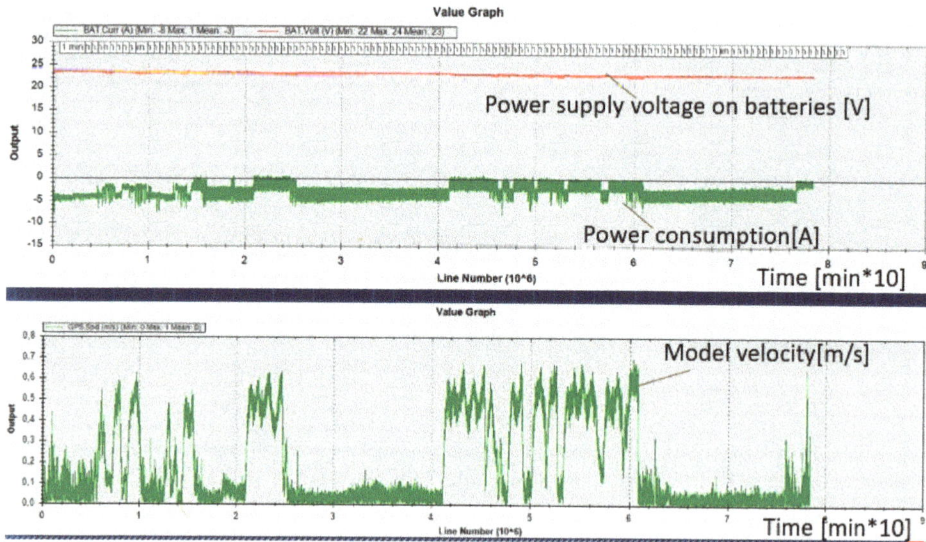

Figure 9. Example recording variables parameters of the ship model: speed (from a GNSS antenna) and power supply of one of the azimuthal thruster motors (from power measurement sensors).

Figure 10. Unmooring maneuver from the platform (Pos. 1 from Figure 8)—photo of the model during unmooring from the platform.

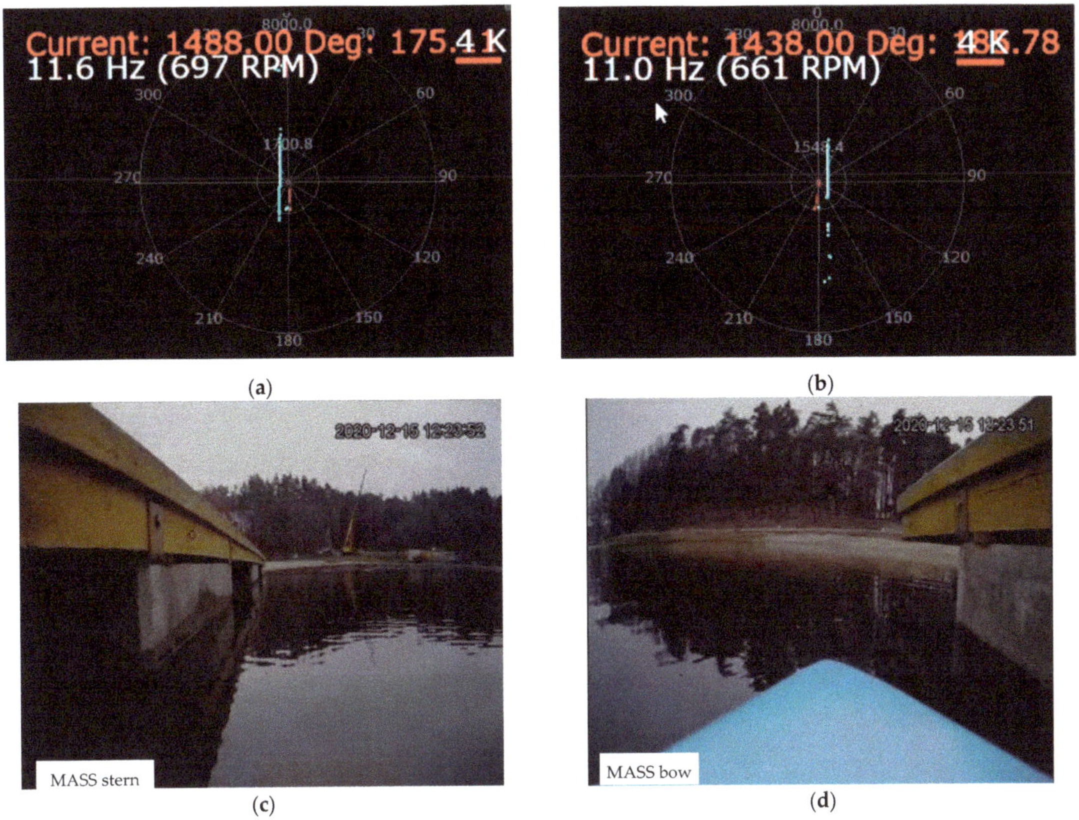

(**a**) (**b**)

(**c**) (**d**)

Figure 11. Unmooring maneuver from the platform (Pos. 1 from Figure 8)—snapshot from lidar and camera screens: (**a**) stern lidar image; (**b**) bow lidar image; (**c**) stern camera image; (**d**) bow camera image.

Figure 12. Model floating towards the obstacle (Pos. 2 from Figure 8)—photo of model floating towards the obstacle.

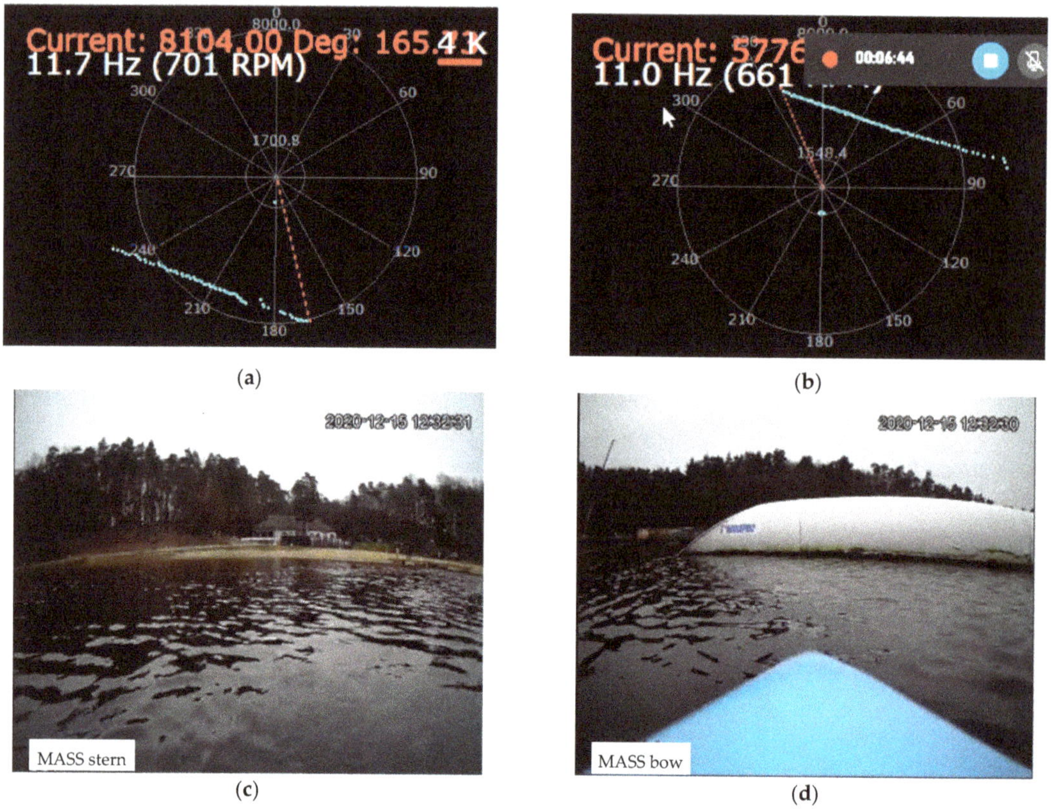

(**a**) (**b**)

(**c**) (**d**)

Figure 13. The model is floating towards the obstacle (Pos. 2 from Figure 8)—snapshot from lidar and camera screens: (**a**) stern lidar image; (**b**) bow lidar image; (**c**) stern camera image; (**d**) bow camera image.

Figure 14. The model floating along a fixed obstacle (Pos. 3 from Figure 8)—a photo of the model floating along the obstacle.

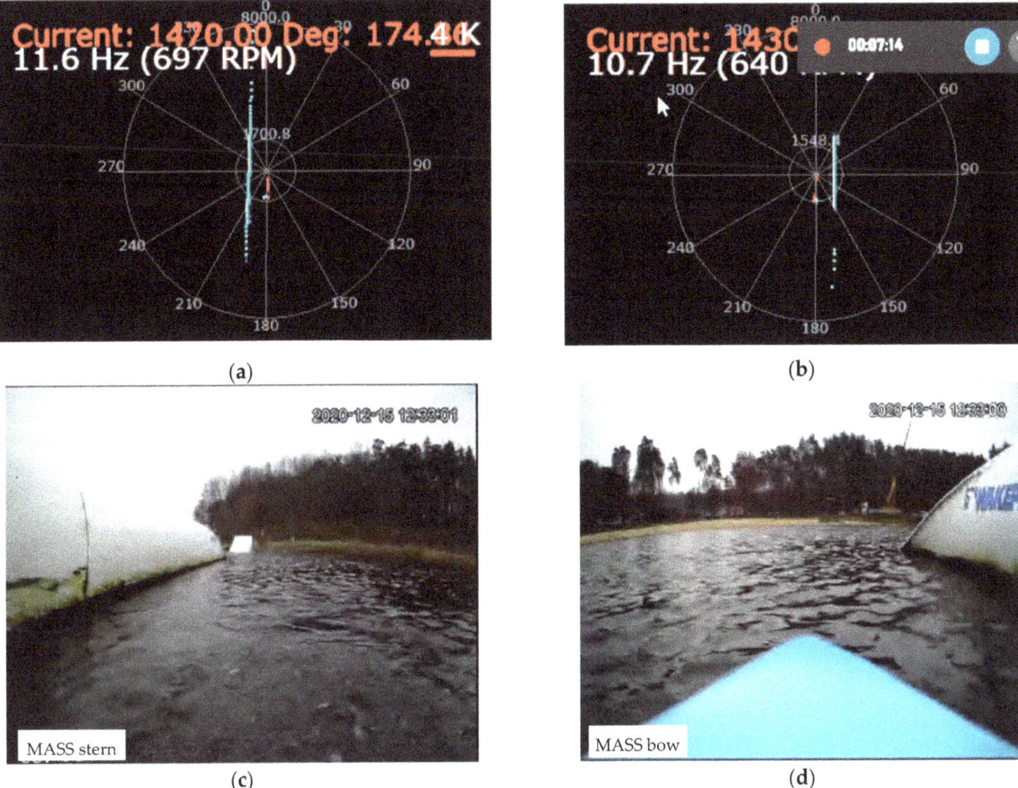

Figure 15. The model floating along a fixed obstacle (Pos. 1 from Figure 8)—snapshot from lidar and camera screens: (**a**) stern lidar image; (**b**) bow lidar image; (**c**) stern camera image; (**d**) bow camera image.

7. Discussion of Test Results

The main objective during the conducted tests was to determine whether the designed and built computer control system including all measurement and navigation sensors will be able to perform all tasks (functions) during autonomous control and whether the applied computer system will have sufficient computing power.

During the tests, the time of realization of individual computer operations and the delays occurring in the system were determined:

- During the software hang of the I/O module, the software Watchdog was 500 ms (can be reduced if necessary).
- The time delay for the I/O module to respond to a command from the PC is between 1.2 and 1.6 ms. To speed up communication, it is possible to change communication from ASCII to binary format. The system is equipped with a CAN bus, which can also be used for future data exchange between the PC and I/O module.
- The conversion time in the I/O module of the analog signals to digital was about 1 µs (12-bit converters). The total conversion time of analog values is, for 5 A/D channels, about 5 µs. Each measurement channel is software averaged and also filtered using a median filter for a sequence of 5 samples. The result is then converted to an output value (mA) and compared to a 5-point calibration curve for each sensor (calibration performed according to a certified current meter with adjustable current load).
- Approximate delay time between issuing the command from the PC to change the speed of the thruster and its execution is about 2–3 ms. The delay is the sum of

communication time—UART, ASCII, decoding of information sent over the I2c bus (400 kHz) to the PWM generator, and the response of the generator itself.

The execution time of all operations (The execution time of all operations (processing, transmission, recording, computation)) was about 0.5 s, and the control loop delays were minimal, and, for the tested ship model, did not significantly affect its control. The model floats at a maximum speed of 1.64 m/s (Table 1), which is a very low speed relative to fast flying drones, for example, and during maneuvers, the model speed can be much lower.

The following conclusions can be drawn from the experimental study:

- All measuring sensors and navigational equipment were working properly and all recorded parameters were transmitted in real time to the control computer installed in the model vessel.
- Lidars were installed to detect obstacles on the water, regardless of the course of the ship in relation to the obstacles (the distance and position of the model ship in relation to the obstacle was determined on the basis of the image from the lidar).
- All recorded parameters from measuring sensors and navigation devices were properly registered in the on-board computer and prepared for transmission to the land station (GCS).
- All computational operations connected with recording and processing parameters from measurement sensors and navigation devices required from 10 to 15% of the total computational power of the on-board computer.
- After changing the broadband network frequency, the interference in the transmission of video from cameras and lidars no longer occurred; after a series of tests, it was concluded that the broadband transmission should be modified and the 5G LTE network should be used.

Based on the experimental study, it is concluded that the designed and built computer control system works properly and is optimal for the built unmanned autonomous vehicle model, and the on-board computer not only performs all computational processes, but still has reserves of computing power necessary for the expansion of the control system.

8. Conclusions

The presented computer system with a complete set of measurement and navigation devices, including the propulsion system of the ship model and the conducted experimental tests, can be considered as a new approach to research on autonomous control systems for unmanned ships. During the conducted tests, it was found that the technical parameters of the computer system were correctly selected for this unmanned ship model and the planned further research, which can also be considered as a new approach (such information is missing in the available literature). During the experimental tests, the computer hardware and the measurement and navigation equipment worked properly, although we anticipate some modifications (broadband transmission), and the results of the tests can be used for the design and construction of the computer control system of the actual unmanned vessel.

The information gathered from the tests and the conclusions drawn allow the development of guidelines for the design of a computer system for a real ship that will travel at a much higher speed (Table 1) than the tested ship model.

Since the hardware tests so far are positive, the next stage of the research will be to test the software for autonomous control of the ship model.

The control software consists of many modules—the autonomous ship must independently implement various maneuvers resulting from different situations arising during the voyage from the starting port to the destination port.

Author Contributions: Design of unmanned ship model, propulsion system, and computer control system, T.S.; design and preparation of computer software, K.Ż.; construction and installation of computer control system, tests of computer system in ship model; A.A.; construction of ship model, M.S.; tests of propulsion system, remote control system, T.S., K.Ż., A.A. and M.S. All authors have read and agreed to the published version of the manuscript.

Funding: The design, construction, and testing of the model of the unmanned vessel was financed entirely from private funds of CONMAR. No external financing.

Institutional Review Board Statement: Not applicable.

Informed Consent Statement: Not applicable.

Data Availability Statement: Not applicable. (All results and data are included in the manuscript.)

Conflicts of Interest: The authors declare no conflict of interest.

References

1. Chauvina, C.; Lardjaneb, S.; Morela, G.; Clostermannc, J.P.; Langard, B. Human and organisational factors in maritime accidents: Analysis of collisions at sea using the HFACS. *Accid. Anal. Prev.* **2013**, *59*, 26–37. [CrossRef] [PubMed]
2. Herdzik, J. Analiza skutków wybranych wypadków na morzu jako zagrożeń utrudniających akcje ratownicze. *Logistyka* **2014**, *4*, 419–429.
3. Cordon, J.R.; Walliser, J.; Mestre, J.M. Human factor: The key element of maritime accidents. In Proceedings of the Atlantic Stakeholder Platform Conference, Brest, France, 29 October 2015. [CrossRef]
4. gCaptain. Rolls-Royce on Drone Ships: It's Not If, It's When. 2016. Available online: https://gcaptain.com/rolls-royce-on-drone-ships-its-not-if-its-when/ (accessed on 15 May 2019).
5. PortalMorski.pl. Pierwszy, w pełni Autonomiczny Prom Rolls-Royce'a. 2018. Available online: https://www.portalmorski.pl/zegluga/41181-pierwszy-w-pelni-autonomiczny-prom-rolls-royce-a (accessed on 15 May 2019).
6. GospodarkaMorska.pl. Kolejny Eksperyment z Udziałem Promu Bezzałogowego Zakończony Sukcesem. 2018. Available online: https://www.gospodarkamorska.pl/Porty,Transport/kolejny-eksperyment-zudzialem-promu-bezzalogowego-zakonczony-sukcesem-html (accessed on 30 May 2019).
7. CTV Seazip 3. 2019. Available online: https://www.seazip.com/our-fleet/ctv-seazip-3/ (accessed on 4 October 2019).
8. DNV-GL. The ReVolt. A New Inspirational Ship Concept. 2019. Available online: https://www.dnvgl.com/technology-innovation/revolt/index.html (accessed on 30 May 2019).
9. Kongsberg New Norwegian Autonomous Shipping Test-Bed Opens. 2017. Available online: https://www.kongsberg.com/maritime/about-us/news-and-media/news-archive/2017/new-norwegianautonomous-shipping-test-bed-opens/ (accessed on 4 October 2019).
10. GospodarkaMorska.pl. Prace przy Yara Birkeland Wstrzymane ze Względu na Pandemię Koronawirusa. 2020. Available online: https://www.gospodarkamorska.pl/stocznie-offshore-prace-przy-yara-birkeland-wstrzymane-ze-wzgledu-na-pandemie-koronawirusa-49143. (accessed on 12 May 2020).
11. Specht, C.; Świtalski, E.; Specht, M. Application of an Autonomous/Unmanned Survey Vessel (ASV/USV) in Bathymetric Measurements. *Pol. Marit. Res.* **2017**, *24*, 36–44. [CrossRef]
12. Schiaretti, M.; Chen, L.; Negenborn, R.R. *Survey on Autonomous Surface Vessels: Part I—A New Detailed Definition of Autonomy Levels, Part II—Categorization of 60 Prototypes and Future Applications*; Computational Logistics. ICCL 2017. Lecture Notes in Computer Science; Springer: Cham, Switzerland, 2017; Volume 10572. [CrossRef]
13. Stateczny, A.; Burdziakowski, P. Universal Autonomous Control and Management System for Multipurpose Unmanned Surface Vessel. *Pol. Marit. Res.* **2019**, *26*, 30–39. [CrossRef]
14. Moreira, L.; Santos, F.P.; Mocanu, A.; Liberato, M.; Pascoal, R.; Guedes Soares, C. Instrumentation Used on Guidance, Control and Navigation of a Ship Model. In Proceedings of the 8th Portuguese Conference on Automatic Control (CONTROLO), Vila Real, Portugal, 21–23 July 2008.
15. Moreira, L.; Guedes Soares, C. Autonomous Ship Model to Perform Manoeuvring Tests. *J. Marit. Res.* **2011**, *8*, 29–46.
16. Perera, L.P.; Moreira, L.; Santos, F.P.; Ferrari, V.; Sutulo, S.; Guedes Soares, C. A Navigation and Control Platform for Real-Time Manoeuvring of Autonomous Ship Models. In Proceedings of the 9th IFAC Conference on Manoeuvring and Control of Marine Craft, The International Federation of Automatic Control, Arenzano, Italy, 19–21 September 2012.
17. Perera, L.P.; Ferrari, V.; Santos, F.P.; Hinostroza, M.A.; Guedes Soares, C. Experimental Evaluations on Ship Autonomous Navigation and Collision Avoidance by Intelligent Guidance. *IEEE J. Ocean. Eng.* **2015**, *40*, 374–387. [CrossRef]
18. Reed, S.; Schmidt, V.E. Providing Nautical Chart awareness to autonomous surface vessel operations. In Proceedings of the OCEANS 2016 MTS/IEEE Monterey, Monterey, CA, USA, 19–23 September 2016.
19. Son, N.S.; Kim, S.Y. On the sea trial test for the validation of an autonomous collision avoidance system of unmanned surface vehicle. In Proceedings of the OCEANS 2018 MTS/IEEE Charleston, Charleston, SC, USA, 22–25 October 2018; pp. 1–5.
20. Lee, J.; Woo, J.; Kim, N. Vision and 2D LiDAR based autonomous surface vehicle docking for identify symbols and dock task in 2016. In Proceedings of the 2017 IEEE Underwater Technology (UT), Busan, South Korea, 21–24 February 2017.
21. Song, H.; Lee, K.; Kim, D.H. Obstacle avoidance system with LiDAR sensor based fuzzy control for an autonomous unmanned ship. In Proceedings of the 2018 Joint 10th International Conference on Soft Computing (SCIS) and Intelligent Systems and 19th International Symposium on Advanced Intelligent Systems (ISIS), Toyama, Japan, 5–8 December 2018.

22. Zaleski, P. Problemy rozwoju statków autonomicznych. In *Akademickie Aktualności Morskie, 3(107), Magazyn Informacyjny Akademii Morskiej w Szczecinie*; Scientific Publishing House of the Maritime University of Szczecin: Szczecin, Poland, 2020.
23. Szelangiewicz, T.; Żelazny, K. Hydrodynamic characteristics of the propulsion system thrusters of the unmanned ship model. *Sci. J. issued Marit. Univ. Szczec.* **2020**, *32*, 136–141.
24. Szczecin Google Maps. Available online: https://www.google.pl/maps/place/Jezioro+G%C5%82%C4%99bokie/@53.4768631,14.4658762,13.92z/data=!4m5!3m4!1s0x47aa0c2d98d84a27:0x9fd8d04bd7861d6!8m2!3d53.4771739!4d14.4759493 (accessed on 30 August 2020).

Article

Assessment of the Propulsion System Operation of the Ships Equipped with the Air Lubrication System

Mariusz Giernalczyk * and Piotr Kaminski

Faculty of Marine Engineering, Gdynia Maritime University, 81-234 Gdynia, Poland; p.kaminski@wm.umg.edu.pl
* Correspondence: m.giernalczyk@wm.umg.edu.pl

Abstract: This paper is an attempt to evaluate the effectiveness of the ship's hull air lubrication system in order to reduce the drag leading to fuel consumption reduction by ships. The available papers and reports were analyzed, in which records of the operation parameters of the propulsion system of ships equipped with this system were presented. These reports clearly show the advantages of using air lubrication system. On the basis of collected operating parameters of the propulsion system the authors performed analysis of operation effectiveness of the Air Lubrication System on the modern passenger ship was. The results of this analysis do not allow for a clearly positive opinion about its effectiveness. Additionally, the conditions that should be met for the system to be more effective and to significantly increase the propulsion efficiency were indicated.

Keywords: emission reduction; air lubrication system; drag reduction; energy efficiency design index

Citation: Giernalczyk, M.; Kaminski, P. Assessment of the Propulsion System Operation of the Ships Equipped with the Air Lubrication System. *Sensors* **2021**, *21*, 1357. https://doi.org/10.3390/s21041357

Academic Editors: Leszek Chybowski, Arkadiusz Tomczak, Maciej Kozak and Jongmyon Kim

Received: 29 December 2020
Accepted: 10 February 2021
Published: 14 February 2021

1. Introduction

The MARPOL Annex VI came into force on 19 May 2005, concerns on the prevention of air pollution by ships. It forced the ship-owners to apply solutions aimed at reducing the emission of harmful substances, such as nitrogen oxides (NOx), sulphur oxides (SOx), carbon oxides (CO), hydrocarbons (HC) and particulate matter (PM) into the atmosphere. This annex did not initially include carbon dioxide emission reductions. However, international institutions including the International Maritime Organization (IMO) have noticed the threat of the greenhouse effect, caused in a large scale by carbon dioxide. In July 2011, the Annex VI of the MARPOL Convention was extended by Chapter IV that aims to reduce greenhouse gases emissions in particular carbon dioxide by ships [1,2].

The reduction of CO_2 emissions is to be achieved by introducing for all newly built vessels greater than 400BRT, the Energy Efficiency Design Index (EEDI) [3]. The EEDI index is defined as the ratio of the amount of CO_2 [g] to the amount of cargo [t] on a specific shipping distance [Mm] and is a specific balance between the social benefit of cargo transport and the negative phenomenon of CO_2 emissions to the atmosphere. It is to be used as a tool to indirect control of CO_2 emissions and to increase the energy efficiency of ships power plants. The EEDI value for the ship is calculated in accordance to the procedure contained in Resolution MEPC.308(73) [4] and must be equal to or lower than the value required for the type and size of the vessel. It is calculated based on the formula presented on Figure 1.

In 2018, the IMO published a preliminary strategy to reduction of the greenhouse gases emissions reduction from ships with the principles of its application [4]. This forced ship-owners to search technological solutions aimed at reducing carbon dioxide emissions and improving sailing efficiency by decreasing fuel consumption. These goals can be achieved, inter alia, by reducing the vessel's hydrodynamic resistance [5–9]. One of the methods to reducing the drag by reducing frictional resistance is insertion of an air layer between the underwater part of ship's hull and water. The air bubbles in this method are used as lubricant and it is called Air Lubrication (AL) [10–12]. AL systems (ALS) are recognized by IMO as category B-1 (Innovative Energy Efficiency Technology) as described in MEPC.1/Circ.815 [4]. This technology significantly lowers the EEDI value, mainly by reducing the components surrounded by the frame in formula (Figure 1).

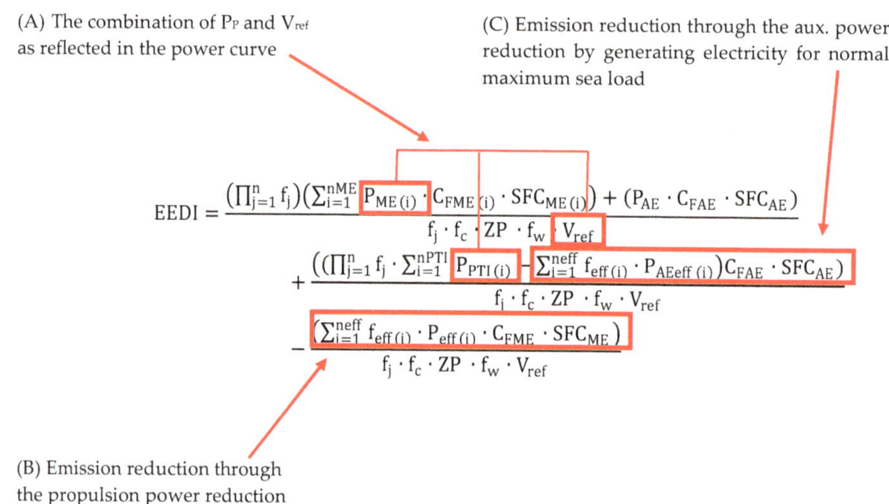

Figure 1. The Energy Efficiency Design Index (EEDI) formula with indicated elements that may affect the emission reduction by the use of ALS.

2. Ship Hull Resistance

The ship moves on the boundary of two fluids-air and water, which counteract movement by causing hydrodynamic and aerodynamic forces that, create movement drag. The total resistance R includes the sum of the aerodynamic resistance R_A and the hydrodynamic resistance R_H. The hydrodynamic resistance R_H is the sum of the components of the frictional resistance R_F and the pressure R_P (wave resistance R_W and viscous pressure resistance R_{VP}). Thus, the total resistance of a ship moving through the water is given by the formula:

$$R = R_F + R_W + R_{VP} + R_A \tag{1}$$

The structure of the total hull resistance is shown on Figure 2.

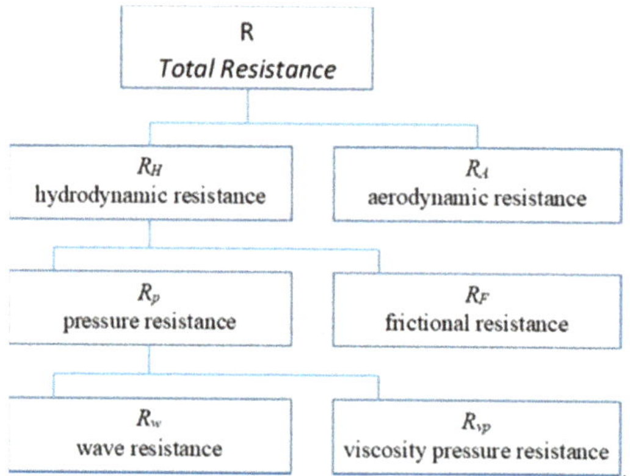

Figure 2. The total hull resistance structure [13].

The distribution of the total resistance components is presented in Figure 3. Both, hydrodynamic and aerodynamic resistance are described by the general resistance forces equation:

$$R = c \cdot \frac{\rho \cdot v^2}{2} \cdot S \quad (2)$$

where, R is the drag force, c the dimensionless drag coefficient, ρ is the fluid density, v is the velocity, S is the hull surface in fluid.

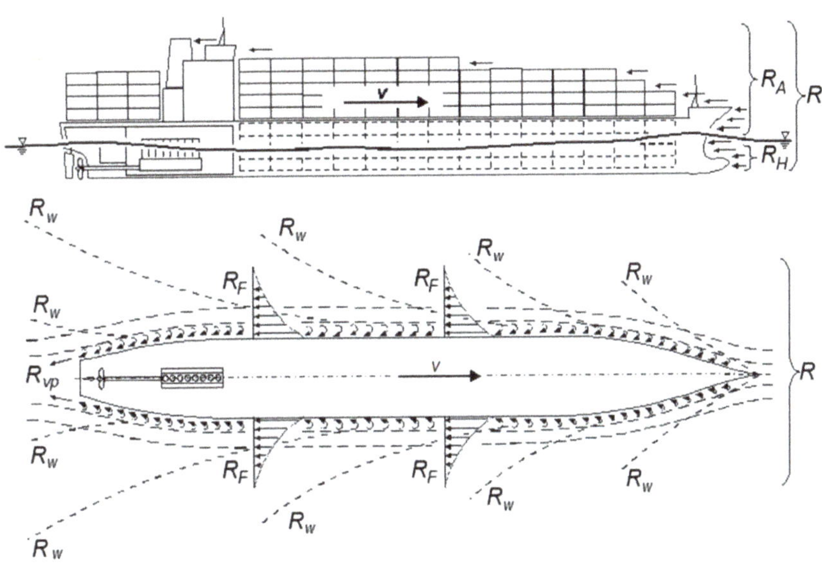

Figure 3. Distribution of the total hull resistance [13].

The individual components of a ship total resistance affect its size to a different extent [14,15]. The pressure R_P (viscous pressure resistance R_{VP} and in particular, wave resistance R_W) and frictional resistance have the greatest share in the structure of total resistance. The shape of the hull-its slenderness, fullness and the speed of the vessel significantly influence the wave resistance R_W. For lower sailing speeds, the average value of the wave resistance is 8 ÷ 25% of the total resistance while at high sailing speeds it may reach the value of 40 ÷ 65% of the total resistance [13]. In order to minimize the resistance associated with sea waves the hull shape is optimized already at the design stage. Designers use computer simulations during design and then models' tests in the ship model basin in order to reduce this resistance to a minimum.

On other hand, the frictional resistance increases with the ship's operation (service) time. It is caused by an increase of roughness of the underwater part of the hull as a result of its overgrowing with seaweed, crustaceans, algae, mollusks and other organisms living in the water. It is estimated that from the moment the ship leaves the dry dock, the daily increase of resistance due to fouling of the hull is 0.2 ÷ 0.5% of the total resistance, although there are lower values for colder waters and higher values for warmer waters riche in flora and fauna [16–19].

To reduce viscosity friction the area of the hull wetted surface needs to be reduced. This can be performed by separating the underwater part of the hull surface from the water with a layer of air [20]. The general term used to describe this phenomenon is called hull "Air Lubrication" (AL).

The remaining components have a smaller influence on the total resistance, although it they may be different in particular ship sailing conditions. A good example can be large container ships, where during the ship's movement in a direction opposite to a very strong wind the containers loaded on board (even up to ten layers) create above water part of the hull resistance (aerodynamic resistance) which is a significant share in the total resistance of the ship.

3. Method to Reduce the Ship's Hull Resistance by Introducing an Air Layer under the Hull

Surface frictional resistance is proportional to the wetted surface of the ship's hull; therefore the ALS works on a simple principle of keeping a layer of air bubbles under the hull [21]. The method of producing and introducing the air layer under the bottom of the passenger ship's hull is presented in Figure 4 [22–24].

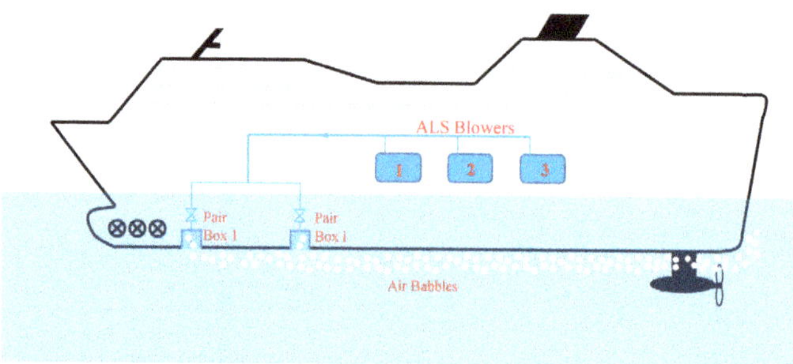

Figure 4. Diagram of the system of introducing the air layer under the hull bottom of the passenger vessel-Air Lubrication System (ALS).

High-capacity blowers are used to generate air bubbles that flow at a constant speed under bottom of the hull. The air bubble outlets are located along the bottom of the hull, symmetrically on both sides of the ship's center line [13,23]. A schematic diagram of the ALS with two blowers, distribution line of compressed air and with air distribution boxes on a large passenger vessel (cruise liner) is presented in Figure 5.

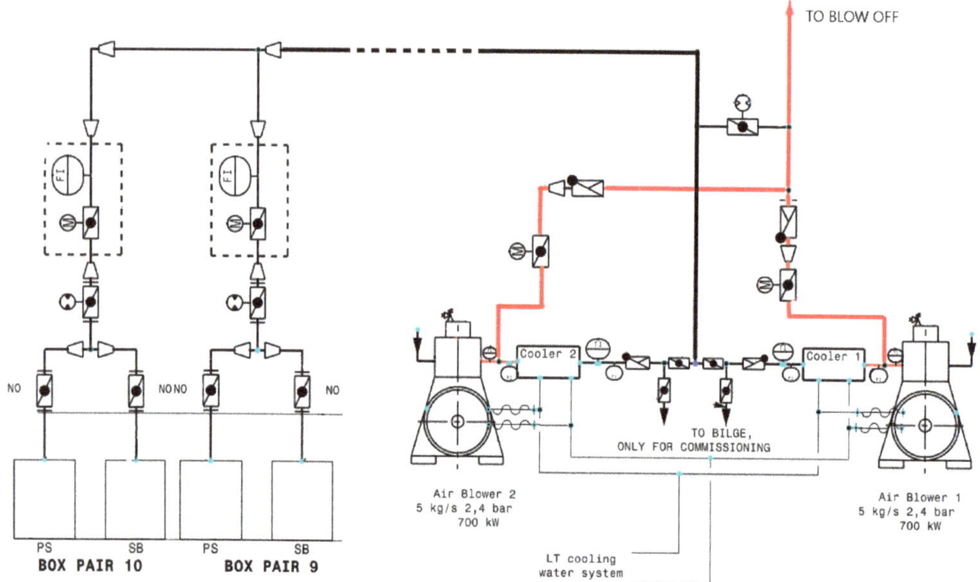

Figure 5. Compressed air supply system (ALS) of the underwater part of the hull on a large cruise liner.

The blowers forced compressed air to 20 distribution boxes (10 pairs) which are a structural element of the ship's hull. Two distribution boxes: one box on the port side and one on the port side are symmetrically connected to one supply subline. The boxes are equipped with corrosion protection (zinc anodes). The compressors run at a constant speed and are controlled by a control system, which can reduce the capacity of one compressor to about 45% of nominal value. This is executed by regulating the air supply with steering wheel with variable angle blades. In this way the energy consumption of the blowers driving motors can be reduced.

The ALS method may be applied at the design stage and built on a new vessel as well as installed on the vessel after a certain period of operation. The introduction of ALS on operated ship is a complicated process and requires comprehensive analyses, calculations, measurements and most often computer simulations [24]. There are several companies specializing in the design and installation of ALS on the vessels and each company calls this system otherwise i.e.: Mitsubishi Co. – Mitsubishi Air Lubrication System (MALS), R&D Engineering – Winged Air Induction Pipe System (WAIP), Samsung Heavy Industries – SAVER System (SAVER Air), Silverstream-Silverstream System, Foreship-Foreship Air Lubrication System (Foreship ALS) and others [5]. The first installation of the ALS called Silverstream System (Addlestone, UK) was applied on a chemical tanker MT Amalienborg with a carrying capacity of 40,000 DWT. This vessel was equipped for propulsion with a low-speed B&W 6S50MC main engine with power of 13,452 BHP [25]. After installing this system on the ship many operation parameters when ALS was ON and OFF were recorded, among others: propulsion system operation parameters and ship speed (on water and GPS), as well as power consumption by blowers, main engine speed, shaft power (torque), fuel consumption by main engines. And additionally, weather conditions (hydrometeorological conditions).

Figure 6 shows the impact of ALS (Silverstream system) operation installed on MT Amalienborg on the changes of the propulsion power (shaft power) and ship speed. The course of the parameters presented in this diagram shows that, while maintaining a constant rpm of main engine, activation of the ALS system causes a decrease in the propulsion power demand and does not significantly affect ship speed, moreover the lower part of the diagram shows the energy consumption by ALS blowers.

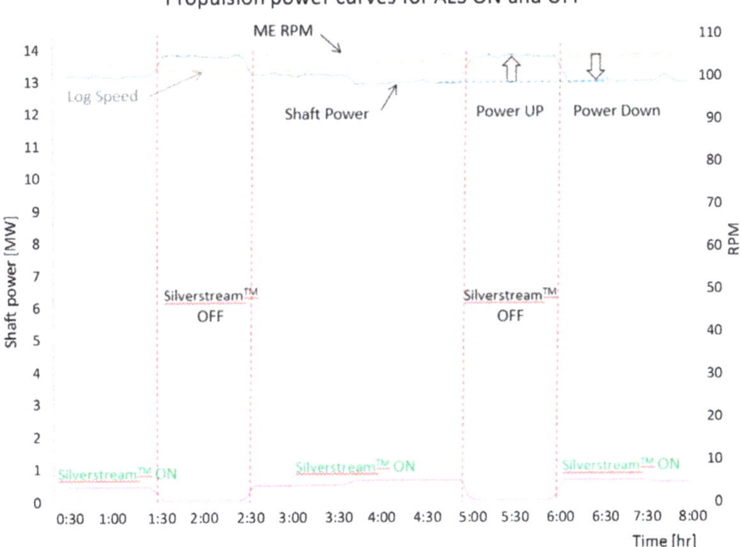

Figure 6. Example of Air Lubrication Effect from Monitoring System on MT Amalienborg [26].

Using the operating parameters of the propulsion system during the sea test of the ALS system the power consumption curve as a function of ship speed (propeller curve) were prepared, shown in Figure 7.

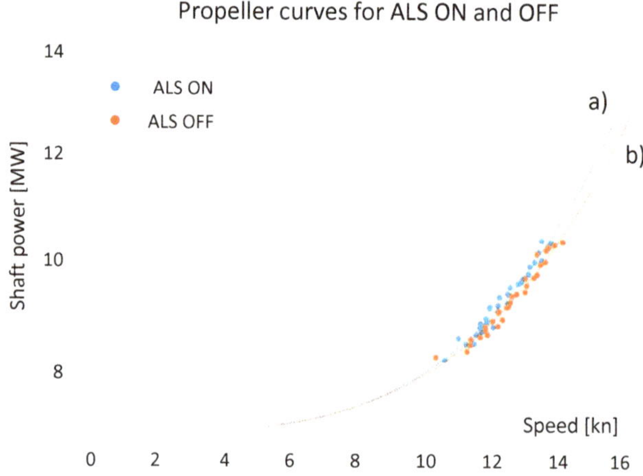

Figure 7. Average Shaft Power for given RPM (Revolution Per Minute) against Speed from MT Amalienborg [26].

The propeller curve for ON and OFF ALS system presented in the diagram P = f (v) shows the benefits of this system operation in the form of lower demand for propulsion power at the same ship speed. This is evidenced by the shift of the propeller curve (a) towards the so-called lighter propeller curve (b).

Measurements results taken on a vessel with ALS were processed by the authors [26,27] and allowed to determine the net energy savings required to propel the vessel, amounting to about: 0.1 ÷ 4.5% (at the sailing speed of 11 ÷ 14 kn). Although these results took into account additional losses related to the blowers drive energy and the resistance of the air distribution boxes, the method of data processing and the obtained results is not precisely explained in the study.

4. Assessment of the ALS Operation

Promising results of operation obtained after installing the ALS on the chemical tanker Amalienborg and other ships encouraged many ship-owners to install this system on their vessels, especially on cruise ships [28–30]. The Silverstream systems were installed, among others on Carnival cruisers (Sapphire Princess, Diamond Princess). On the other hand, the Foreship ALS were installed among others on Royal Caribbean International's large, modern passenger ships. The list of cruise ships equipped with the ALS and delivered up to 2019 is shown in Table 1 [5].

Table 1. The list of cruise vessels with Air Lubrication System delivered 2015–2019.

Year	Vessel Name	Type	System
2015	Quantum of the Sea	Cruise	Foreship
2016	AIDAprima	Cruise	MALS
2017	AIDAperla	Cruise	MALS
2017	Norwegian Joy	Cruise	Silverstream
2018	Diamond Princess	Cruise	Silverstream

The ALS system has been installed on one of the large cruise liners since the ship was put into service. To generate air with the required parameters, two single-stage centrifugal blowers integrated with the gearbox, driven by a motor with power of 700 kW each and a capacity of 5 kg/s at an overpressure of 1.4 bar, were installed on the ship.

On similar passenger vessels, the ALS was installed after some of operation time. Ship-owner decided to install three blowers similar to the ones on the previous cruiser. These vessels are equipped with diesel-electric propulsion system, consisting of six engines driving the main generators with the capacity of 12600 kW each, and two auxiliary generators with a capacity of 2500 kW each. Distribution of the generated electric power is shown in Figure 8.

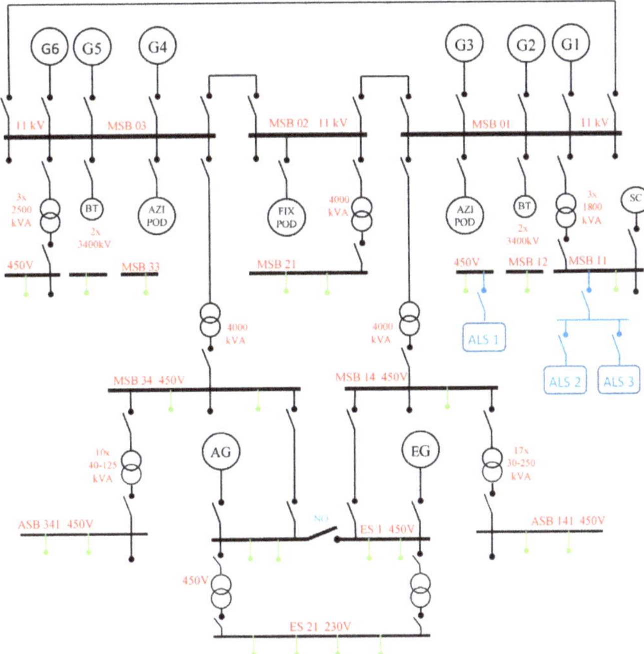

Figure 8. Ae example of simplified diagram of power distribution on a passenger ship with diesel-electric propulsion system.

Electricity is generated by main generators (G1÷G6) and auxiliary generators (AG+EG). The highest voltage current (11 kV) is directed through the Main Bus Bars (MBB) to supply three gondola propellers (2 × AZIPOD + 1 × FIXPOD) and four bow thrusters (BT). Main Switchboard Bus bars (MSB) with a voltage of 450 V power most of the machines and devices in the engine room, including ALS blowers, while the receivers with the lowest power are supplied with 230 V.

After starting the vessel operation with ALS installed, modified propulsion system was tested. Operating parameters of the propulsion system were recorded with the ALS ON and OFF. The system was turned on for a period of 2 ÷ 3 h, and the parameters were recorded before turning on, during operation and after turning off the system. Due to the relatively short time intervals (30 ÷ 60 min) between the recording of parameters it was assumed that the weather conditions were constant. Table 1 shows the recorded and calculated parameters such as energy consumption for propulsion, energy consumption for the other needs of the vessel, the power used by the ALS blowers, fuel consumption of generators engines, etc.

Based on the data from Table 1, the variability of selected parameters is shown in Figure 9. It presented the impact of the ALS operation on the change of the propeller power

PP (Propeller Power/Shaft Power), vessel speed and the power used by blowers ALSP (Air Lubrication System Power). It also took into account the summary power SPP (Summary Propulsion Power) used for the ship propulsion and to drive ALS blowers. The list of selected i.e., measured and calculated parameters is presented in Table 2.

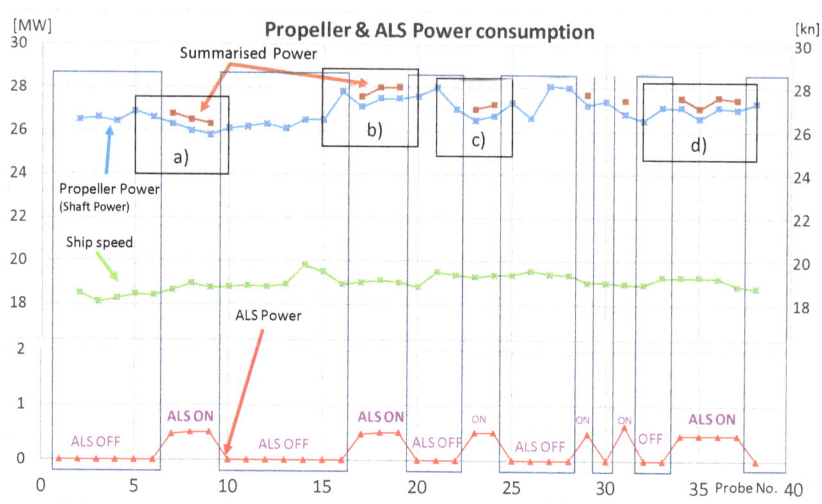

Figure 9. Air Lubrication Effect from Monitoring System on the large cruise liner.

Table 2. Calculated parameters of propulsion system, based on the operating parameters recorded during passenger vessel sailing with ON and OFF ALS.

Probe No.	ALS State	ALS Power Consumption	[1] Propeller Power Distribution	[2] Fuel Consumption Difference	Fuel Consumption Difference	[3] Propeller Power Difference	[4] Summary Propulsion Power Difference	[5] Average Speed Difference
-	-	[MW]	[%]	[kg/h]	[%]	[MW]	[MW]	[kn]
-	-	A	B	C	D	E	F	G
6	off		69%					
7	on	0.47	68%	97	0,8	0.3	−0.17	0.3
9	on	0.51	66%				−0.31	
10	off		69%	−107	-0.9	−0.3	−0.30	0.1
16	off		71%				−1.30	
17	on	0.47	70%	125	1.0	0,7	0.23	0.1
19	on	0.51	69%				−0.51	
20	off		70%	−68	-0.6	−0.1	−0.10	−0.2
22	off		71%	0		1.0	1.00	
23	on	0.51	68%	75	0,6	0.5	−0.01	−0.1
24	on	0,51	69%				−0.71	
25	off		69%	299	2.4	−0.6	−0.60	0.0
26	off		78%				−2.10	
27	on	0.35	77%	237	1.7	−0.3	−0.65	−0.4
28	off		72%				−0.10	
29	on	0.48	69%	20	0.2	0.8	0.32	−0.4
30	on		71%	−101	−0.8	−0.2	−0.20	0.0
31	on	0.63	69%	146	1.2	0.6	−0.03	−0.1
32	off		68%	−95	−0.8	0.3	0.30	−0.1
33	off		71%				−0.60	
34	on	0.46	69%	189	1.5	0.0	−0.46	0.0
37	on	0.46	68%				−0.36	
38	off		70%	−102	−0.8	−0.3	−0.30	−0.1

[1] Propeller Power distribution-the part of total power produced on the vessel used for propulsion purposes. [2] Fuel consumption difference-difference of fuel consumption when ALS is ON or OFF. [3] Propeller Power difference-difference of propeller power consumption before and after start ALS. [4] Summary Propulsion Power difference-difference of propulsion purpose power consumption before and after start ALS. [5] Average speed difference-difference of ship speed before and after start ALS.

The course (track) of parameter variability shown in the diagram (Figure 8) is general and does not allow for a detailed analysis of the system operation in particular periods, i.e., with ALS on and off. Therefore, on the Figure 10 are presented fragments of the diagram where the ALS was started and stopped in detailed.

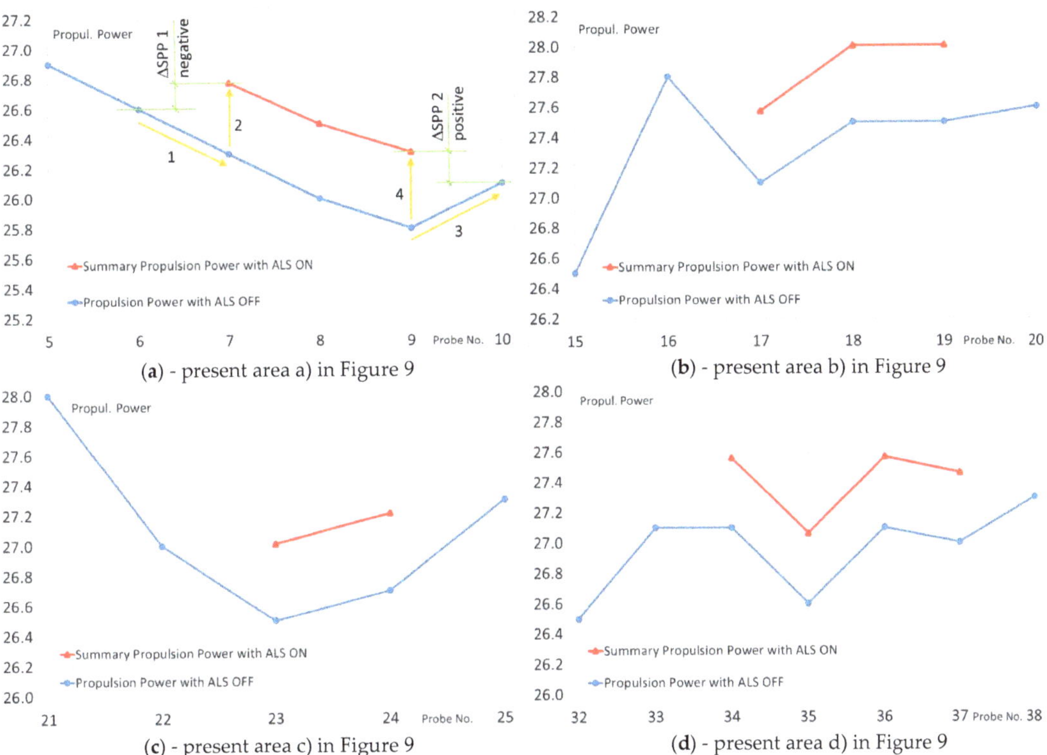

Figure 10. Details of changes in consumption of Propulsion Power and Summary Propulsion Power with ALS (Air Lubrication System) switched ON and ALS switched OFF.

As shown in Figure 10a after switching ON the ALS at measuring point (probe 7) if compared to point 6 there is a decrease (vector 1) of propeller power (PP) consumption (AZIPOD'S + FIXPOD) but at the same time there appeared a demand for energy to drive blowers ALSP (vector 2). Comparing summary power of SPP before and after starting the ALS, a slight increase of the value ΔSPP1 (+0.3 MW) is observed, with a minimal increase of ship's speed (0.3 kn) (the line of the ship's speed in Figure 9). Between the measurement points 9 (probe 9) and 10 (Figure 10a) as a result of the ALS OFF there is a change in the SPP power distribution, due to the lack of power demand for the ALS blowers drive ALSP (line 3) and increase of propeller power (PP) (line 4) with the simultaneous lack of vessel speed changes. This causes only a slight decrease of demand for the summary power ΔSPP2 (0.3 MW).

Similar changes in the distribution of the summary power SPP can be observed when switching the ALS ON and OFF shown on Figure 10b–d.

In Figure 10b is observed a decrease of the summary power SPP is observed demand when ALS is turned ON (transition from point 16 to 17), at constant ship speed. However, after switching the ALS OFF (transition from point 19 to 20) there is a slight increase in propeller power PP and a simultaneous minimal decrease in ship speed.

Figure 10c presents the operating parameters of the propulsion system before (point 22) and after switching the ALS ON (point 23). Switching the ALS into operation does not increase the summary power SPP for propulsion (the power transmitted to the PP ship propulsion decreases, but the ALSP blower propulsion power appears with the same value, at a minimal decrease in ship speed (0.1 kn).

In Figure 10d it is observed that after the ALS is turned ON into operation (points 33 to 34) the propeller power PP demand does not change but the summary power SPP for propulsion increases by the ALSP value (ALS blowers drive power). At the next measurement, points (34÷36) there are fluctuations in the summary power SPP demand without changing of ALSP at the same speed of the ship. When the ALS is turned OFF (transition from point 37 to 38) the propeller power PP demand increases but it is less than the power consumed by the blowers ALSP. At the same time, the speed of the ship drops slightly by about 0.2 kn.

Only at measurement points, 17 and 29 (Figure 9) there are visible slight benefits of switching the ALS ON are visible in the form of decrease the summary propulsion power SPP at unchanged ship speed. Only these points confirm the assumption that ALS reduces fuel consumption for the ship propulsion. Other data do not confirm this assumption.

Moreover, when observing the fuel consumption (Table 1, column D) it can be noticed that switching ALS ON causes an increase or decrease of fuel consumption by approx. −0.8% (positive effect of ALS activation) to 1.7% (negative effect) respectively. In addition, when switching ALS OFF an increase or decrease approx. −0.9% to 2.4 can be observed respectively. It is accompanied by a minimal change in the speed of the ship.

Additionally, based on the collected data, the propulsion power in the ship speed curve (propeller characteristic) was drawn. It was done for propulsion with ALS ON and OFF and presented by points and trend lines in Figure 11. This diagram also confirms that the ALS operation does not increase efficiency of propulsion system. This is due to the location of the points, especially the lack of clear separation between operating points for ALS ON and OFF (like for the MT Amalienborg in Figure 7).

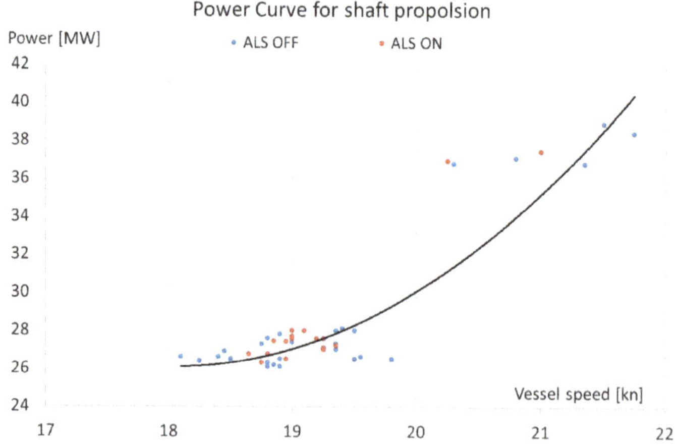

Figure 11. Average Shaft Power for given RPM against Speed from on cruise liner with diesel-electric propulsion.

5. Discussion

This paper presents a more critical assessment of the operation of the ALS system than in the presented and available reports and publications. It should be noted that the presented analyses of the ALS (Silverstream) installed on MT Amalienborg show that the benefits (savings) resulting from its use are about 4.5% at the vessel's speed of 14 kn.

However, the attention should be paid to the fact that the benefits of using this system at speeds below 14 kn are doubtful. The net savings dropped to the amount of 3.2% at 13 kn, 1.7% at 12 kn. Taking into account the fact that the vessel does not always sail at the maximum design speed (14.5 kn for MT Amalienborg) and take into considering the investment and service costs the benefits of using this system seem to be questionable.

On the other hand, on the basis of the observations, analyses and records of operating parameters for the passenger vessel with diesel-electric propulsion system, it can be concluded that the activation of the ALS resulted in a reduction in the propeller power (AZIPOD'S + FIXPOD) demand by 0.1 ÷ 0, 4 MW, while maintaining practically the same ship speed. If taking into account the fact that the power consumption of the activated ALS blowers was 0.34 ÷ 0.63 MW, it appears that the decrease of propeller power demand is balanced by the increase of power consumption by the operated blowers. Practically the fuel consumption of a ship does not change substantially. It can therefore be concluded that the operation of ALS did not improve the ship's propulsion efficiency. This is confirmed by the power curve shown in Figure 11. Contrary to the propeller curve for the chemical tanker MT Amalienborg (Figure 7), switching the ALS ON did not shift curve towards the "light propeller", the propulsion system operating points for a passenger vessel with diesel-electric propulsion were dispersed (Figure 11) and did not clearly divide the propulsion curve with ALS ON (lighter propeller) and OFF (heavier propeller).

6. Conclusions

Based on the conducted analyses of the available literature and records of operating data, it can be stated that:

- the benefits of ALS use seem doubtful (only at the ship design stage, the application of this system improves the EEDI value, which is interesting for ship designers and ship-owners).
- the use of the ALS for the entire ship's speed range is not beneficial, there are minimum and maximum speeds beyond which the use of the system does not give the assumed savings.
- equipment included in the structure of the ALS, including main blowers, require high investment outlays and high operating costs.
- maintaining the same size and evenly distributed air bubbles under the hull surface is a difficult task. Changing the diameter of the air bubbles significantly affects their distribution under the hull and may significantly reduce the effect of reducing the ship's drag. Although features such as protruding ridges on the edges of the hull can help maintain the air layer, but these elements contribute to increased drag and stability of the ship, especially in heavy seas.
- it is difficult to counteract the effect of air bubbles being sucked in by the propeller, causing noise and vibration and leading to a reduction of the propeller efficiency [31].

The authors' observations included in this article, which consist in a rather critical approach to the use of the ALS on ships, are proved in reality. It seems that in recent years, the interest in using this system on newly designed and built units has decreased, and if it appears, it is usually dictated by the need to obtain the recommended EEDI value.

Author Contributions: Conceptualization, M.G.; methodology, M.G.; formal analysis, M.G. and P.K.; investigation, P.K.; data analyzed, P.K.; writing—original draft, M.G and P.K.; writing—review and editing, P.K. and M.G. All authors have read and agreed to the published version of the manuscript.

Funding: This research received no external funding.

Institutional Review Board Statement: Not applicable.

Informed Consent Statement: Not applicable.

Data Availability Statement: Not applicable.

Conflicts of Interest: The authors declare no conflict of interest.

Abbreviations

The following abbreviations are used in this manuscript:

NOx	Nitrogen Oxides
SOx	Sulphur Oxides
CO	Carbon Oxides
HC	Hydrocarbons
PM	Particular Matter
IMO	International Maritime Organization
MARPOL	MARine POLution convention
EEDI	Energy Efficiency Design Index
ALS	Air Lubrication System
R	Total Resistance
R_P	Pressure Resistance
R_W	Wave Resistance
R_F	Frictional Resistance
R_A	Aerodynamic Resistance
R_H	Hydrodynamic Resistance
R_{VP}	Viscous Pressure Resistance
c	Dimensionless Drag Coefficient
ρ	Fluid Density
v	Velocity
S	Surface of the ship's hull
AL	Air Lubrication
MALS	Mitsubishi Air Lubrication System
WAIP	Winged Air Induction Pipe System-R&D Engineering
SAVER Air	SAVER System-Samsung Heavy Industries
BHP	Brake Horse Power
GPS	Global Positioning System
MT	Motor Tanker
kn	Knots
MBB	Main Bus Bars
AZIPOD	Azimuthal stern thruster
FIXPOD	Fixed gondola thruster
BT	Bow Thruster
PP	Propeller Power
ALSP	Air Lubricating system Power
SPP	Summary Propulsion Power

References

1. Giernalczyk, M.; Krefft, J. Metody ograniczania zużycia paliwa przez statki morskie zmierzające do obniżenia emisji szko-dliwych substancji do atmosfery. *Logistyka* **2015**, *4*, 7461–7466.
2. Bell, M.; Deye, K.; Fitzpatrick, N.; Wilson, C.; Chase, A.; Lewis, C.; Bates, G. *Reducing the Maritime Sector's Contribution to Climate Change and Air Pollution*; A Report for the Department for Transport: London, UK, 2018.
3. Resolution MEPC.308(73) 2018 Guidelines on the Method of Calculation of the Attained Energy Efficiency Design Index (EEDI) for New Ships. Available online: https://wwwcdn.imo.org/localresources/en/KnowledgeCentre/IndexofIMOResolutions/MEPCDocuments/MEPC.308(73).pdf (accessed on 21 December 2020).
4. MEPC.1/Circ.815 17. Guidance on Treatment of Innovative Energy Efficiency Technologies for Calculation and Verification of the Attained EEDI. Available online: https://wwwcdn.imo.org/localresources/en/OurWork/Environment/Documents/.
5. Available online: Circ-815.pdf (accessed on 21 December 2020).
6. Air Lubrication Technology. ABS. 1 April. Available online: https://ww2.eagle.org/content/dam/eagle/advisories-and-debriefsAir%20Lubrication%20Technology.pdf (accessed on 21 December 2020).
7. Giernalczyk, M.; Górski, Z.; Krefft, J. Metody zmniejszania oporów kadłuba w celu ograniczenia zużycia paliwa przez statki. *Logistyka* **2015**, *4*, 7453–7460.
8. Park, S.H.; Lee, I. Optimization of drag reduction effect of air lubrication for a tanker model. *Int. J. Nav. Arch. Ocean Eng.* **2018**, *10*, 427–438. [CrossRef]
9. Ahmadzadehtalatapeh, M.; Mousavi, M. A review on the Drag Reduction Methods of the Ship Hulls for Improving the Hydrodynamic Performance. *Int. J. Marit. Technol.* **2015**, *4*, 51. Available online: http://ijmt.ir/browse.php?a_code=A105411&sid=1&slc_lang=en (accessed on 21 December 2020).

10. Perlin, M.; Dowling, D.R.; Ceccio, S.L. Freeman Scholar Review: Passive and Active Skin-Friction Drag Reduction in Turbulent Boundary Layers. *J. Fluids Eng.* **2016**, *138*, 091104. [CrossRef]
11. Foeth, E.J. Decreasing Frictional Resistance by Air Lubrication. Proceedings of the 20th International HISWA Symposium on Yacht Design and Yacht Construction. Amsterdam, Netherlands, 2008, Session 6. Available online: https://repository.tudelft.nl/islandora/object/uuid:4a50369c0a494813863628cd9d2e3016?collection=research (accessed on 21 December 2020).
12. Trial success for Silverstream Air Lubrication. The Motorship – Insight for Marine Technology Professionals. Available online: https://www.motorship.com/news101/engines-and-propulsion/trial-success-for-silverstream-air-lubrication (accessed on 21 December 2020).
13. Silberschmidt, N.; Tasker, D.; Pappas, T.; Johannesson, J. Silverstream®System–Air Lubrication Performance Verification and Design Development. Available online: https://conferences.ncl.ac.uk/media/sites/conferencewebsites/scc2016/1.3.2.pdf (accessed on 21 December 2020).
14. Namura develops NCF for ship energy saving. Japan Ship Exporters' Association. SEA Japan. 2001; No. 287; pp. 4–5.
15. Hashim, A.; Yaakob, O.B.; Koh, K.K.; Ismail, N.; Ahmed, Y.M. Review of Micro-bubble Ship Resistance Reduction Methods and the Mechanisms that Affect the Skin Friction on Drag Reduction from 1999 to 2015. *J. Teknol.* **2015**, *74*. [CrossRef]
16. Murai, Y. Frictional drag reduction by bubble injection. *Exp. Fluids* **2014**, *55*, 1–28. [CrossRef]
17. Park, H.J.; Tasaka, Y.; Murai, Y. Bubbly drag reduction accompanied by void wave generation inside turbulent boundary layers. *Exp. Fluids* **2018**, *59*. [CrossRef]
18. Tanaka, T.; Park, H.J.; Tasaka, Y.; Murai, Y. Spontaneous and artificial void wave propagation beneath a flat-bottom model ship. *Ocean Eng.* **2020**, *214*, 107850. [CrossRef]
19. Elbing, B.R.; Winkel, E.S.; Lay, K.A.; Ceccio, S.L.; Dowling, D.R.; Perlin, M. Bubble-induced skin-friction drag reduction and the abrupt transition to air-layer drag reduction. *J. Fluid Mech.* **2008**, *612*, 201–236. [CrossRef]
20. Latorre, R. Ship hull drag reduction using bottom air injection. *Ocean Eng.* **1997**, *24*, 161–175. [CrossRef]
21. Ceccio, S.L. Friction Drag Reduction of External Flows with Bubble and Gas Injection. *Annu. Rev. Fluid Mech.* **2010**, *42*, 183–203. [CrossRef]
22. Hoang, C.L.; Toda, Y.; Sanada, Y. Full scale experiment for frictional resistance reduction using air lubrication method. In Proceedings of the Nineteenth International Offshore and Polar Engineering Conference, Osaka, Japan, 21–26 July 2009; pp. 812–817.
23. Górski, Z.; Giernalczyk, M. Siłownie okrętowe, Część I, Podstawy napędu i energetyki okrętowej. In *Basics of Ship Propulsion, Part I, Basic Principles of ship Propulsion and ship Power Engineering*; Wydawnictwo Akademii Morskiej w Gdyni: Gdynia, Poland, 2014.
24. Cucinotta, F.; Guglielmino, E.; Sfravara, F. An experimental comparison between different artificial air cavity designs for a planing hull. *Ocean Eng.* **2017**, *140*, 233–243. [CrossRef]
25. Makiharju, S.A.; Perlin, M.; Ceccio, S.L. On the energy economics of air lubrication drag reduction. *Int. J. Nav. Archit. Ocean Eng.* **2012**, *4*, 412–422. [CrossRef]
26. Mizokami, S.; Kawakita, C.; Kodan, Y.; Takano, S.; Higasa, S.; Shigenaga, R. Experimental Study of Air Lubrication Method and Verification of Effects on Actual Hull by Means of Sea Trial. *Mitsubishi Heavy Ind. Tech. Rev.* **2010**, *47*, 41–47.
27. Yousefi, R.; Shafaghat, R.; Shakeri, M. High-speed planing hull drag reduction using tunnels. *Ocean Eng.* **2014**, *84*, 54–60. [CrossRef]
28. He, Y.; Song, B.; Dong, H. Multi-objective optimization design for the multi-bubble pressure cabin in BWB underwater glider. *Int. J. Nav. Arch. Ocean Eng.* **2018**, *10*, 439–449. [CrossRef]
29. Quadvlieg, F. *Maine Propulsion and Fuel Economy*; Maritime Research Institute Netherland: Amsterdam, The Netherlands, 2009.
30. Busch, J.; Barthlott, W.; Brede, M.; Terlau, W.; Mail, M. Bionics and green technology in maritime shipping: An assessment of the effect of Salvinia air-layer hull coatings for drag and fuel reduction. *Philos. Trans. R. Soc. A Math. Phys. Eng. Sci.* **2019**, *377*, 20180263. [CrossRef] [PubMed]
31. Kawakita, C. Study on Marine Propeller Running in Bubbly Flow. In Proceedings of the Third International Symposium on Marine Propulsors. SMP'13, Launceston, Tasmania, Australia, 5–8 May 2013; pp. 405–411.

MDPI
St. Alban-Anlage 66
4052 Basel
Switzerland
Tel. +41 61 683 77 34
Fax +41 61 302 89 18
www.mdpi.com

Sensors Editorial Office
E-mail: sensors@mdpi.com
www.mdpi.com/journal/sensors

www.ingramcontent.com/pod-product-compliance
Lightning Source LLC
LaVergne TN
LVHW071535170726
843515LV00008B/2146

* 9 7 8 3 0 3 6 5 2 3 0 9 5 *